P9-APK-214

DATE DUE		
APR 0 9 1985		
APR 3 0 1985		
MAR 3 1993 Renewed		
5/25/93		
JAN 2 3 1994		
JUN 0 9 2003		
JUN 1 0 2004		

COMMUNICATIONS
FOR A MOBILE SOCIETY

COMMUNICATIONS
FOR A MOBILE SOCIETY

An Assessment of New Technology

Edited by **Raymond Bowers**
Alfred M. Lee
Cary Hershey

and

Philip L. Bereano   Erwin A. Blackstone   Sara Edmondson
Jeffrey Frey   Kurt L. Hanslowe   Heidi Kargman
Arnim Meyburg   Mark V. Nadel   T. J. Pempel
Eric Shott   Harold Ware   and others

SAGE PUBLICATIONS / BEVERLY HILLS / LONDON

This book was prepared with the support of NSF Grant ERP 74-20555. However, any opinions, findings, conclusions and/or recommendations herein are those of the authors and do not necessarily reflect the views of NSF.

For information address:

SAGE PUBLICATIONS INC.
275 South Beverly Drive
Beverly Hills, California 90212

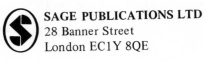

SAGE PUBLICATIONS LTD
28 Banner Street
London EC1Y 8QE

Printed in the United States of America

Library of Congress Cataloging in Publication Data
Main entry under title:

Communications for a mobile society.

Includes bibliographical references and index.
1. Mobile communication systems. I. Bowers, Raymond, 1927-
TK6570.M6C65   621.38'041   77-28119
ISBN 0-8039-0960-8

FIRST PRINTING

Contents

Acknowledgments

The research we have conducted has involved a high level of collaboration and interaction between members of the project. While principal authors are designated for each chapter—reflecting our division of responsibility—the way in which topics have been treated and the conclusions drawn in each chapter have been substantially influenced by other members of the project. The sovereignty of authors had to yield to the requirements of collaborative research. Moreover, the frequency with which individuals are listed as chapter authors may not be a fair measure of their contribution to the overall project. For example, Philip L. Bereano was a major contributor to the planning of the project and its execution in the earlier phases, but his departure in 1975 for the University of Washington allowed only minor involvement in the writing. Similarly, new responsibilities at Cornell prevented Douglas Van Houweling, another important early contributor, from participating in the writing phase. Three of the authors—Beth Baldwin, Robert Glanville, and Howard Hammerman—took part only in the research described in the chapters they coauthor.

Many persons have provided us with valuable assistance and counsel. It is impossible to acknowledge their contributions in a manner that fairly reflects the important help that each of them gave us. Yet we must record, even if inadequately, our gratitude for their contributions.

During the planning phase of the research, four members of the Cornell faculty—Alfred Kahn, Walter Ku, Simpson Linke, and the late Neil Brice—served as consultants to the project. In carrying out the study, many undergraduate and graduate students at Cornell participated as research assistants. We especially wish to thank the following: Edward Baldwin, Daniel Boehnen, Eleanor Brown, William Bush, Alan Christenfeld, Edward Edelson, Barry Eisenberg, Jennifer Helbraun, Jeffrey Hill, Gary Klein, Steve Kratzer, Michael Martin, Catherine Minuse, Barbara Robins, Christine Russell, Edward Sinick, Sue Smith, Thomas Sofo, and Brian Wengenroth. In addition to students, members of the staff at

Cornell provided important research assistance: Betsy Sachter did extensive library research and Michael Kaplan was a key person in our attempt to develop computerized methods that would be useful for executing assessments.

Many persons outside of Cornell provided us with data, helped us gather new information, gave us advice, and criticized our written material—all of which was important in improving the quality of our work. Among such persons were: Sidney Aronson of the City University of New York, Gary Bowman of Temple University, Smiley Ashton and Phil Travers of the Law Enforcement Assistance Administration, Richard Larson and Keith Stevenson of the Massachusetts Institute of Technology, Clifford Orloff of the University of California at Berkeley, Arthur Griffiths and Larry Rose of the Public Health Service, Robert Sohn of the California Institute of Technology, Donal Kavanaugh of the Associated Public Safety Communications Officers, William Elder of the American Trucking Association, Dale Hatfield and Carlos Roberts of the FCC, Richard Tell of EPA, and Val Williams of NABER. Mizumo Kunitoshi of the National Diet Library provided valuable bibliographic help in our work on Japan.

Members of industrial organizations have given us similar help. Such organizations included operating companies within the Bell System, Bell Telephone Laboratories, Motorola, Inc., and RCA. For assistance in our work on trade with Japan, we must thank Ito Sadao, Komura Mitsuru, and Watanabe Masanobu, all of Nippon Telegraph and Telephone, who provided detailed information during multiple interviews. We assumed that the comments expressed by those outside of Cornell were the views of particular individuals and did not necessarily reflect the policies or positions of parent organizations. In any case, industrial perspectives were diverse even within a single organization.

Throughout this study, we benefited from the counsel of an external advisory committee that conducted a full review of the project on two occasions and provided other kinds of help in the intervening period. The members were Nicholas de Martino of the Washington Communications Video Center, Sebastian Lasher of the FCC, Kerns Powers of RCA, Roger Sutliff of the Public Service Commission of New York State, and Robert Yin of the Rand Corporation.

Most of the research described in this volume was supported by funds provided by the National Science Foundation (Grant ERP 74-20555). A grant from the Cogar Foundation was used to initiate the project. Funds of the Cornell University Program on Science, Technology, and Society derived from private foundations and industry were also used to execute the research. We are indebted to the trustees and program officers of these organizations for their financial support. Special thanks are due to the staff of the National Science Foundation. Joseph Coates, now at the Office of Technology Assessment, helped us in many ways during the planning of the project. And Patrick Johnson, the program officer for the period of the grant, deserves particular recognition for his sustained understanding, assistance, and encouragement.

The members of the Science, Technology, and Society Program at Cornell have provided valuable encouragement for and criticism of our work; the contributions of Mark Sagoff and Neil Orloff are worthy of special mention. Dianne Ferriss, who edited the material in this volume, has greatly improved its clarity, *precision, and style; we thank her for this very important contribution to our work. Finally, all of us who participated in this project owe special thanks to Janet Epstein, the project secretary in the last and critical year of our activities. Her extraordinary secretarial and organizational skills were essential to the completion of the project. She deserves our gratitude and the respect that we have for her.*

Our list of acknowledgments is long; it reflects the extraordinary degree to which we depended upon people who were not principal members of the research team to make our project a success. This was an irrefutable finding of our research.

Introduction

Technologies are rapidly maturing that could provide widespread communication capabilities to mobile units such as automobiles and other vehicles; even the portable hand-held telephone is now feasible. During the past 35 years, many important but specialized applications of mobile communications have been developed for police functions, safety services, and commercial purposes. But these uses represent only a fraction of the potential. It now appears that we are at the threshold of not only a major expansion within these specialized areas, but also in the development of a broader range of uses for commercial, public, and private purposes. Taken together, these uses of land mobile communications have the potential for extensive and significant impacts on our society. This volume is an attempt to appraise some of these consequences.

We live in a mobile society; in 1974 there were approximately 105 million automobiles and 25 million trucks and buses in this country, and most of the autos and trucks were equipped with standard radio receiving units for entertainment purposes. The recent acceleration of Citizens Band radio sales, as well as that of other forms of land mobile communication, raises the question whether deployment of two-way devices could ultimately become as common. Seen in a broader perspective, eliminating the wire leash from the telephone–the most prevalent two-way communication device–provides a new dimension of freedom to the use of electrical communications. The consequences could be profound.

Our work has been motivated by an increasing concern with problems that result from technologies being deployed without sufficient regard for their potential social impacts. Similar concerns have been expressed in an extensive literature,[1] and public interest and other citizen groups have utilized political action and litigation to force consideration of these problems. New legislation, such as the National Environmental Policy Act of 1969 and the Technology Assessment Act of 1972, has provided a statutory basis for focusing attention on

[11]

the consequences of technological growth, requiring consideration of the potential impacts that transcend a particular technology's intended and primary purposes.

Appraisals of the potential costs and benefits of new technologies have occurred throughout history. When introducing the concept of "Technology Assessment,"[2] the proposers did not imply their goals were new, but rather argued that greater effort should be devoted to making the process more systematic and comprehensive. Although few thought that assessing the future impact of technology could be reduced to a science, or even a craft with universally accepted methods, the early literature did assert that present methods and processes could be greatly improved. Hence, the rubric of "technology assessment" has come to be associated with a conscious attempt to utilize more disciplined techniques, while striving for the goal of a holistic analysis. Ideally, technology assessment should present an overview of the implications of a new technology that can provide a context for more specialized studies.

Thus, we are dealing with an area of analysis that is in an embryonic state. Although many are enthusiastic about its prospects, others are pessimistic. And even among the enthusiasts, there is no consensus about methods and occasional disagreement about goals. The breadth of examination that distinguishes technology assessment from other modes, such as economic studies, presents special intellectual and organizational problems, since in most cases it implies the necessity of multidisciplinary teams.

The research strategy of the Cornell group has had a strong "experimental" rationale. We have been conducting a series of specific assessments involving telecommunications, regarding each of them as an experiment for exploring the limits of what can be achieved in this area. While each of the technologies studied has particular policy implications, we are also concerned with those features of the analysis that are generalizable to other assessments. The project on mobile telecommunications to be described in this volume is the third specific case that we have studied; the two earlier cases involved microwave solid-state devices and video telephones.[3] In each successive case, we have attempted to broaden the range of impacts considered and to undertake a more penetrating analysis.

This, then, is the context within which we undertook the research to be described in this volume. Our examination of the extent to which social impacts can be anticipated, analyzed, and incorporated into the relevant decision-making and regulatory processes is in the mode of the so-called "technology-initiated" assessment. As a result, the starting point is a proposal for the development of a particular technology, and from this one attempts to foresee a wide range of sociological and technological impacts that could ensue. This approach contrasts with "problem-initiated" assessments that start with some social problem, such as the improvement of health services or education, and explore a wide variety of technologies for means to alleviate the problem.[4] Because many public policy issues have their origin in the effects produced by a specific technology, we felt justified in undertaking a technology-initiated approach to our assessment.

Major work on the project was underway by the spring of 1974, after an initial 18-month planning period that began with a review of developments in communication technology, possible social impacts, and relevant policy questions. We explored a range of emerging technologies, including wideband switched systems, satellite communications, videotapes, and so forth. From a large list, we eventually selected "Mobile Communications" as the subject for intensive study, having concluded this was a topic of increasing importance that was receiving too little attention by policy-oriented research groups.

There are currently several means of expanding communications to mobile units being proposed. One of the most far-reaching proposals, and the central focus of our assessment, is the "cellular system," in which a large service area is divided into small cells, each covered by low-power transmitter-receivers that can be connected to a central station by a landline or high-frequency directed electromagnetic link. As the mobile unit moves from cell to cell, a monitoring system would relay this information back to a central computer and switching structure so that communication via the appropriate local transmitter-receiver could be maintained. This advanced technology would have the capacity to provide service for a very large number of users, and it might, in the long run, make possible virtually universal deployment of mobile devices on the scale of the ordinary telephone.

At the present time the Federal Communications Commission is favoring the use of FM cellular technology as a way to accommodate widespread demand for more mobile services, but the use of other techniques has been suggested that may also increase subscriber capacity. One alternative approach is the use of narrow-band digitized voice systems, and methods are being developed that may utilize computer techniques to lower present spectrum requirements for conveying both analog and digital information and thereby accommodate more users. The exploitation of satellite techniques has also been proposed.

Some argue that the cellular method may not be needed because the alternatives may ultimately be more economical, resulting in a controversy over whether this advanced technology should actually be deployed, and if so, when and in what form. But there is no question that this system will be capable of accommodating a very large number of users. Thus, we can proceed with our analysis on the assumption that, if a high level of deployment of land mobile devices is a desirable social goal over the next two or three decades, we have the technical means to bring it about. Moreover, as will be discussed in subsequent chapters, economic factors are becoming steadily more favorable.

Because the subject of our work is telecommunications, we are dealing with an industry that has a special structure and is subject to a high level of state and federal regulation. Therefore our assessment has had to be concerned with the apparatus, substance, and process of governmental regulation. There was a need to examine the basis on which governmental regulatory decisions were made and also to explore the range of considerations that did not enter into the relevant decisions. Aware that it would be insufficient merely to point out that a certain set of implications was not considered in the regulatory process, we sought to carry out enough analysis to indicate the feasibility and advantage of extending regulatory considerations into previously neglected areas. We have also considered the implications of the present industrial structure, especially the relative absence of market mechanisms for determining investment in new technologies and the distribution of the costs of those technologies when deployed. Thus our work has involved not only the assessment of a particular technology, but also an assessment of the regulatory process associated with the telecommunications industry.

There are several reasons why we have chosen to focus our efforts on the cellular system. In the first place, plans for its deployment are more advanced than for the other sophisticated techniques being proposed. Consequently, much more information is available on the technical and economic aspects of the cellular technology than for the other methods. Furthermore, as mentioned earlier, recent regulatory decisions by the FCC have given this technology an advantageous start. Two large-scale trial systems have been authorized for Chicago and Baltimore-Washington.

Nevertheless, it should be stressed that, although some of our analysis is based on the special characteristics of cellular technology, much of it deals with general properties common to a majority of the advanced methods being proposed in the land mobile field. As a result, most of our analysis and conclusions are relevant to various means of increasing the utilization of mobile communications and are not limited to cellular technology alone. While concentrating on the latter technology, we have also examined less versatile and cheaper competing technologies that can provide alternative ways of communicating with mobile units, such as paging devices and more widespread use of "conventional" two-way radio systems.

Having designed an assessment as broad in scope as our resources would allow, we hope that our work will result in an improved understanding of:

the probable course of the technology's development;

potential uses, particularly in the private and commercial sectors;

economic and political aspects of regulation;

issues of spectrum allocation;

impact on public safety services;

the possibility of new work patterns;

potential legal problems, including privacy, liability, and individual freedom;

energy and transportation issues related to mobile communications; and

certain international issues, with emphasis upon the special case of Japan.

In appraising the results of our work, we ask the reader to keep in mind two interrelated purposes that underlie its design. One purpose was to conduct a broad assessment in the specific area of land mobile communications. The second was to carry out the analysis in such a way that an example might be provided for conducting future studies in related areas; for this reason, in the final chapter we shall discuss the various techniques and methods used in the work. Our emphasis will be on prescriptions and procedures for organizing the research, rather than the development of formal methodologies that can be applied in a relatively direct way to other studies. It will be clear after examining the text that, although formal methodologies have influenced our modes of analysis, no single technique or set of formal methodologies is applicable to all technology assessments. A critical step in the designing of any new assessment involves the careful "tailoring" of procedures, techniques, and research methods to fit the job at hand.

The material in this volume is divided into five parts, each of which presents a specific aspect of our assessment. In Part I, we review the historical, technical, and economic factors that provide the background and context for the subsequent analysis. Part II examines uses of mobile communications within particular sectors of the economy, beginning with current patterns of use but extending the analysis to include changes in those patterns that might accompany deployment of new technology. Implications of the technology's evolution that are not restricted to particular sectors are considered in Part III. Part IV contains an analysis of the future structure of the industry, along with some general speculations on the future of land mobile communications and issues associated with public policy and planning. Finally, in Part V, we provide some reflections on the assessment that we have conducted by reviewing the organization and methods used and by making some observations we hope will be useful to others undertaking a similar task.

NOTES

1. See, for example, an extensive set of essays in Philip L. Bereano, *Technology as a Social and Political Phenomenon* (New York: John Wiley and Sons, Inc., 1976).

2. National Academy of Sciences, "Technology: Processes of Assessment and Choice," Report prepared for the U.S. House of Representatives, Committee on Science and Astronautics (Washington, D.C.: U.S. Government Printing Office, 1969); National Academy of Engineering, "A Study of Technology Assessment," report prepared for the U.S. House of Representatives, Committee on Science and Astronautics (Washington, D.C.: U.S. Government Printing Office, 1969); Library of Congress, Science Policy Research Division, "Technical Information for Congress" (Washington, D.C.: U.S. Government Printing Office, 1969). Also, F. Hetman, *Society and the Assessment of Technology* (Paris: OECD, 1973).

3. Jeffrey Frey and Raymond Bowers, *Advances in Electronics and Electron Physics*, vol. 38 (New York: Academic Press, 1975), p. 147; Jeffrey Frey and Raymond Bowers, "What's Ahead for Microwaves," *IEEE Spectrum* 9 (March 1972):41-47; Raymond Bowers and Jeffrey Frey, "Technology Assessment and Microwave Diodes," *Scientific American* 226, no. 2 (February 1972):13-21; Edward M. Dickson and Raymond Bowers, *The Video Telephone* (New York: Praeger Publishers, 1973).

4. See National Academy of Sciences, "Technology: Processes of Assessment," p. 123, and National Academy of Engineering, "A Study of Technology Assessment," p. 15.

Part I
BACKGROUND
Setting
the Stage

Chapter 1
LAND MOBILE COMMUNICATIONS
The Historical Roots

Heidi Kargman

INTRODUCTION

The extent to which telecommunications influences daily life is practically immeasurable. Television sets, radios, telephones, and more recently Citizens Band radios have had a profound effect on how we live, work, and play. Modern communication technologies have permeated the organizations within our social structure. In conjunction with other technologies, especially those of transportation, they have had a major role in shaping contemporary American society.

It is not surprising that many new communication technologies have been deployed and adopted as fast as advances in electronics and industrial organization permitted, since throughout history the size and stability of nations have been affected by the coordination and sophistication of their communication networks. Natural geological boundaries often became political and cultural boundaries in societies where communication required direct person-to-person interaction.

Even though more than 90 percent of American homes and businesses are presently equipped with telephones, the demand for more versatile communication systems continues. The Citizens Band radio explosion of the mid-1970s is only the most recent example of this trend; the growth of business data transmission systems is another. Now we are at the threshold of a major new development in the technology of land mobile communications, exemplified by the proposed cellular mobile telephone system.

The changes brought about by innovations in communications have not always been immediately apparent; the effects were only gradually realized and appreciated from a historical perspective. In this chapter we will consider some dimensions of the history of modern communication in order to set the stage for our subsequent assessment of the significance of advances in mobile communication technology. Our historical review will be highly selective, emphasizing the interaction of electrical communications with mobility, their role in business and the delivery of public services, and the need to coordinate growing and competing demands for communication services. Also, we shall restrict our discussion to the American experience.[1]

THE EMERGENCE OF ELECTRICAL COMMUNICATION

Various means to convey information between distant points emerged as part of the evolution of American society. During the first century of the American colonial period, special arrangements were required to send a letter or message. Within a town or village, messengers were hired to deliver notes and return with a reply. Letters were often carried between towns by ship captains, sailors, and other regular travelers. Once at their destination, they would leave the pouch of letters at the local tavern, trusting that eventually the letters would reach the addressees. This means of communication was haphazard at best; some letters took months to reach their intended readers, if they arrived at all.

Several colonial postal services were organized during the eighteenth century. Although these systems were more reliable than the informal methods that preceded them, communication between distant parts of the country was still very slow. After the Revolutionary War, many felt that the United States was too large to be centrally governed with existing communications. President Washington sought to dispel this uneasiness in his 1796 farewell address: "Is there a doubt whether a common government can embrace so large a sphere? Let experience solve it. To listen to mere speculation in such cases were [sic] criminal . . . it is worth a fair and full experiment."[2]

The United States, of course, remained a single union, and the fledgling postal service continued to expand within it. By the standards of the day, communication and transportation in the young nation improved steadily. For example, by 1798 letters took only forty-four days to travel between Philadelphia and Nashville, and thirty-two days to go from Philadelphia to Lexington, Kentucky.[3] In general, however, communication of news and the transport of people and goods was a slow and ambitious venture, limited by the speed of horses, wagons, and boats, so that localism and regionalism remained strong.[4]

Encouraged by the success of visual signaling in Europe, a system of semaphores was set up between Martha's Vineyard and Boston in 1800 to transmit shipping news to the busy Massachusetts port. Each semaphore was a tall building with removable appendages or shuttered windows that could, in combination, represent any letter in the alphabet. Such structures were erected on hilltops approximately one to two miles apart, and each required a person on duty to receive the message from the preceding semaphore and to relay it on to the next. Although messages could be transmitted one hundred miles or more in

several minutes, the system was extremely expensive, used manpower ineffici-
ently, and could be rendered inoperable by fog or bad weather. As a result,
semaphore systems were not widely adopted in the United States, despite their
usefulness.[5]

The principle of the electromagnetic telegraph was demonstrated by Samuel
F. B. Morse in 1835, and as early as 1844 the first commercially successful
telegraph line was established between Washington, D.C., and Baltimore. In that
year, the line was used to transmit news of Polk's nomination at the Democratic
National Convention.[6] This was a revolutionary change in communication; never
before had instantaneous communication between very distant points been
possible. New telegraph companies rapidly overcame public skepticism of the
gadget, and there were more than fifty telegraph companies in the United States
by the time Western Union was founded in 1851.[7] This growth occurred in spite
of the high rates charged for sending telegraph messages (ten words sent from
Boston to Springfield, Massachusetts, cost 15¢, 25¢ to New York City; each
additional world was 2¢).[8]

As the telegraph network expanded, the posted letter and the telegraph
message became complementary rather than competing modes of communication. When actual transmission of the original written message or privacy of the
message was important, letters were sent. However, if rapid transmission of less
personal correspondence was necessary, the messages were telegraphed.[9] In
addition to the press and news reporting services, which made extensive use of
the telegraph system, much business was transacted by telegraph; many early
transmissions were for placing business orders, scheduling deliveries, and the like.

The early interaction between the railroad and the telegraph is especially
important to our assessment of mobile communication, as it provides, to some
degree, a preview of the later use of CB and dispatch radio by the trucking
industry. Development of telegraphic dispatching procedures for train scheduling
was a major factor in the rapid growth of the rail network during the second half
of the nineteenth century, because it replaced older methods that could not
prevent long and frequent delays of the trains.[10] Although use of the telegraph
met with some resistance from railroad engineers when first introduced in 1851,
its advantages over the old system were considerable, and it soon became the
standard method of dispatching.

Trains provided quick and relatively inexpensive access to the frontier regions
of the United States, facilitating increased spatial mobility and redistribution of
population.[11] Prior to the laying of the rails, migration and long distance travel
of more than one hundred miles—even by boat—were undertaken by only a very
small proportion of the population. The telegraph, by aiding the growth and
expansion of the railroads, was thus a significant element in the settlement and
development of the western frontier regions.

Because railroad dispatchers could only communicate with each other from
the train stations along the line, the need for movable communication devices
that would allow emergency transmission from any point to the nearest dispatch
office soon became apparent. Yet it was not until the Civil War that the
telegraph itself became mobile.

The field telegraphs of the Civil War were movable instruments used to establish temporary telegraph stations linking adjacent troops during battle. In 1861, the newly organized United States Signal Corps and the Army of the Potomac demonstrated the practicality of the first portable field telegraph stations. As other Union armies began establishing field telegraph wagon trains, the Signal Corps continued to improve the new communication system, and by 1862 it was possible to transmit messages even while the telegraph line was being reeled out.[12] Temporary telegraph stations were placed almost anywhere that could be reached by lines. Perhaps the precursor of modern aerial reconnaissance can be traced to the Civil War when telegraphic communication was established between a hot-air balloon and the ground to guide artillery direction.[13]

As the telegraph network expanded to other parts of the nation and the world, it became a more familiar element of American life. After the completion of the transcontinental telegraph line in 1861, there was no longer doubt about the outcome of the "full and fair experiment" that Washington had called for a century before. Improvements in communications and transportation technology had extended the reach of a "common government."[14]

Although widespread initial mistrust of the telegraph had been overcome, suspicion of other new gadgets and inventions remained. Many were as skeptical of the telephone as they had been of the telegraph. Soon after Alexander Graham Bell developed the telephone in 1876, newspaper accounts referred to the device as a new toy or novelty. Few recognized the extraordinary possibilities of this personal new communication medium, since the telegraph was commonly perceived to adequately satisfy existing needs for rapid communication.

In demonstrations that Bell and his assistant Watson arranged to publicize the telephone, its potential as an entertainment device was emphasized. Newspaper reviews of these events predicted that musical recitals and Sunday sermons might be transmitted through the telephone. Not long after, the first "telephone-newspapers" were established in the U.S. that kept residents of both urban and rural areas up-to-date on the latest news events.[15] This "radio" notion of the telephone was short-lived but laid the foundation for the quick acceptance of broadcast radio programming during the early 1900s.

One of the first to recognize the potential of the new instrument was E. T. Holmes, who ran a burglar alarm service in Boston. Little more than a year after the first intelligible words were transmitted over wires, Holmes ordered several telephones from A. G. Bell and formed the Telephone Dispatch Company to provide messenger and express services for his burglar alarm customers.[16] The earliest telephone systems required that a separate wire link each pair of telephones in the network. To connect ten subscribers so that any two could converse meant that 45 lines had to be installed, connecting each of the ten locations with the nine others. Obviously, this method becomes unwieldy as the number of subscribers is increased; in rounded figures, it would take 5,000 lines to interconnect 100 subscribers, and 500,000 lines for 1,000 subscribers. The number of individual lines required is approximately equal to one-half the square of the number of subscribers. Recognizing this problem, Holmes had the phone lines run only between his clients' establishments and his own office, where he installed the first switchboard in Boston, thereby permitting customers to converse with one another without having a direct wire connection between their telephones.[17] By 1878, the first regular telephone exchange was operating in New Haven, Connecticut.[18]

Uses of the telephone multiplied rapidly. Because the dispatching of instructions is a major means of coordination in a geographically dispersed organization, the telephone soon assumed the same importance that the telegraph had had for coordinating train schedules. Police departments were among the first organizations to extensively utilize telephones. In 1877 precinct stations of the Albany Police Department were connected with the mayor's office; the first police call box was placed on a Chicago street in 1880; and by 1917 Detroit policemen were instructed to park their autos near the telephone booths for speedy dispatch. Instead of patrolling the neighborhood, these officers waited for each set of orders to be issued by telephone; thus the telephone substantially changed the mode of police operation.[19]

Services similar to modern telemedicine projects were attempted via the early telephone network. In 1910, heart sounds were reproduced with a microphone and transmitted audibly, albeit with much distortion, over a commercial telephone line in London.[20] Though limited, the success of this experiment encouraged physicians and engineers to continue refining the technique.

Again, as in the case of telegraphy, high initial cost of the service did not prevent rapid growth. In 1880, a call from a pay phone cost 15¢ and charges for subscriber services were around $150 for 1,000 messages and $240 for 2,500 calls. That is equivalent today to a charge of 50¢ for a pay phone call and $500 for the lower subscriber service.[21] Despite high cost, poor transmission quality, and unreliable service, however, the number of telephones in the U.S. increased rapidly. By the end of 1880—four years after Bell's invention—there were 47,900 telephones in the United States.[22] This burgeoning continued through the rest of the nineteenth century; it is estimated that there were 1,356,000 telephones in the U.S. in 1900. The Bell System tightened its grip on the telephone market early in the twentieth century, and although the total number of telephones in the U.S. increased steadily from 1900 to 1930, the number of non-Bell System phones remained relatively stable after 1915.[23] By 1930, the total number of telephones had reached 20,000,000.

A number of technological developments contributed significantly to the early growth and present-day ubiquity of the telephone. We have already mentioned the development of the telephone switchboard. Automatic switching was another important advancement and evolved from the development of the dial telephone in 1889 by A. B. Strowger.[24] Delay in the advent of long-distance telephony resulted from attenuation—or the weakening of the transmission signal—in the telephone lines. A practical solution to this problem was not found until about 1900 when substantial improvements were made through the use of loading coils and attempts at amplification by electromechanical methods.[25] As the popularity of the telephone increased, the web of telephone lines and poles that appeared became a much-noted eyesore. The open wire lines were also dangerous; hurricanes and sleet storms toppled poles, froze or severed lines, and often caused serious physical damage in addition to interrupting telephone service—difficulties that motivated the efforts to use buried cables.[26]

The telephone differed functionally from the telegraph mainly in its simplicity of operation. Its capacity for voice transmission—which eliminated the need for a Morse operator—and two-way communication permitted exertion of more personal control and flexibility of use than was possible with either the telegraph or post. Before 1900 casual visits during the morning hours were common. As the telephone became more popular, the personal visit was gradually replaced by a "telephone visit." While some were pleased to be relieved of unannounced intrusions, others felt that the phones were contributing to a growing social alienation.[27] In many rural areas, the telephone afforded considerable relief from loneliness.[28]

Business activity in the expanding urban centers was also notably affected by the telephone. For instance, retail businesses were enabled to extend their marketing areas into the developing suburban areas, and the telephone provided the means for coordinating the large, geographically dispersed corporations that began to appear late in the nineteenth century.[29]

COMMUNICATION BY RADIO

By the time Guglielmo Marconi sent and received the first wireless telegraph signal in 1895, the use of electrical communication had already grown substantially. In contrast to the telephone, a need for wireless communication had been widely felt prior to its invention, and it was put to immediate use.

Early work in radio communication was preoccupied primarily with meeting two needs: communicating with and linking wireline systems to remote mobile units, such as ships and airplanes, and communicating with less accessible locations, such as islands, where laying wires was not practical. The absence of transmission cables established the wireless as the first truly mobile electrical communication technology, and its earliest applications demonstrated its potential importance in providing emergency assistance to remote and isolated areas. Among the first installations were included the establishment of wireless communication with a lightship off the coast of Fire Island, New York, in 1899, and the linking of five Hawaiian islands by radiotelegraph by 1901.[30] The first wireless communication with balloons occurred in 1908, and two years later followed messages to heavier-than-air craft.[31]

The usefulness of radio communication was especially evident in maritime communication. As early as 1901 the U.S. Navy established a wireless system, which practically replaced homing pigeons as the means of ship-to-shore communication. At the 1903 Berlin conference on international frequency allocation, two frequencies were reserved for such uses and were intended to be shared by all maritime operations. What is thought to be the first distress call on record came from a Navy relief ship in 1905. The ship "Titanic" used the wireless to summon help when it collided with an iceberg on its maiden voyage in 1912.[32]

While several demonstrations showed that voice transmission by radio was possible,[33] it remained barely practical until the invention of the audion (or vacuum tube) in 1912. Improvements in the audion permitted another trial in April 1915, when speech was transmitted 220 miles from Montauk, New York, to Wilmington, Delaware. By August, the human voice had been successfully transmitted overseas—from Washington, D.C., to Paris and Honolulu.[34] But these were only one-way transmissions; trials with two-way systems were conducted as early as 1915, and the first ship-to-shore two-way radio conversation occurred in 1922, between Deal Beach, New Jersey, and the S.S. "America."[35]

Like the wireless telegraph, early radiotelephone services were intended to provide links where wires ordinarily could not be used, primarily across large bodies of water. In February 1926, the first two-way radiotelephone conversation between the United States and England was held,[36] and in the following year a commercial radiotelephone service between the U.S. and Europe was available.[37] By December 1929, a high-seas public radiotelephone service was operating between the ship "Leviathan" and the U.S. that enabled phone calls between the ship and any Bell subscriber.[38]

The development of land mobile radiotelephone service was a relatively unimportant goal for early Bell System planners and engineers. Land mobile technological development was undertaken primarily by police departments rather than telephone companies. The first automobile radio receiver used in 1928 was the work of two patrolmen for the Detroit Police Department. Not until 1932, however, was the first vehicular mobile transmitter[39] licensed to the Bayonne, New Jersey, police.[40]

EARLY USES OF MOBILE COMMUNICATION TECHNOLOGIES

Although the earliest developments of radio and mobile communication technologies were aimed at specific applications, they were quickly adapted to suit the needs of professional organizations in a variety of fields. This section will focus on the earliest users of land mobile communications, the police, and briefly discuss its effect on their practices.

The sinking of the "Titanic" drew worldwide attention to the effectiveness of the SOS distress call and led Detroit Police Commissioner William P. Rutledge to suggest that his department develop a radio system for police communication. The proposal was approved in 1916 after the New York City police department had successfully used telegraphic spark transmitters to communicate with their harbor patrol. From 1920 to 1922 the Detroit police department experimented with radio communication under a variety of licenses, including the humorous call letters "KOP." Police radio was not yet recognized as a separate radio service, however, and in 1921 experimental station W8BNE had to provide entertainment between police calls.[41]

Partly because the earliest police radio experiments produced few technical breakthroughs, commercial broadcasting stations worked closely with the police. Bulletins were sometimes broadcast to solicit assistance in locating a lost child or a stolen automobile, and the success of these early transmissions indicated the potential benefits of police and public radio. When commercial radio stations were used to communicate intercity police business, the responding agency broadcast their reply from a different radio station originating in their own city.[42] The New York City police department used special radio receiving equipment designed by Bell Laboratories that signaled the precinct radio operator just before a police bulletin was broadcast on municipal station WNYC.[43]

Meanwhile, in Detroit, work continued on designing an automobile radio receiver for police use. Detroit patrolmen Kenneth Cox and Bob Batts succeeded in building such a receiver in 1928, and early trials confirmed its utility in apprehending burglars, car thieves, and other legal offenders.[44] During the next two years, the Cleveland and Indianapolis police departments and the Michigan state police installed radio systems similar to the one designed by Cox and Batts. As police radio was more widely used, it became evident that there was a need to respond to orders and transmit field reports from the mobile unit to the precinct. Effort was directed toward developing a two-way mobile radio system, and in 1932 the first mobile transmitter was licensed to the Bayonne, New Jersey, police department. By the following year the system was successfully in operation.[45]

These early police radio systems mark the beginning of land mobile radio. Of the many technical changes and modifications made in land mobile systems since then, one of the most significant was the switch from AM to FM radio systems. The first two-way mobile FM radio system was installed by the Connecticut state police in the late 1930s. Some of its advantages over AM radio were the ability to maintain communication even when signals were very weak, the high level of receiver sensitivity, and the superiority of its noise squelch system.[46]

One of the most often noted factors contributing to the need for better police communications was the increasing popularity of the automobile. Although automobiles were in use by police before electrical communication (in Indianapolis in 1908, for example), it was clear that some method of quickly dispatching patrol cars was badly needed. Getaway cars from robberies were often impossible to track if a police car was not alerted and on its way in minutes.

The "telephone booth" system of communication used by the Detroit police department in 1917 did speed up dispatching, but at the same time it changed the nature of patrol. Policemen did not exercise the personal judgment and autonomy that was characteristic of the officer on the "beat"—they waited until instructions were received by telephone.[47] A further change occurred with the implementation of land mobile systems; automobile patrols were now freed to roam areas outside their precinct but could still remain under the control of a central office. This led to internal reorganization of police departments and the creation of a communications branch that coordinated the mobile units of the force.[48]

RAPID GROWTH AND THE EMERGENCE OF REGULATION

In 1927, amid the confusion caused by the growing number of unregulated commercial and amateur radio stations in the U.S., Congress authorized the establishment of a new federal agency, the Federal Radio Commission (FRC). Among its assigned responsibilities, the FRC was authorized to classify radio stations, regulate radio service classes, assign radio frequencies, and regulate equipment specifications for all nonfederally owned radio stations.[49] The President of the United States, as commander-in-chief of the Armed Forces, had already been given similar responsibilities with respect to federally owned stations by the Radio Act of 1912, but it was felt that a new agency was needed to regulate activities in the private sector.

By 1930, police radio stations existed in 29 cities in the United States, and there was every indication that this number would continue to increase. In April of that year, the FRC adopted the first organized scheme for police radio broadcasting and allocated eight frequencies strictly for police use. According to the plan, patrols were signaled from headquarters by radio but the responses were transmitted by telephone.[50]

The FRC made its first reference to two-way mobile radio in its 1932 annual report. Noting the increase in police effectiveness and flexibility that mobile radios could effect, the report also cautioned that "no consideration can be given to the authorization of such a service until [a] sufficient number of frequencies becomes available."[51] Thus, as early as 1932 spectrum scarcity was identified as an important constraint, one that still plagues regulators and broadcasters today.

In 1934, the newly organized Federal Communications Commission reported that on the 11 frequencies that had been made available for police radio there were 194 municipal police radio stations, 58 state police radio stations, and a total of 5,260 police cars equipped with radio receivers.[52] A need for additional spectrum became evident during hearings held in 1936, and the following year the FCC issued Order No. 19 allocating 29 VHF channels (30.58-39.9 MHz) to law enforcement agencies.[53] The use of mobile radio by police continued to increase during the 1930s and gradually spread to fire departments and other emergency services. Utilization of the very high frequencies around 30 MHz and the development of FM significantly augmented the value of these systems.[54] Several power companies, for example, became interested in using mobile communications. In 1938, the Central Hudson Gas and Electric Company of New York installed a two-way land mobile system, and Detroit Edison followed soon after. Power utilities were allowed to operate radio systems under special emergency licenses.

Police radios were classified as an emergency service in 1939. Hundreds of new radio stations were beginning service each year, and by 1940 there were 1,027 licensees in police radio, with 6,290 stations, and 76 licensees in special emergency services, with 452 stations.[55] The phenomenal growth of two-way land mobile radio since 1940 is startling; although there were only a few thousand licensed mobile transmitters in 1940, there were 86,000 by 1948, 695,000 in 1958, and 1,390,000 in 1963.[56] These figures do not include one-way automobile radios and transistor radios, devices that are also certainly a form of land mobile communication.

The diffusion of the standard car radio also occurred in the same period. Because of the popularity of commercial radios during the 1930s, a number of radio manufacturers began installing radio receivers in automobiles. However, the average citizen was not trained in the use of the radio as police officers were, and, as Henry Petrakis notes, "agitation developed in a number of states for legislation prohibiting the use of radios in automobiles on the basis that they constituted a hazardous distraction. In the course of a coroner's investigation into the reason for auto accidents, questions were asked whether the auto contained a radio in operation at that time."[57] Over the years, that fear has gradually abated. Today the automobile radio is a common item; 72 percent of 1974 American-made cars were equipped with factory-installed radios, 7 percent with stereo tape players, and thousands of other car owners have had these devices installed after purchase of their automobiles.[58]

During the 1940s, the number of different sectors requesting spectrum for two-way mobile radio increased dramatically. Report No. 13 (1945) by the Radio Technical Planning Board of the FCC recommended mobile radio spectrum allocations for the following categories of users: police; fire departments; forestry and conservation; electric, gas, and water utilities (as "special services"); transit utilities, including buses and streetcars; special emergency services; miscellaneous radio services, including motion pictures and relay press; and railroad radio. Docket No. 6651, released May 25, 1945, contained the first definition of land mobile radio, proposing the following service classes: railroad, bus, truck, taxi, and three categories of common carrier radio—general, urban, and highway. Not until the *General Mobile Radio* (1949) proceeding, however, did the FCC officially recognize mobile radio as a new class of service.[59] This classification finally permitted land mobile radio to be made available to small businesses through an allocation to the Radio Common Carriers, and the institution of an automobile telephone service was authorized.

Development of the automobile telephone dates back to 1940 when New York Telephone made the first connection to vehicles in establishing an emergency radiotelephone service.[60] In 1946, the first overseas telephone call was made from a moving vehicle—St. Louis to Honolulu.[61] And by 1948, a fully automatic radiotelephone system was in operation in Richmond, Indiana.[62]

The need for reliable modern communications for military use during World War II contributed to the shift in emphasis from AM to FM mobile radio broadcasting. Additionally, as a result of their military training, many veterans were familiar with the installation, maintenance, and operation of radio equipment. Their availability as trained technicians during the postwar era facilitated the rapid development and growth of radio that occurred in that period. A World War II spin-off of major significance for the land mobile market was the walkie-talkie, which was widely used by taxis and quickly became a popular item with the public.[63]

Introduction of the transistor by Bell Laboratories in 1948 allowed the manufacture of smaller and more portable units with increased performance reliability. Although attention has been given to producing more compact equipment and improving the signal-to-noise ratio, the principal technical emphasis in the development of mobile radio since 1940 has been on improving spectrum utilization. Measures adopted have included the use of special codes to shorten message lengths, channel splitting, and other technical innovations.[64] During the past decade, however, advances in microelectronics, large-scale integrated circuitry, and also electronic switching and computer technology have contributed to the production of smaller, less expensive, and more reliable radio units.

Recognizing the need for a mobile communication service for general citizen use, the FCC in 1949 created the Citizens Band Radio Service (CB); but, because of restrictions on transmission type and the high cost of CB equipment, initial growth of the CB was extremely limited. In 1958, to stimulate growth, the FCC allocated twenty-three additional channels for CB use in the 26.96- to 27.26-MHz band. Equipment used to broadcast in this range being much less expensive than that required for the original service, the number of licensed CB stations rose correspondingly;[65] the number of licensees grew from 49,000 in 1959 to nearly a million by the early seventies. A continuing increase in popularity of the transistor radio, walkie-talkie, and CB radio has since prepared the way for the explosive growth in the use of the Citizens Band during the mid-seventies, a phenomenon that will be discussed in detail in a later chapter.

The Citizens Band radio involves a broadcast party line and is not suited for communication of a confidential nature. Compared to CB, the mobile telephone provides considerable privacy, in addition to offering interconnection with a vast number of nonsubscribers to the mobile telephone service—namely all telephone subscribers. The trend toward greater reliance on communication and the demand for more sophisticated systems is reflected in the changing attitudes toward mobile telephone service. In the early sixties, even as mobile telephone systems were spreading across the country, the high cost of the mobile equipment and the poor quality of the service were considered to be the main factors limiting mobile telephone growth.[66] Yet today, in major American urban centers, the shortage of available radiotelephone channels is the primary limitation, since demand has surpassed service supply despite the persistence of poor quality service.

TAXONOMY OF USES AND SERVICES

The increased diversity and magnitude of land mobile communications discussed in this chapter has required that the spectrum allocated for their use be continually increased as well. As will be discussed more fully in the next chapter, the FCC had to supplement the early allocations below 50 MHz with new allotments near 150 MHz and 450 MHz.

Current uses of land mobile communications are so diverse and multiple in character that it is not possible to define a set of functions for them that are mutually exclusive. Nevertheless, in subsequent chapters we will find it helpful to make the following partial distinctions concerning *mode of use:*

paging: a basic function consisting of signaling a person that there is a message waiting to be received by some other means of communication.

dynamic routing: the capability of redirecting personnel and vehicles after they have left their base of operation, thus reducing response time of units deployed in the field.

vehicle monitoring: keeping track of the location of vehicles in the field in order to provide more efficient deployment.

emergency beaconing: notification of a base station in the case of emergency by specialized one-way beacon systems or by accessories to two-way systems. Such systems can involve automatic or manually activated signals.

data transmission: provision of data received on mobile printout devices to facilitate transactions.

conversation: the verbal exchange of information for business or personal objectives requiring participation by two or more individuals.

A set of telecommunications *services* have evolved that facilitate the modes of use described above. These services have different characteristics and the technology required for providing them also differs. The principal categories are:

paging service:[67] the provision of one-way communications that permit an extremely brief message to be transmitted to a recipient. Frequently, this message is no more than a signal to the recipient to call a prearranged number, though simple voice messages are possible. Some paging services involve interconnection with the wireline telephone system, permitting direct access to mobile units from a telephone.

dispatch service: service providing two-way communication frequently used by taxis and ambulances, as well as various units of commercial organizations. The messages are brief and often consist of instructions concerning deployment. A few channels can accommodate a very large number of users.

mobile telephone service: a means of communication with characteristics comparable to those of the fixed telephone. Interconnection with the fixed telephone system is an essential feature. The conversation can be extended and have a higher degree of privacy than is possible with a dispatch service.

Citizens Band service: service based on a broadcast system of limited range. Use of the system is not restricted to a particular class of users. Essentially, it provides a very large number of broadcasters with a "party line."

While each of these services can be used to perform more than one of the functions listed earlier, they are not equally efficient or convenient for any given use. The amount of spectrum required per user varies greatly from one service to another, and the nature of the technology required to provide the service also varies. In the next chapter we shall review the technological means for providing these services.

SOME GENERAL CONCLUSIONS

In this chapter we have traced the evolution of land mobile communication from the invention of primitive devices useful for specialized purposes to the deployment of systems able to provide a wide range of services. We have noted that with each major communication advance of special relevance to our present assessment—the telegraph, telephone, and land mobile radio—high initial cost of the service did not prevent an impressive growth rate for the new technologies shortly after their introduction. Even though the early devices were not especially convenient to use, many found that the increased speed of communication more than compensated for any inconvenience and costliness. As happens with many important new inventions, only a few could see the potential of most of these particular technologies; many more asked, "who needs it?" With each innovation there was a period of resistance to the new technology, during which the existing communication systems were considered to adequately satisfy current communication demands. After the initial skepticism had been overcome, however, popular acceptance of the technology proceeded with surprising rapidity.

Another point worth noting concerns the interactive and synergistic character of the new technologies. Rather than replacing the posted letter, for example, the early telegraph was used to meet demands for speedy communication that the letter could not satisfy. The railroad and the telegraph were complementary: the telegraph fostered westward expansion of the railroad, and the train tracks provided a convenient right-of-way for the stringing of telegraph lines. Historically, there has been this special interaction between communication and transportation. Increased use of the automobile in crime and its prevention provided the impetus for the police force to develop land mobile radio. Furthermore, communication technologies have been used steadily for coordination and control in the business and public sectors, as well as for personal purposes.

In light of this history, an investigation of any new communication system might consider the following themes: that few originally saw the potential of the new technology; that the technologies were adopted in spite of high initial costs and poor service quality; that they often interacted in a complementary manner with other new and existing technologies; and that coordination and control has consistently been a major function of each of these communication systems. This is not to imply that the impact of communication technology on our society has been beneficial in every instance, nor that proposals for new telecommunication technology will necessarily serve positive social purposes and therefore should be immediately adopted. In some cases, the answer to the question "who needs it?" might be "almost nobody."

This historical introduction is intended to provide perspective for our subsequent assessment of the potential social impact of new developments in land mobile communication technology. Past experience with the telephone, telegraph, and radio provides significant clues concerning the potential impact of the newly proposed technologies. If history is our guide, the impacts will be significant and opinions will differ as to whether, on balance, they will affect our lives in a positive or negative way.

Finally, we should note that the rapid growth in the number of land mobile devices is a relatively recent phenomenon, occurring in the period since World War II. Despite this growth, however, they are still relatively uncommon compared with the telephone, radio, and television. Yet the potential for massive growth in the future is real; in many respects, the area of land mobile communications has been charted and explored, but it is still sparsely populated.

NOTES

1. For a more comprehensive history of modern communication, see Anthony R. Michaelis, *From Semaphore to Satellite* (Geneva: International Telecommunications Union, 1965).

2. Wayne E. Fuller, *The American Mail* (Chicago: University of Chicago Press, 1973), p. 80.

3. Ibid.

4. Carl Bridenbaugh, *Cities in Revolt* (New York: Knopf, 1955), p. 55; Charles Frederick Carter, *When Railroads Were New* (New York: H. Holt & Co., 1909), pp. 33-36.

5. Two significant exceptions to this were the visual signaling by the railroads, which was in use until very recently, and the colorful signal flags used by sea-going vessels. For additional examples of early semaphore and visual signaling systems, see Michaelis, *From Semaphore to Satellite.*

6. Joseph Newman, ed., *Wiring the World* (Washington, D.C.: U.S. News and World Report, 1971); and Malcolm Macdonald Willey and Stuart A. Rice, *Communication Agencies and Social Life* (New York: McGraw-Hill, Inc., 1933), pp. 122-23.

7. Clayton C. Mau, "The Early History of the Telegraph in the United States" (Ph.D. dissertation, Cornell University, 1930), p. 185.

8. New York and Boston Telegraph Association, *Tariff of Charges* (New York: Chatterton and Crist, 1946), p. 10.

9. Willey and Rice, *Communication Agencies* , p. 123.

10. Ibid.; and Carter, *When Railroads Were New,* pp. 103-5.

11. Thomas C. Cochran, "The Social Impact of the Railroad," in *The Railroad and the Space Program: An Exploration in Historical Analogy,* ed. Bruce Mazlish (Cambridge, Mass.: MIT Press, 1965).

12. J. Willard Brown, *The Signal Corps U.S.A. in the War of the Rebellion* (Boston: U.S. Veteran Signal Corps Association, 1896), p. 177.

13. John Emmet O'Brien, *Telegraphing in Battle* (Scranton, Pa.: The Raeder Press, 1910), pp. 82-83; and E. W. Chapin, "The Past and Future Techniques of Vehicular Communications," *IRE Transactions on Vehicular Communication* VC-9, no. 2 (August 1960):6.

14. Newman, *Wiring the World.*

15. Sidney H. Aronson, "The Sociology of the Telephone," *International Journal of Comparative Sociology* 12, no. 3 (September 1971):160.

16. M. D. Fagan, ed., *A History of Engineering and Science in the Bell System* (Bell Telephone Laboratories, Inc., 1975), p. 20; and Frederick Leland Rhodes, *Beginnings of Telephony* (New York: Harper and Brothers, 1929), p. 147.

17. Rhodes, *Beginnings of Telephony.*

18. Fagan, *A History of Engineering.*

19. Daniel E. Noble, "The History of Land-Mobile Radio Communications," *Proceedings of the IRE* 50, no. 5 (May 1962):1405.

20. H. A. Frederick and H. F. Dodge, " 'The Stethophone,' An Electrical Stethoscope," *The Bell System Technical Journal,* October 1924, p. 531.

21. John Brooks, *Telephone* (New York: Harper & Row, 1976), p. 81; and U.S. Department of Commerce, Bureau of the Census, *Historical Statistics of the United States; Supplement to the Annual Statistical Abstracts* (Washington, D.C.: U.S. Government Printing Office, 1976).

22. Newman, *Wiring the World,* p. 181.

23. Willey and Rice, *Communication Agencies.*

24. Fagan, *A History of Engineering,* pp. 484, 544-54.

25. Ibid., pp. 232-56.

26. Ibid., pp. 225-27.

27. Robert Straughton Lynd and Helen Merrell Lynd, *Middletown* (New York: Harcourt, Brace & Co., 1929), p. 275.

28. Arthur Pound, *The Telephone Idea* (New York: Greenery Publisher, 1926).

29. Willey and Rice, *Communication Agencies.*

30. Newman, *Wiring the World.*

31. Chapin, "Past and Future Techniques," p. 6.

32. Newman, *Wiring the World;* and Fagan, *A History of Engineering,* p. 383.

33. John F. Mason, "Dr. Ernst Alexanderson: A Pioneer to Remember," *Electronic Design* 23, no. 13 (June 21, 1975):68-72.

34. R. A. Heising, "Radio Extension Links to the Telephone System," *Bell System Technical Journal* 19 (October 1940):611-46; and Fagan, *A History of Engineering.*

35. Newman, *Wiring the World;* and W. Wilson and L. Espenschied, "Radio Telephone Service to Ships at Sea," *Bell System Technical Journal* 9 (July 1930): pp. 407-28.

36. Heising, "Radio Extension Links," p. 611.

37. Newman, *Wiring the World.*

38. Wilson and Espenschied, "Radio Telephone Service."

39. Noble, "History of Land-Mobile."

40. Motorola, Inc., "Reply Comments," Formal submission to FCC Docket 18262, Motorola, Inc., Schaumburg, Ill., July 20, 1972, Appendix A.

41. Noble, "History of Land-Mobile"; Chapin, "The Past and Future Techniques"; Motorola, Inc., "Reply Comments."

42. Motorola, Inc., "Reply Comments."

43. S. E. Anderson, "Radio Signalling System for the New York Police Department," *Bell System Technical Journal,* October 1926, p. 529.

44. Noble, "History of Land Mobile."

45. Ibid.; Motorola, Inc., "Reply Comments."

46. Noble, "History of Land-Mobile."

47. Ibid.

48. Chapin, "The Past and Future Techniques."

49. U.S. Congress, *Public Law no. 632,* 69th Congress, 2nd sess., February 23, 1927.

50. Motorola, Inc., "Reply Comments," p. A-1; and Stephen E. Parker, "From 8BNE to WCK: An Early History of the Detroit Police Department Radio System," *APCO Bulletin,* Special Edition, February 1971.

51. Motorola, Inc., "Reply Comments," p. A-3.

52. Ibid.

53. Noble, "History of Land-Mobile"; and Marce Eleccion, "Electronics in Law Enforcement," *IEEE Spectrum* 10, no. 2 (February 1973):33.

54. Chapin, "The Past and Future Techniques."

55. Motorola, Inc., "Reply Comments."

56. Noble, "History of Land-Mobile."

57. Harry M. Petrakis, *The Founder's Touch: The Life of Paul Galvin of Motorola* (New York: McGraw-Hill, 1965), p. 100.

58. Motor Vehicle Manufacturers Association, *1975 Automobile Facts and Figures* (Detroit, Mich.: Motor Vehicle Manufacturers Association, 1975), p. 19.

59. FCC Docket No. 8658, April 1949; and Motorola, Inc., "Reply Comments."

60. David Talley, "A Prognosis of Mobile Telephone Communications," *IRE Transactions on Vehicular Communication* VC-11, no. 1 (August 1962):27.

61. Newman, *Wiring the World.*

62. Ramsey McDonald, " 'Dial Direct' Automatic Radiotelephone System," *IRE Transactions on Vehicular Communication* PGVC-11 (July 1958):80.

63. Motorola, Inc., "Reply Comments."

64. Newman, *Wiring the World;* and Motorola, Inc., "Reply Comments."

65. Carlos Roberts, "Two-way Radio Communications Systems for Use by the General Public" (Master's thesis, University of Colorado, 1975).

66. Talley, *Prognosis of Mobile,* p. 36.

67. The reader should note that the nomenclature is far from satisfactory because words such as "paging" and "dispatch" are frequently used to refer to a use, a service, or a technology. The context in which the word is used is often the only guide to clarifying the reference.

Chapter 2
TECHNOLOGIES FOR LAND MOBILE COMMUNICATIONS 900-MHz Systems

Jeffrey Frey
Alfred M. Lee

INTRODUCTION

In this chapter we shall examine the technologies that are capable of providing the services previously described. While we shall review some general principles that apply to devices and systems irrespective of the frequency range in which they operate, most of our attention will be focused on technologies suited for the 900-MHz band. We emphasize such systems because recent FCC decisions now permit a rapid growth of land mobile services in the band. Briefly, in Docket 18262 the FCC allocated 115 MHz to land mobile use, quadrupling the amount of spectrum available for that use. Forty MHz were allocated to high-capacity common user systems employing cellular technology, 30 MHz to dispatch use employing current technologies, and 45 MHz were held in reserve. These FCC decisions will be discussed in detail in the next chapter.

We shall concentrate on system technologies; for reasons of brevity, little attention is given to emerging devices and component technologies (e.g., microprocessors, charge-coupled devices, alternative transmission techniques). At the outset, however, we wish to emphasize that advances in the latter will play an important role in reducing costs and stimulating mobile radio growth.

Various radio technologies have been proposed or are available for sending messages to and receiving messages from dispersed, mobile units. These technologies differ in the nature and quality of service they provide. A simple taxonomy of mobile communication technologies can be based on the following categories: one-way paging, conventional, multichannel trunked, cellular. Citizens Band radio is an additional category that is currently less suited for professional and business uses. Other methods are being proposed that would significantly increase the capacity of land mobile systems, but we shall restrict our discussion to those systems described in Docket 18262. The nature of the systems and the services they make possible are as follows.

ONE-WAY PAGING SYSTEMS.   A central unit (base station) is used to transmit a signal to a portable unit to inform that unit's possessor that someone wants to communicate with him. The signal received by a mobile unit can activate a tone or light and might include a brief verbal message. Some paging systems employ automatic interconnection with the fixed telephone system, giving a person the ability to initiate a "page" from a standard telephone. There is no provision for communication from the paged unit back to the base station; the recipient of the signal or page uses the standard telephone in order to respond. The operation of one-way paging systems is schematically illustrated in Figure 2.1.

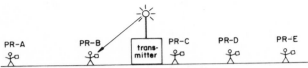

Figure 2.1: *Functional diagram of one-way paging system. The paging message is actually transmitted to all page receivers (PRs), but a signal code triggers only one unit, e.g., (B), to "accept" the message.*

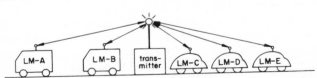

Figure 2.2: *Functional diagram of simplest conventional system. A human dispatcher indicates by voice the land mobile (LM) unit he wishes to contact. All LMs may overhear messages between any LM and dispatcher; communication among LMs usually proceeds through dispatcher.*

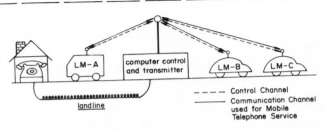

Figure 2.3: *Functional diagram of multichannel trunked system used for mobile telephone service. If unit A wishes to communicate with a fixed telephone, it requests and is assigned a free channel; it communicates over that channel plus landline to the fixed telephone. Communication between mobile units can be achieved by an analogous process.*

CONVENTIONAL SYSTEMS. Conventional systems are two-way broadcast systems that are usually operated in a simplex mode: they have one voice path available for use in the communication, so that signals cannot be transmitted and received at the same time. A conventional system is schematically illustrated in Figure 2.2. Many mobile units from one or more organizations are usually assigned a single frequency; if more than one mobile unit wants to transmit, they must join a queue and wait until the channel is free. Conventional systems are widely used for dispatching delivery vehicles and taxis, aiding emergency services, and directing field operations and personnel.

MULTICHANNEL TRUNKED SYSTEMS. Multichannel trunked systems (MCTS) allow users access to several two-way channels; users can search among this "bank" of channels for a free one when they want to communicate. An MCTS permits a more efficient use of allocated channels than would be possible if each channel were dedicated to a single organization or user. Waiting time can be greatly reduced or, alternatively, the number of users that can be accommodated per channel can be increased. Trunked systems can be used to provide all the services currently available on conventional systems, including dispatch services, although no such "all-service," systems are authorized now. The use of a multichannel trunked system in a mobile telephone application is schematically illustrated in Figure 2.3.

CELLULAR SYSTEMS. In cellular systems a large service area is divided into smaller areas, or cells. Within each cell, a subsystem functionally similar to a multichannel trunked system is operated; the base station transmitters in adjacent cells operate on different sets of frequencies to avoid cochannel interference. A possible arrangement of cells for the Philadelphia area is shown in Figure 2.4. Seven different sets of frequencies are reused many times in nonadjacent cells to cover the limited geographical area—multiple reuse in a market area is made possible by limiting transmitter power and antenna height and by maintaining adequate distance between base stations. The fundamental advantage of the cellular approach is that more users can be accommodated than with any other mobile communication technology, as long as the users are not concentrated in one cell most of the time. However, cellular systems require an elaborate technological design.

In addition to the basic radio component of MCTS, a cellular system requires facilities for identifying the cell that contains an activated mobile unit, and for automatically switching duplex channel frequencies and transmitter stations as the mobile moves from cell to cell. When a caller wants to contact a mobile unit, a search is conducted by the central processing computer for the location of the called mobile unit. Once the unit's cell is identified, a signal is transmitted through wirelines to the appropriate base station; a channel in the proper cell is then assigned by the trunking computer, and the message is radioed to the mobile unit. This operation is illustrated in Figure 2.5.

CITIZENS BAND. Citizens Band, the most ubiquitous mobile communication technology, has many features similar to those of conventional systems. A

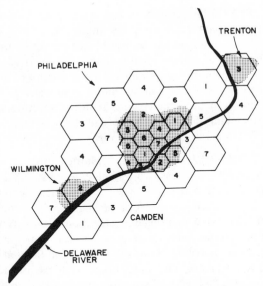

Figure 2.4: *A possible cellular system for the Philadelphia area. Seven sets of frequencies are used, with the set in use in each cell identified by the number in the cell.*

SOURCE: Adapted from Bell Laboratories, "High-Capacity Mobile Telephone Service Technical Report," Unpublished document, Bell Laboratories, Holmdel, N.J., December 1971, figure 5-3.

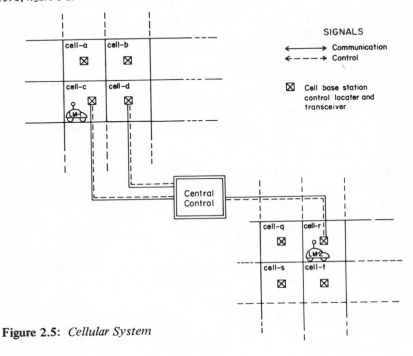

Figure 2.5: *Cellular System*

total of 40 channels is now available, although one of these is supposed to be restricted to emergency use. While CB technology can be used for business functions, the delays often incurred (for example, by spectrum congestion), and the lack of privacy have made CB a relatively inefficient means for performing these functions. Consequently, CB is primarily used for private and entertainment purposes, for which wait time, privacy, and interference are less important parameters.

Before turning to a detailed technical analysis of these systems, it is desirable to consider some technical features that are important to several of the systems and that will later aid us in differentiating between and comparing the various systems under discussion.

GENERAL TECHNICAL ASPECTS OF MOBILE COMMUNICATIONS

All systems must be so designed as to take into account the technical factors that determine particular qualities of message transmission. The technical determinants of transmission range, interference potential, error detection (for digital systems), waiting time to use a system, spectral efficiency, and privacy must all be considered.

Range of Service

The range of radio services depends upon many factors, including transmitter power, antenna height, receiver sensitivity, ambient noise, terrain, antenna gain, and carrier frequency. For instance, the signal strength at a given distance from a high-frequency transmitter will usually be less than that of a much lower frequency transmitter with equal output power and antenna gain. This is true for all of the new 900-MHz systems under consideration. The range of a base station can be extended by the use of repeaters, which can also be used to amplify and relay signals from relatively weak mobile units back to base stations; relaying via the base makes mobile-to-mobile communication possible with a high degree of reception probability. However, direct mobile-to-mobile communication is usually very difficult across a large market area because of the limited power and antenna height of mobile transmitters.

Message Type and Interference

Interference potential and error detection are dependent upon the type of message and the transmission system used. Four possible combinations for mobile communications are:

1. analog signals on analog transmission systems
2. digital signals on analog transmission systems
3. analog signals on digital transmission systems
4. digital signals on digital transmission systems

Analog messages are the signal type most commonly transmitted over radio channels. This message format includes voice and music and is currently used in systems operated by public safety and service organizations, such as police and fire departments, and by businesses involving roving units, such as taxicab companies and service firms. Transmission of data, such as printed or video displayed text, is becoming an increasingly popular alternative, however. Automatic reception and decoding of digital information by the mobile unit displays allows the user to perform other necessary tasks at the same time that fairly detailed messages are received and stored for later examination. Both analog and digital information can be transmitted over all of the proposed 900-MHz systems.

Voice and data messages can be relayed by either of two methods. The predominant transmission system type is currently analog, where voice or data signals are linearly modulated according to a time-varying, continuous wave form. At least one study has shown that the most efficient way to send voice messages over analog channels is by narrow-band frequency modulation with channel spacing of 20 to 30 kHz.[1]

Transmission of digital information over analog channels requires the use of modulation methods that convert input data to streams of pulses, either of differing frequencies transmitted at uniform intervals or of uniform frequencies transmitted at varying intervals. In the first case a receiver must sense the order of frequencies transmitted; in the second, the phase difference or time delay between pulses must be measured. That information must then be decoded so that data can be presented to the mobile user on a video screen or printer. It has been argued that channel occupancy would be much shorter per communication and spectral utilization improved if message types were digital rather than vocal over analog transmission systems.[2]

An alternative way to send voice and data messages is by using a digital transmission system designed to relay only a particular, limited set of codes. Signals received on digital systems are compared and interpreted only according to this set. Analog signals must be converted to digital form when using a digital transmission system and then must be decoded by the receiver to recreate the audio message. Digital transmission systems can tolerate a lower signal-to-transmission noise ratio but may in return require more bandwidth than analog systems. Some authorities argue that digital voice transmissions offer inferior performance.

All mobile communication systems are sometimes subject to interference that causes delays in the arrival time of signals, producing shadow signals, or that changes—through the Doppler effect—the frequencies of signals received. Voice messages can easily be repeated if they are garbled or when errors are detected. In order to minimize errors in digital messages due to reception irregularities, two methods are commonly used: digital messages are often automatically repeated numerous times and compared in the receiver; also, error detecting or correcting codes are used that insert checking data into specific places in a pulse stream. Both message redundancy and the transmission of additional checking data reduce the spectral efficiency of sending data messages over analog mobile radio links.

As reflected by Docket 18262, the FCC currently favors the use of FM cellular technology as a means of accommodating widespread demand for more mobile services, but other techniques that may also increase subscriber capacities have been suggested. For example, narrow-band digitized voice systems are an alternative approach; such systems are now feasible because of recent advances in waveform coding, multiplexing, and solid-state technology. Several companies currently build hand-held portables utilizing digital voice techniques for military applications, but this equipment is not yet "type approved" by the FCC for general public use. Technology that utilizes computation to lower present bandwidth requirements for conveying both analog and digital information is also being developed. Linear predictive coding techniques make intelligible communication of speech possible over digital systems, with lower bandwidth requirements than current systems. Yet some authorities argue that the quality of voice transmission over synthesizing speech systems will be inferior to that utilizing full-width telephone channels.

While digital voice equipment is now more expensive than FM equipment, it is likely that the cost differential will decrease. Since new digital techniques are expected to reduce bandwidth requirements even further than presently possible, it has been argued that the installation of digital transmission systems would allow new advances to be more easily accommodated. Such technological alternatives will have to be considered as the evolution of land mobile systems proceeds.

Coding: Privacy and Shared Systems

Privacy of conversations may be enhanced in various ways, although absolute security from eavesdropping is impossible to achieve—the privacy attainable is relative and partial when one base station serves a large fleet of mobile units. If a station is assigned only one or a few frequencies, all of the mobiles sharing a channel would normally be able to hear a call to any one of them. However, relatively inexpensive units are now available to impress signaling codes on the transmitted message. These codes are sensed by special units in the mobile receivers, and if the transmitted code corresponds to the code assigned to a given receiver, that receiver will automatically activate its loudspeaker to present the call. This signaling device increases selective calling only within the system; anyone with a noncoded receiver tuned to the assigned frequency or with a scanning set can overhear all transmissions.

Privacy can also be enhanced by the use of scramblers. These devices utilize various techniques, including frequency inversion and tone masking, to prevent unauthorized eavesdroppers from understanding the transmission. Normally only base stations and mobiles equipped with the proper devices may encode and decode these communications. Even with these sophisticated electronic components, however, absolute privacy cannot be ensured; a determined eavesdropper with adequate knowledge and technical resources can decode virtually any transmission. Scrambler systems vary in price according to the complexity of the modulation techniques employed. An economical system will cost about $600 for the base station and $300 for each mobile. More secure systems are available that cost over $1,000 for base stations and $600 for each mobile.

We have described above how coding methods can be used to allow a base station to call individually one of the mobile units in the fleet that it serves. The use of coding is therefore advantageous in systems operated by large organizations with many mobile units. But coding can also be beneficial for shared systems and community repeaters in which several organizations, possibly too small to operate mobile systems on their own, join together to share a channel and the cost of transmission equipment. Community repeaters are set up by the "community"—organizations licensed to use a frequency pair—in order to increase range and improve quality of service. While each user of these systems must be licensed to operate on the repeater's frequency, this communal action allows participation with a reduced capital expenditure per user.

Spectral Efficiency

In comparing mobile communication systems, it is useful to have a measure of how efficiently each system uses the spectrum allotted to it. Admittedly, comparisons between diverse systems offering markedly different services, such as one-way paging and mobile telephone service, can be very misleading; thus, choice of a single spectral efficiency measure is controversial. Nevertheless, measures of spectral efficiency do give an indication of the number of prospective customers that may be served by a given type of technology. Possible measures of spectral efficiency include:

A. mobiles/megahertz
B. mobiles/channel
C. Erlangs/megahertz
D. Erlangs/channel
E. Erlangs/megahertz/square mile
F. mobiles/megahertz/square mile
    If the width of a channel in MHz is known, (A) and (B), and similarly (C) and (D), become equivalent measures.

(A) and (B) are adequate gauges of spectral efficiency if the systems under comparison have similar specifications for service parameters. For instance, various tone-voice paging systems can be compared if maximum message length is restricted by technological design. But the use of (A) and (B) has been criticized[3] because the mobiles of two systems being compared may not generate the same traffic. Mobiles in one system, for example, may be used much less than those in another due to pricing schemes that penalize long messages.

Instead of using indicators that reflect only the number of mobile units that can be accommodated on a channel, we may choose measures that reflect the amount of speaking that the channel can support; measures (C) and (D) are suitable for this purpose. The traditional unit of telephone traffic is the Erlang,[4] a dimensionless quantity defined as follows:

$$\text{Traffic Intensity} = \frac{t_d \times n_c}{60} \text{ Erlangs} \tag{1}$$

where t_d = average call duration in minutes,
n_c = average number of calls/hour.

A channel carries one Erlang of traffic if it is occupied 100 percent of the time. Alternatively, a user making three calls per hour, lasting 10 minutes each, will generate 0.5 Erlangs of traffic. Since the Erlangs/MHz or Erlang/channel measures of spectral use compensate for the possibility of unequal traffic generation by individual mobiles in different systems, it has been argued that these are superior measures of spectral use.[5]

The addition of the "geographic area" term in (E) and (F) allows inclusion of the effects of frequency reuse within relatively small areas and is particularly appropriate for cellular systems. Since no one measure of spectral efficiency is adequate for all of the diverse systems we are considering, we shall use several of these measures, choosing those that are appropriate for a particular context in our subsequent technical discussion.

Service Characteristics and Wait Times

Other comparisons between diverse services, such as paging and mobile telephone, are difficult because of the great differences in the nature of these services. For instance, paging is only a one-way system, and this is an unquantifiable "quality restriction." Also, paging customers, sent much shorter messages than mobile telephone customers are, so that the traffic intensity per paging unit is much smaller than that per mobile telephone. Another difference is that paging customers are not conscious of a wait time in receiving their messages. The mobile telephone user, on the other hand, would probably be annoyed if required to wait several minutes for a dial tone. In fact, one of the reasons for the FCC action in Docket 18262 was in response to complaints by present mobile telephone users who often wait one or two minutes for a "dial tone" or free channel.

We can compare the differences in the amount of time each subscriber must wait for free equipment or a free channel for various types of services. A measure of this delay, or waiting time, is related to the Blocking Probability, B, which is the probability that a caller would find all channels occupied. The theory of probability has been used to derive relationships among blocking probability, number of channels, and traffic intensity. This derivation makes use of experiential evidence on the nature of telephone calls, such as distribution in time of calls placed, calling rate, holding time variation, and customer behavior upon encountering occupied facilities.[6] The resulting theory yields formulas for the probability of losing a call as a function of the number of channels and traffic intensity. Alternatively, these formulas can be used in engineering studies to tabulate the total traffic intensity that can be accommodated as a function of the number of channels (or trunks) and blocking probability. A graphical presentation of typical results is shown in Figure 2.6.

Approximate answers to traffic problems can be obtained without recourse to this elaborate analysis based on probability theory. One can use a nonprobabilistic "worst case" approach in which each subscriber is assumed to experience, on the average, a maximum possible wait time for which a service is designed. Using this approach, N, the number of subscribers that can be accommodated per channel, is

$$N \cong \rho\,(W,\tau) \times \frac{1}{\text{fraction of time sub-}} , \tag{2}$$
$$\text{scriber is using unit}$$

where ρ is the carried load, W is the maximum wait time, and τ is the duration of each call. Various organizations use different expressions for estimating ρ, but when W/τ is much greater than 1, ρ is roughly equal to 1 Erlang.[7]

Communication systems engineers have, through experience, arrived at values for acceptable service parameters for various types of communications, some of which have become virtual standards for the industry. Typical values are summarized in Table 2.1. We shall use the acceptable performance data from this table, the appropriate formulas above, Figure 2.6, and other information to compare the spectral efficiency of various mobile communication systems.

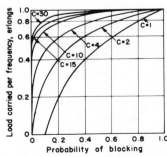

Figure 2.6: *Radio traffic carried as a function of blocking probability for various degrees of trunking. (C = frequency channels per mobile.)*

SOURCE: Adapted from H. Staras and L. Schiff, "Spectrum Conservation in the Land Mobile Radio Services," **IEEE Spectrum 8,** no. 7 (July 1971), figure 4.

Table 2.1: *Acceptable Service Parameters for Mobile Communication Service*

	Paging (BHU−2 pages)		Dispatch	Improved Mobile Telephone Service	Mobile Telephone Service (goal)
	5-digit tone only	10-second message			
Traffic intensity (Erlangs)	0.0002	0.005	0.004	0.03	0.03
Average message length (seconds)	0.42	10	30	140	140
Blocking probability	a	b	0.67	0.65c	0.02

SOURCE: Adapted from Richard N. Lane, "Spectral and Economic Efficiencies of Land Mobile Radio Systems," **IEEE Transactions on Vehicular Technology** VT-22 (November 1973):93.

a. A five-digit, tone only signal may require a wait time of 300 seconds.
b. The wait time for a 10-second message may be 300 seconds.
c. Bell Laboratories, "High-Capacity Mobile Telephone Service Technical Report," Unpublished document, Bell Laboratories, Holmdel, N.J., December 1971. The figure of 0.65 is too high to provide completely satisfactory service, but it is tolerated in current service.

THE TECHNOLOGY OF NONCELLULAR SYSTEMS

Conventional Radio Systems

Conventional radio technology is used in most of the systems operated by public and private organizations. The FCC categorizes users of conventional systems, called "private land mobile systems," into several groups including: industrial (plumbers, contractors, etc.); public safety (police, fire, etc.); and land transportation (trucking firms, railroads, taxicabs, etc.). This categorization reflects priority differences in the assignment of spectrum. Allocation of spectrum is also affected by channel-loading characteristics (i.e., the number of mobile units to be served by each channel).

Each organization with a land mobile authorization is assigned one or more channels on which to transmit and receive messages; transmissions are usually simplex and use FM broadcast techniques. The technology of conventional base stations and mobile units differs little in complexity from that of a commercial FM radio station. Anyone equipped with a receiver tunable to the right frequencies can overhear the communications and, in fact, "scanner" receivers are available for under $200 that automatically scan all the public-service frequencies and stop at any one on which a transmission is being made.

In the past, growing demand for mobile radio services has been met by increased allocations of spectrum to those services. As the technology developed, operation at higher frequencies became economically feasible; allocation of

Table 2.2: *FCC Loading Guidelines for Channel Allocation*

Service	Mobiles/Channel
Transportation	160
Business	100
Emergency	60

SOURCE: "Nature of Inquiry," Docket 18262, **FCC Reports**, 2nd series, vol. 14, Appendix, p. 26.

higher bands became necessary as utilization of lower bands for a variety of purposes prevented expansion of mobile services at the lower frequencies. Consequently, the FCC set aside frequencies in the ranges of 150 MHz and 450 MHz to complement the original allocations near 27.4, 35.9, and 43 MHz. These allocations are commonly described in the following way: "low band" refers to frequencies below 50 MHz, "high band" to frequencies near 150 MHz, and "UHF band" (for ultrahigh frequency) to frequencies near 450 MHz. In these three bands, the total spectrum allocated to public safety, industrial, and transportation in use is 39 MHz. An additional allocation of more than 3 MHz is dedicated to domestic public radio (radiotelephone) service. This allocation is divided between wireline telephone companies and radio common carriers who provide mobile telephone and paging services through interconnection with the wired network.[8]

Prior to the reallocations specified in Docket 18262 (which will be discussed in the next chapter), the provision of spectrum for conventional systems was not adequate to meet the demand for such services with an acceptable level of service quality. Since there has been little interservice sharing of blocks of channels among widely differing organizations, spectrum congestion has been severe in certain geographic regions, particularly urban areas. In Docket 18262, the FCC allocated an additional 30 MHz in the 900-MHz band for conventional systems to be shared with multichanneled trunked systems. Table 2.2 shows the loading guidelines specified for the various types of mobile services based on five-channel systems. Thus, a community that operates 180 police cars would, by these guidelines, be allocated three radio channels for its police services.

The various bands afford differing technological advantages and disadvantages.[9] The low band is subject to interference from faraway stations on the same frequency because of skip—the reflection of radio signals from the ionosphere. Radio waves in higher frequency bands, including the 900-MHz band, pass through the ionosphere and are therefore not subject to skip. Receivers for the low band are also sensitive to interference from automobile ignition systems and other sources of impulse radiation. In some cases signals in the higher bands need not be as strong as those in the low bands for an equally intelligible signal to be received. This consideration is most important in noisy (e.g., urban) environments. Also, signals in the UHF band are much more subject to reflections from stationary objects than are the lower frequency signals; it is considerably more difficult to design receiving equipment and to place antenna arrays to assure fadefree reception in this band than it is for the lower bands. Nine hundred-MHz conventional systems will display many of the same technical advantages and disadvantages as UHF (450-MHz) systems.

Currently marketed conventional mobile radio systems that respond to individual user needs are available at a wide range of prices. For instances, E. F. Johnson markets a simple, one-channel system rated at 25 watts that costs approximately $700 for base stations and about $600 for mobile units. Johnson also markets a more elaborate four-channel system that is tone activated and automatically scans channels for an incoming signal. Mobile units cost $1,200-$1,400; base stations are $200 more. Motorola manufactures a similar four-channel system. A 100-watt base station will cost around $2,200, with single frequency mobile units selling for $1,200. Motorola also markets a simpler system in which a 60-watt base station costs $1,600, while mobile units cost $600. Further analysis of costs will be found in Chapter 4.

One-Way Paging

The use of one-way paging (OWP) has grown rapidly over the last decade. Since high channel-loading factors are possible because of the service characteristics of OWP systems, such growth has been accommodated by allocating just a few frequencies in the standard land mobile radio bands. Message lengths are relatively short, users are unaware of any waiting time involved, and no response from the paged receiver is required. Thus, little spectrum is used and receiving equipment can be simple. The FCC has set aside a total of 13 channels for business use, four in the high band and nine in the UHF band.[10]

Various types of messages can be sent by OWPs. Even if no verbal capacity is provided paging units can offer more than one (prearranged) message by using either continuous or alternating tones, or by illuminating combinations of lamps. Pagers that vibrate, instead of beep or illuminate, are also available, as are pagers that can have their signaling devices turned off. In the latter units, a memory circuit "remembers" a page that has been received while the signaling component is turned off, and later signals the user when the unit is reactivated. Most domestic paging systems use either a one- or two-message tone/lamp system and offer localized, simplex operation.

The control equipment at the central station can be manual or automatic and may vary in complexity depending upon the level of automation desired. In a manual system, the person initiating a page usually telephones the dispatcher, relays a message, and identifies the intended recipient. The dispatcher then transmits a signal to the particular pager, indicating a prearranged message. Nonverbal paging messages may be broadcast twice, at perhaps one-second intervals, to reduce reception errors due to weak signal, interference, or other factors.

With automatic systems, the person initiating the page dials a special telephone exchange to obtain access to the paging system, and then dials an additional number to specify the particular unit desired. The most sophisticated systems envisaged are tone-voice and allow the page's initiator to include a short verbal message with the page. To send a page, the base station control equipment would first search a catalog to find the access code of the desired paging unit and then radio a burst of either analog (tone) or digital (pulsed) information that includes the access code and the message. In a voice system, the access code (analog or digital) is followed by the vocal message. If the radio transmitter is being used when the call is received, this information would be stored and transmitted when the channel becomes free. These high-level control systems, which may utilize minicomputers, can easily handle automatic page rebroadcast and out-of-queue priority messages.

Typical service characteristics can be identified by examining several currently available paging systems. Motorola's "Metro Pageboy" is a completely automatic digital system that can provide service for up to 100,000 subscribers per central unit. Each subscriber can be alerted by either a steady or pulsating beep, providing a capability for two prearranged messages. Metro Pageboy receivers are versatile: in addition to receiving two different possible audible messages, the units contain a memory, which enables the user to delay reception of "disrupting" tones until a more convenient time. The units are small; they typically occupy only 5.35 cubic inches (4.8″ × 1.4″ × 0.8″) and weigh 4.2 ounces.

An interesting feature included in each unit is a small, interchangeable coding key that retains a specific paging code (equivalent to a telephone number). Each key can be easily removed from its pager and reprogrammed, using a programming device similar in size and appearance to a desk-top calculator. A typical Metro Pageboy receiver costs around $300; in comparison, a simple two-tone nonvoice pager can be obtained for under $200.

Pagers with voice reception capabilities have increased message versatility with little additional cost, size, or weight. A typical voice pager (Sonar Radio SPL 518) occupies 11 cubic inches (4.2″ × 2.4″ × 1.3″) and weighs seven ounces. It costs about the same as the Motorola Metro Pageboy. However, voice reception systems accommodate fewer subscribers per channel than tone only paging systems.

Because of the extremely high loading factors associated with paging systems, expected growth in the demand for these services can probably be accommodated at frequencies lower than 900 MHz. It is conceivable, however, that if demand for paging grows more rapidly than anticipated, allocations in the 900-MHz band may be desirable. In the long run, some economies—both financial and spectral—might result from interconnection of paging systems with emerging mobile communication systems.

Multichannel Trunked Systems

Multichannel trunked systems are currently being used to provide mobile telephone service. These systems, unlike conventional ones, are intended to allow users easy access to a number of channels. The advantages of these newer systems is that they should provide

> the user a higher grade of service than is possible in comparable loaded non-trunked systems by reducing the amount of time he must wait for a channel and/or reducing the probability that his call will be blocked. . . . [They are] particularly suitable for serving different types of users on the same group of channels without interference. With today's single-channel systems it is generally desirable to put similar types of users on the same group of channels without interference. In the trunked system, different types of users can be intermixed more readily as they operate essentially independently of each other, the computer assigning channels on demand.[11]

Furthermore, when a user is making a call, "no one else on the system can listen or interrupt during normal operation."[12] The implication here is that only standard, coded transceivers will be authorized for use on trunked systems at 900 MHz; calling or listening on a channel is possible, however, if a person illegally operates an unauthorized transmitter, or uses a scanning receiver.

Trunked systems have become increasingly automated during the last decade. The early domestic mobile telephone service offered by AT&T (i.e., MTS) was a totally manual low-band trunked system. Anyone desiring to make a call had to switch among several channels to find a free one, and then signal a mobile service operator by pressing a "push-to-talk" button; the mobile service operator would manually complete the call. This was a time-consuming process. The waiting time to receive a call could also be long, since each mobile unit could be reached on only one designated channel.

Substantial design changes have been introduced that improve the quality of mobile telephone service over trunked systems. MTS has now been largely supplanted by "improved" MTS, or IMTS. This service exists in several forms: IMTS-MJ, a high-band system; IMTS-MK, a UHF-band system; and IMTS-B, which incorporates several control and switching refinements. All IMTS systems incorporate automatic search for free channels and automatic interconnection of the mobile signal with the wired system. IMTS-B has more sophisticated control circuits, which increase the handling efficiency of calls to mobiles and also automatically connect prearranged land-to-mobile calls when a mobile is outside of its usual service area.

Eleven channels are allocated in the high band for telephone service provided by the wireline companies, along with seven more for a similar service provided by radio common carriers (RCCs). These channels are spaced 30 kHz apart. In the UHF band, the wireline companies are allotted 12 channels and the RCCs 11, with a channel spacing of 25 kHz. The total amount of spectrum allotted to the mobile telephone service in the low, high, and UHF bands is more than three MHz.

The more automated multichannel trunked systems require control circuits in both the base station and the mobile equipment, in addition to their radio components. The base station radio equipment of trunked systems could be virtually the same as that needed for conventional systems, except that multiple transmission and reception capabilities are needed. Since several signals are broadcast on adjacent frequencies from a single antenna, a well-designed antenna configuration is a difficult yet important requirement.

The base station control equipment of a trunked system must, of course, be much more sophisticated than that required for a single-channel conventional system. For calls going to mobile units, base equipment must signal a particular mobile that a call is being routed to it, locate a free channel, and then switch both transmitting and receiving frequencies to that channel. For mobile-originated calls, the control equipment must sense which channel is unoccupied, complete the call, and instruct all of the other mobiles to move to a new free channel. Since much of the system operation is monitored and guided by the base equipment, a flexible signaling format is required to pass necessary control information and instructions between the base and mobile units. Some trunked systems now in operation in the U.S. use a tone signaling format, with five tones that can convey a wide variety of control messages; the five tones are sent out in streams of varying length, with variable intervals between stream transmissions. Both mobiles and base stations are able to sense not only the frequency of each tone sent, but also its place in a sequence of tones and the time intervals between tones. The numerous combinations of tone bursts make a large number of control messages possible.

When a base station finds a free channel, it transmits a single-frequency tone over that channel. All mobile units sense that tone and lock onto that channel, awaiting an incoming or outgoing call. If this channel should become occupied, the base station will find a new free channel and "tone signal" all mobiles to switch accordingly. A different audio frequency is used to indicate that all mobiles should be alert for an incoming call. The base then transmits a particular mobile unit access code, represented by strings of these two tones presented alternately; the number of such tones indicates which mobile has been called and the appropriate unit responds. Connect and disconnect signals are sent to the base control equipment using other tones and sequences of tones.[13] Base station control equipment must be designed to generate, transmit, receive, and interpret these tone signals.

Mobile units for new trunked systems will also be much more sophisticated than paging or conventional units because they incorporate control circuits that differentiate among the various tones, count the number of tones in alternating-tone bursts, and direct the radio system to change channels. In addition, sequences of tones must be generated to relay control information back to the base station. The mobile units of the more advanced trunked systems use custom-designed large-scale integrated circuits (LSI) to perform these control functions.

The control and the radio equipment described above are incorporated into currently available mobile units of advanced systems used for mobile telephone service. A typical UHF-band unit, Motorola's Pulsar T1407, produces an output power of 16 to 20 watts, occupies a volume of 11" X 22" X 5", and weighs about 40 pounds. A complete UHF Pulsar radio unit with control head (which includes a telephone-type handset and dial) costs around $2,500. If a UHF-band trunked system mobile unit without dialing capability were offered, it would probably cost about 20 percent less.

Current service on an IMTS trunked system is almost indistinguishable from that available with the wireline system, except that the former has extended waiting times for dial tones and problems with fading reception. If the blocking probability of these systems were lower, the already available features (i.e., automatic free-channel selection, automatic interconnection, and full duplex operation) would supply a degree of convenience equivalent to that of the wired telephone network.

Before turning to a description of the technology associated with cellular systems, it is desirable to compare the spectral efficiency and user capacity of the systems already discussed. We do this for two reasons: first, such a comparison is relevant to significant policy questions concerning spectrum allocation for these systems; and second, this analysis will introduce the concept of spectrum reuse, which is fundamental to the achievement of high spectral efficiency and high user capacity by means of cellular technology.

Spectral Efficiency and User Capacity

The three systems thus far examined can offer various services that have different technological and service parameters (as displayed in Table 2.1). In order to compare spectral efficiencies, we shall make a number of illustrative calculations assuming 25-kHz channel widths and 10 MHz of allocated spectrum for each system.

Assuming a blocking probability of approximately 65 percent, each simplex channel of a conventional system can carry about 0.65 Erlangs.[14] Since Table 2.1 indicates that each dispatch unit generates 0.004 Erlangs of traffic, conventional systems can accommodate on the order of 160 mobiles per channel, or about 6,400 mobile units per MHz. This figure can be compared to FCC loading guidelines at 900 MHz of 100 mobiles/channel for conventional business services and 160 mobiles/channel for transportation services. A geographical measure of spectral efficiency for a radial service coverage area of 20 miles would be five mobiles/MHz/mi^2. This quantity would be even lower if one allowed for the fact that the channels used in this service zone are not reusable in the immediately surrounding area.

Less stringent service parameters and technologically limited message lengths permit paging systems to offer the most spectrally efficient service. A simple example will illustrate the importance of brevity in allowed voice transmissions and simplicity of transmitted information for conserving spectrum. The Metro

Pageboy system requires 165 milliseconds to signal a five-digit identification number using coded tones. Each page signal is separated from the next by 45 milliseconds. Therefore, each five-digit paging signal requires 210 milliseconds for transmission. Redundant transmission, if desired for reliability, doubles this total transmission time per message to only 420 milliseconds. This short transmission time implies that 3,600/0.42, or more than 8,500 five-digit pages, can be sent in an hour over one channel. The large maximum acceptable wait time of 300 seconds (Table 2.1) allows designers to load many pagers on each channel, contributing further to the efficiency of such systems.

Blocking probabilities for paging systems can be very high, since busy-hour wait times can be as much as 300 seconds. In these cases, an approximate "worst-case" calculation of the number of subscribers that can be accommodated on a single channel can be made with the use of equation (2). Since wait time (W) is very large in comparison to call duration (τ), ρ (the carried load) will be approximately equal to 1. A nonredundant tone-voice paging system designed for two ten-second messages in the busy hour and a wait time of 300 seconds can load only

$$N = 1 \times \frac{3,600}{2 \times 10} = 180$$

subscribers on a single channel. In contrast, a five-digit, tone only paging system requiring 0.42 seconds for redundant transmission and handling two pages per hour can accommodate

$$N = 1 \times \frac{3,600}{2 \times 0.42} = 4,285$$

units on each channel. The geographical spectral efficiency of a five-tone paging system with a 20-mile range and the service parameters listed above will be 136 mobiles/MHz/mi^2. A paging system with a ten-second message capability will have a geographical efficiency of 5.7 mobiles/MHz/mi^2.

Trunked systems increase spectral efficiency by making use of otherwise "dead time" on dedicated channels. While any service now offered with conventional technology could employ trunking, only mobile telephone service is actually trunked. Because mobile telephone service is operated in duplex fashion, 200 full telephone channels can be accommodated in ten MHz. Extrapolating from Figure 2.6, we estimate that this large number of channels could handle about 200 Erlangs with a designated blocking probability of about 0.65; that is, present mobile telephone systems provide about 20 Erlangs/MHz. Since about 0.03 Erlangs of traffic are generated by each subscriber, this system can accommodate 666 mobiles per MHz, or approximately 33 mobiles per channel. For systems with a 20-mile range, the geographical spectral efficiency is 0.5 mobiles/MHz/mi^2 for a blocking probability of 0.65.

A trunked mobile telephone system offering the same quality of service as the wireline telephone system would provide a maximum blocking probability of

0.02. This figure is a design goal for future mobile telephone services. Such a system with 200 50-kHz channels could accommodate about 180 Erlangs of traffic within ten MHz, or about 30 mobiles/channel. A system with this low blocking probability and a 20-mile radius would have a geographic spectral efficiency of 0.48 mobiles/MHz/mi^2.

Trunked systems at 900 MHz will offer dispatch service. At 900 MHz, as at UHF, such service will use a two-frequency simplex system in which the base station and the mobile each use a separate 25-kHz channel. The system is not duplex because the mobile's receiver is turned off while its transmitter is on. Thus 900-MHz dispatch service would have 200 channels in ten MHz. With an 0.65 blocking probability, 200 trunked channels could handle about 200 Erlangs or 20 Erlangs/MHz. Since each dispatch unit generates only 0.004 Erlangs of traffic, a trunked dispatch system can accommodate 5,000 mobiles/MHz or 250 mobiles/channel. A geographical measure of spectral efficiency for this service is 4.0 mobiles/MHz/mi^2.

Table 2.3: *Spectral Efficiency Measures for Noncellular Mobile Communication Services*

System	One-Way Paging (2 pages/BHU)		Conventional	Trunked		
Service	tone only (5 digit)	15-second message	Dispatch	IMTS present	IMTS desired	Dispatch
Linkage	simplex	simplex	simplex	duplex	duplex	2-frequency simplex
Channel width (kHz)	25	25	25	2 X 25	2 X 25	2 X 25
Total spectrum considered (MHz)	10	10	10	10	10	10
Blocking Probability	a	a	0.65	0.65	0.02	0.65
Mobiles/MHz	171,400	7,200	6,400	666	600	5,000
Mobiles/channel	4,285	180	160	33	30	250b
Erlangs/MHz	40	40	26	20	18	20
Erlangs/channel	1	1	0.65	1	0.9	1
Mobiles/MHz/mi^2	136	5.7	5	0.5	0.46	4

a. The waiting time may be as large as 300 seconds.
b. A channel is defined here as the 2 X 25 kHz-frequency pair.

Measures of spectral efficiency are summarized in Table 2.3. As one might expect, the more limited service offerings are the most efficient ones. In other words, there is an inverse relationship between system flexibility (i.e., variable service parameters) and spectral efficiency.

TECHNOLOGY OF THE CELLULAR SYSTEM

In the cellular system, a large service area is divided into small cells, each of which has a low-power base station. These fixed base stations can be interconnected through a central station by means of a landline or directional radio or optical link. Communication between mobiles is carried out in the following manner: the signal from one mobile is conveyed by radio link to a cell base station and relayed by landline to the central station; the central station then routes the signal to the cell base station of the receiving mobile; and, finally, this base station establishes a radio link with the receiving unit. Cellular systems will be fully interconnected with the wired telephone system, making high quality telephone service between mobile and fixed units possible (see Figure 2.5). Subsequently, we shall refer to the interconnected fixed base and central station components of the system as "base equipment."

The cellular approach has been proposed as a means of providing various mobile services (e.g., dispatch, mobile telephony) for very large numbers of users. The previous discussion of spectral efficiency indicated that providing mobile telephony would normally require a great deal of spectrum per subscriber to supply the high quality of service characteristic of wireline telephony. Cellular systems exploit a channel reuse feature that allows many more subscribers to be accommodated by each channel, without changing the blocking probability, than was previously possible within a given market area. Consequently, these systems can offer high quality mobile telephone service with lower spectrum requirements per user; efficient dispatch service can be provided as well.

Frequency reuse is accomplished by assigning one set of radio channels to mobile communicators in one geographic area (or "cell") and assigning a different set of channels to communicators in the immediately adjacent cells. Then, the original set of channels can be reassigned to cells sufficiently removed from the area of utilization. This channel assignment and reassignment scheme permits cells that maintain a required separation distance to use identical sets of radio channels without producing unacceptable interference. A sample assignment scheme is illustrated in Figure 2.4.

Frequency reuse is now widely applied in various situations. For instance, the VHF/TV allocations in the New York-Washington-Philadelphia area are as follows:

Coverage Area	TV Channels Assigned						
New York	2	4	5	7	9	11	13
Philadelphia		3		6		10	12
Washington, D.C.		4	5	7	9		

Channels 4, 5, 7, and 9 form a set that is initially used in New York City and then reused in Washington, D.C.; the set is not used in Philadelphia, however, since it lies between the other two cities. If stations in Philadelphia used these channels, they would be subject to interference from New York or Washington stations.

Frequency reuse is only feasible if channel users in any cell or area can be guaranteed that they will not experience interference because the same channel is being used in other cells. Such interference can be controlled by ensuring that the more powerful base stations, using the same channel sets, are separated by an adequate distance. This distance is on the order of 18 miles for four-mile radius cells for the 900-MHz system.

Optimistic estimates indicate that cellular systems may require one-way channel bandwidths as narrow as 25kHz.[15] Since the regulation of transmitter frequency may not be precise, each channel must be separated by an additional five kHz, resulting in a total one-way channel bandwidth of 30 kHz. Thus, these optimistic estimates imply that high quality mobile telephone service could be provided with cellular systems using 60 kHz duplex channels. If the entire 40 MHz allocated in Docket 18262 is fully utilized, $(40 \times 10^6)/(60 \times 10^3)$ or 667 duplex channels will be available for mobile telephone service with this channel width. Without channel reuse, a 667-channel system could accommodate only about 20,000 subscribers with a blocking probability of 0.02. We will show later that the same number of channels could theoretically be used to provide a much larger number of subscribers with high quality mobile telephone service in a single service area using a cellular approach.

The radio equipment of both the base and mobile stations of cellular systems will involve fairly standard technology, similar in principle to that used in trunked systems at 900 MHz. The use of frequency modulation will encourage static-free reception and will also allow receivers to be built that quite easily "capture" a signal at a particular frequency, and "reject" undesired strong signals at nearby frequencies. An important difference between radio components built for use at 900 MHz and those built for use at lower frequencies is cost. For instance, transistors that operate at 900 MHz, particularly those that will produce powerful broadcast signals, are now considerably more expensive than those designed for lower frequencies. Engineering costs are also generally greater at the higher frequencies.

In Motorola's proposed cellular systems, the radiated power of the mobile units will be adjusted by the base equipment so that transmission levels can be increased when reception difficulties exist or decreased when reception is clear, allowing battery power in portable units to be conserved. Such adjustments could also correct for reception difficulties resulting from excessively weak signals at the cell station or prevent offset cell interference in many cases. Because effective broadcast power of the mobile unit will be much less than that of the cell stations, additional radio equipment may be necessary to ensure that mobile transmissions are accurately received by the base stations. Fixed satellite receivers could be placed in problem areas, such as in canyons, between large buildings, or ahead of significant obstructions to the radio path, in order to improve coverage. These supplemental stations require only reception capabilities, since received messages may be relayed to the appropriate base station by wire link, thereby eliminating the need for radio transmitters at these sites.

The control circuits of cellular base and mobile units must be more sophisticated than those required for automated trunked systems. Since cellular systems will offer mobile telephone service, they will employ trunking and will require equipment similar to that described in the last section. These circuits will monitor and allocate channels upon demand to complete calls to or from mobile units. In addition, the base equipment will also require control circuits to monitor and coordinate the operation of the numerous individual low-power base stations characteristic of cellular systems.

The base equipment must rapidly perform several tasks to begin a mobile-initiated call. Although several approaches are possible, we will use the system shown in Figure 2.5 for illustration. When LM-1 wishes to place a call, a request signal is received at cell station C. The base equipment must find and assign a free duplex channel and then ask the mobile to adjust to those authorized transmission and reception frequencies. These tasks are carried out while the mobile waits for a dial tone. Service quality will therefore be partially influenced by the speed with which the base control circuits can perform these tasks.

For calls directed to mobile units, the base equipment must first inform the mobile that a call is being routed to it by wide area page; the mobile is then located, and a free duplex channel found and assigned to that cell. Referring again to Figure 2.5, when LM-2's number is dialed, a signal is broadcast by all cell stations over a special set of paging channels[16] asking LM-2 to identify itself. Because all mobile units are ready to receive a call when not in use, they are all tuned to those paging channels. LM-2 broadcasts a digital signal on one of the signaling channels overheard by a cell station. This digital response signal is used by the base equipment to locate the desired unit.

Some possible methods of locating cellular mobile units include:

1. *Trilateration.* The location of a mobile may be determined if its distance from each of three or more fixed points (or cell stations) is known. These distances can be estimated by measuring the delay time required for a transmitted pulse (or response signal) to arrive at each fixed point ("pulse delay ranging"), or by measuring the shift in phase of the transmitted tone ("phase ranging"). While this technique might locate a mobile unit quite precisely, it is very costly in terms of both spectrum and equipment.

2. *Triangulation.* The location of a mobile may be determined by measuring its direction relative to several fixed sites. Although this system was mentioned in early AT&T proposals for a high capacity MTS, the complex and expensive equipment required for triangulation makes its application doubtful.

3. *Signal strength sensing.* A mobile is considered to be "closest" (for the purposes of radio linking) to the base station that receives the strongest signal from it. This method does not determine location as precisely as the use of trilateration or triangulation, but engineering costs are much lower and little or no extra spectrum is required.

Determination of location through signal strength sensing will require a fairly complex procedure in urban environments, since buildings and other features of the terrain will cause signal strength to vary continuously as the unit moves. The sensors must also be able to compensate for adjustments and variations in broadcast power of mobile units. Schemes for using signal strength sensing can be tested and implemented primarily using software that is contained in the control circuits. However, the other locational methods mentioned above would require extra hardware (e.g., special receivers and signal-processing equipment, and/or scanning antennas) for testing, as well as for implementation.

Once the cell location of LM-2 is known, the control circuits in the base equipment assign a free duplex channel (from the set authorized for that cell) to LM-2. This assignment information may be sent on yet another channel that all cell stations use or on the already identified paging channel. In the latter case, fewer overall channels are used to set up calls, but subscriber capacity is limited since the paging channel must handle both paging and set-up information. Once its channel assignments are received, the mobile unit automatically tunes its transmitter and receiver frequencies, and voice communication can begin with LM-2.

In cellular systems, adjacent cells cannot use identical sets of transmitter and receiver frequencies due to possible interference, a restriction that creates the need for additional base control equipment. As a mobile crosses a cell boundary, a new duplex channel must quickly be assigned so that uninterrupted communication can be maintained. The base equipment "hands off" the call from the base station in the user's original cell to that in the cell being entered. Components needed for the hand-off process, particularly the monitoring system that relays location information back to the base control equipment and the switching circuits, will be important determinants of service quality.

Several additional schemes have been proposed for designing a more efficient system, that is, one that uses the least possible spectrum for the least cost. These suggestions include: dynamic channel allocation, in which the set of channels assigned to particular cells can shrink or grow according to how the demand for channels is distributed throughout the system; deterministic location, in which the strength of signals received is compared to the known behavior of radio signals within the system in order to locate the mobile unit more easily; use of a single channel or special combination of channels for transmission of paging and channel assignment information; and various methods for sending a specialized signal simultaneously with the voice signal to facilitate signal strength sensing. These proposals are intended to further conserve spectrum and improve service performance, and these advantages must be balanced against considerations of cost, complexity, and reliability. During the initial, introductory period of cellular development, however, it does not appear likely that spectrum and service quality considerations will necessitate use of any of these schemes.

It is clear that the memory, computation, and switching requirements of the base control equipment are demanding. Channel assignment and hand-off processes require continuous storage and updating of channels requested and channels in use. The computer facilities necessary to control and operate a cellular system must be quite large, but need not be exceptional; they may be centralized or decentralized, concentrated at a central switching office (as in Figure 2.5), dispersed to individual cell stations, or grouped as satellite computers each serving several cells. Current stored-program electronic switching systems, such as AT&T's ESS units, can be converted to handle the logic and switching functions that are required. If computation were decentralized, the computers required in each cell could be smaller than an ESS.

The control circuits in mobile units are not as complex as those in the base control circuits, because most cellular system operations are monitored and directed by the base equipment. Each mobile unit must contain a small computer that will decipher various digital signals received from the base equipment, act upon transmitted information or instructions, and send appropriate digital responses. These actions include: answering a page, changing transmitter power, requesting a channel, and dialing a number. The most distinctive feature of these mobile units is that they must be able to change both transmission and reception frequencies while in use, when a unit moves to a new cell, and this switch must be executed rapidly enough to be imperceptible to the user.

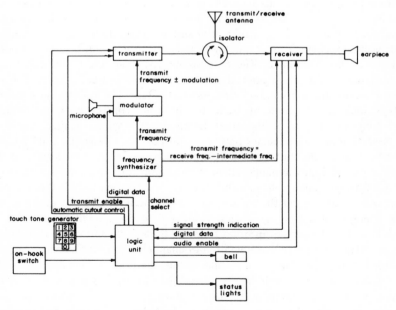

Figure 2.7: *Block diagram of a cellular mobile telephone unit. Digital data from the receiver is interpreted by the logic unit (a small computer) that instructs the frequency synthesizer to adjust, transmit, and receive frequencies. The logic unit also interprets signals from the touch-tone pad (number the user wishes to call) and the on-hook switch (user desires a channel to be assigned). The modulator impresses the voice signal from the microphone, plus any digital data that must be transmitted to central control, on the transmitted signal. The isolator prevents the transmitted and received signals from interfering with each other.*

The logic unit of the mobile telephone need not be any more complex than that of a programmable scientific hand-held calculator. The logic circuit will receive its inputs either from base equipment signals or from the touch keyboard pad of the unit itself, and its power can be drawn from the mobile unit. Without including a display capability, the logic unit should not need any more space than that required for two or three integrated circuit chips. In addition, when custom designed and built in quantity, the logic unit should be relatively inexpensive. Our conception of the structure of a cellular mobile telephone unit is shown in Figure 2.7.

Adequate power supply for car-mounted mobile units will not be a problem since the electrical system of the vehicle can provide more than sufficient energy. However, suggestions have been made that the hand-held or portable telephone is now feasible. A working prototype of such a unit has been constructed by Motorola, and plans are being made to introduce portable service into Motorola's trial system. Also, development of Japanese portable units is now receiving serious attention at NTT. These smaller units might present special problems because battery power must be provided to run the unit for at least a working day, allowing for daily (probably overnight) recharging. Batteries could contribute the major fraction of weight and volume to these portables.

Although energy is consumed at a much smaller rate by the receiver and logic unit than by the transmitter, the receiver unit ordinarily must be continuously active. If the receiver consumes 25 mw when waiting to receive a call, it will require 300 mwh for a twelve-hour day. A portable telephone that has a 5-watt transmitter stage operating at 33 percent efficiency (i.e., requires an input of 15 w to produce a 5 w radio output) and that operates according to the service parameters of Table 2.1 (or about 20 minutes in a twelve-hour day) will require its batteries to supply 5.3 wh a day. The logic unit can be ignored, as its energy consumption will be negligible compared to that of the transmitter. Thus, the total battery capability must be on the order of 6 wh. Since three rechargeable penlite-size nickel-cadmium batteries in a pocket calculator can supply roughly 0.5 wh, about 36 NiCad penlite cells would be needed to operate a portable unit for twelve hours without recharging. This battery pack would occupy approximately 155 cubic centimeters and weigh between 500 and 900 grams.

It should be emphasized that cellular system designs are still considered experimental. System trials have been proposed for the near future, however. AT&T has designed a system for the Chicago area, and Motorola has designed one for Baltimore-Washington. NTT has also scheduled a system trial for the Tokyo area. The technology actually deployed to realize a completely satisfactory system may well differ in detail from that described here, but any system must follow the general spectrum-use, control, and locational-method guidelines we have summarized. In addition, since there is not yet much experience with radio propagation in the 900-MHz band, particularly in urban areas, the FCC is now allowing system builders considerable flexibility in the specification of the detailed technical configuration of cellular systems.

Spectral Efficiency and Subscriber Capacity

The number of subscribers that can be accommodated by a cellular system will depend upon patterns of deployment, geographic distribution of users, and the service quality constraints that are imposed. Thus the number of users will be determined by the nature of the service provided. For example, dispatch service can accommodate many more units than mobile telephone service in the same bandwidth. The subsequent discussion of spectral efficiency and sub-scriber capacity will deal first with telephone service.

In general, the maximum capacity of a cellular system, assuming a uniform distribution of users, can be roughly calculated by using the formula:

$$S = \frac{N_c \times N_f \times D}{BHU}, \tag{3}$$

where S = the maximum number of subscribers for a given level
of service;
N_c = number of cells in a service area;
N_f = number of channels per cell;
D = channel occupancy factor, which traffic engineers
estimate to lie between 0.7 and 0.8—we shall use
0.75 in our calculations;
BHU = busy-hour usage—i.e., the fraction of time in
which unit is in use during the busiest hour
of system use, or traffic intensity per unit
in the busiest hour.

We shall use Chicago as an example in our calculations, with a total service area of 1,300 square miles.

The number of cells in any service area, N_c, will be roughly equal to the size of the service area divided by the area of each cell. Cellular proposals have suggested that the service area be divided into hexagonal cells of "radius," r, each having an area of $2.6 \times r^2$. Therefore, in Chicago a four-mile radius cell system would have N_c equal to $1,300/(2.6 \times 4^2)$ or 31. It is important to note that N_c would increase in our example if the Chicago market were expanded to include portions of southern Wisconsin or western Indiana, or if cell radii were reduced. Such changes would ultimately have a large effect upon S, the sub-scriber capacity.

The number of channels per cell, N_f, can be calculated from the following relation:

$$N_f = \frac{\text{(allocated spectrum)}}{\text{(channel width)} \times \text{(no. of different cells in a set)}}.$$

While opinions differ as to the optimum duplex channel width, we shall use 60 kHz, as mentioned previously.[17] Further, several frequency assignment schemes, yielding different numbers of cells in a set, are feasible. For example, two possible schemes are illustrated in Figure 2.8. We will assume in our calculation use of a seven-cell repeating pattern endorsed by AT&T. (It should be noted that the use of a four-cell design would yield a larger N_f if other parameters remain unchanged.) Assuming, then, that the full 40 MHz of spectrum allocated in Docket 18262 is utilized, N_f would be

$$\frac{(40 \times 10^6)}{(60 \times 10^3) \times (7)}$$

or about 95.

BHU (traffic intensity at the peak hour) is an experimental, rather than a calculated, parameter. To deliver a constant service quality, an increase in BHU would necessitate enlarging the switching network and offering more user channels. We shall use the BHU value of 0.03, the desired service goal for cellular service indicated in Table 2.1.

Thus, our hypothetical cellular system has the following design specifications: (a) 1,300-square-mile coverage with 31 four-mile radius cells; (b) 7-cell frequency assignment pattern; (c) 60-kHz channel bandwidth with 40 MHz of spectrum utilized; and (d) a BHU equal to 0.03. Such a cellular system for the Chicago market could accommodate about

$$S = \frac{31 \times 95 \times 0.75}{0.03} = 73,600 \text{ mobile telephone subscribers.}$$

An alternative estimate of subscriber capacity can be derived by utilizing blocking probability, B. With a blocking probability of only 0.02 percent—the

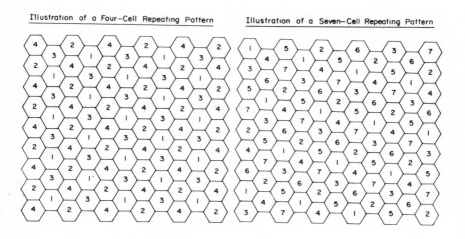

Figure 2.8: *Two possible frequency assignment schemes*

SOURCE: Adapted from Dale N. Hatfield, "Recent Trends in Land Mobile Radio Systems," draft paper, Office of Telecommunications, Boulder, Colo., June 1974, pp. 21-22.

telephone network standard—Erlang statistics for telephone network usage[18] indicate that the 95 channels in each cell could accommodate 81 Erlangs of traffic. Therefore, the 31-cell system we have used to approximate the Chicago market could handle a total of 2,511 Erlangs of traffic. Table 2.1 indicates that each mobile telephone will generate 0.03 Erlangs of traffic, so our hypothetical cellular system could offer mobile telephone service to about 83,700 subscribers. This estimate, which assumes a spatially uniform distribution of users, is similar to that obtained in the previous nonprobabilistic calculation. Since one cannot assume a uniform distribution of users, though, the actual subscriber capacity would presumably be lower.

A major advantage of the cellular system is its capability of handling greatly increased numbers of mobile telephone subscribers, in comparison with current systems. The mobile telephone system now operating in Chicago is saturated with 1,170 subscribers. A cellular system with four-mile radius cells, together with the increase in allocated spectrum through Docket 18262, allows a potential increase to more than 60 times the present capacity; if one-mile radius cells were used, the total capacity would be over a thousand times the number of subscribers that can be accommodated now. The total capacity increases inversely with the square of the cell radius, an important relationship when one considers the ultimate possibility that mobile communication devices will become widespread consumer items. However, we should emphasize that the technological advances and cost reductions necessary to make such small cell sizes practical are substantial.

Telecommunications policy is frequently concerned with satisfying future demands for mobile communication services in ways that conserve spectrum. Spectral efficiency measures of trunked systems have already been calculated and are summarized in Table 2.3. We shall estimate the efficiency measures of cellular systems, first utilizing the full 40-MHz allocation and then using a 10-MHz allocation for comparative purposes.

Our hypothetical Chicago cellular system will accommodate almost 2,100 mobiles/MHz. This means that loading for such a system will be about 126 mobiles/channel. Since 2,511 Erlangs of traffic could be supported, this implies 63 Erlangs/MHz and 3.8 Erlangs/channel. The geographical spectral efficiency would be 1.6 mobiles/MHz/mi^2. These measures are summarized in Table 2.4, along with efficiency estimates considering a spectrum utilization of only 10 MHz. Efficiency measures for trunked IMTS are also included in this table for purposes of comparison. As one can see, the seven-cell system is more than twice as efficient in terms of mobiles/MHz/mi^2 than current IMTS would be if it offered similar service. A four-cell system would be even more efficient, providing about 2.5 mobiles/MHz/mi^2 if 40 MHz were used.

Cellular systems will be allowed to offer dispatch service. Under present rules, however, they are prohibited from offering the "all-call" option, in which all units can be contacted at once. Our spectral efficiency calculations will consider the utilization of ten MHz. Assuming the use of 60-kHz two-frequency simplex dispatch channels, our ten-MHz Chicago system could accommodate close to 140,000 users, or about 14,000 mobiles/MHz. Given that about 560 Erlangs of

Table 2.4: *Approximate Spectral Efficiency Measures for Trunked and Cellular Telephone Services*

System	Trunked			Cellular (7-cell scheme)		
Service	IMTS present	IMTS desired	Dispatch	MTS	MTS	Dispatch
Linkage	duplex	duplex	2-frequency simplex	duplex	duplex	2-frequency simplex
Channel width (kHz)	2 × 25	2 × 25	2 × 25	2 × 30	2 × 30	2 × 30
Total spectrum considered (MHz)	10	10	10	10	40	10
Blocking Probability	0.65	0.02	0.65	0.02	0.02	0.65
Mobiles/MHz	666	600	5,000	1,680	2,100	14,000
Mobiles/channel	33	30	250	101	126	840
Erlangs/MHz	20	18	20	50	63	56
Erlangs/channel[a]	1.0	0.9	1.0	3.0	3.8	3.4
Mobiles/MHz/mi^2 for 1,300 mi^2 service area[b]	0.5	0.46	4	1.3	1.6	11

a. Note that the Erlangs/channel exceed unity for cellular systems because of small cell frequency reuse within markets.
b. Trunked figures are adjusted for larger service area.

traffic could be supported, this implies 56 Erlangs/MHz and 3.4 Erlangs/channel. The geographical spectral efficiency would be 11, or almost three times the efficiency of trunked dispatch service (assuming no all-call).

FUTURE DEVELOPMENTS IN LAND MOBILE COMMUNICATION TECHNOLOGY

Over the past decade, advances in electronics have facilitated the development of new devices at a reasonable cost and the improvement of older devices. Developments in land mobile communication technology are no exception. For instance, conventional system users have been able to share channels conveniently, thereby allowing system costs to be shared. Pagers are now available that will receive several preprogrammed messages, which increases the flexibility of these devices. The increased sophistication found in trunked systems as designs have progressed from standard MTS to IMTS have also made these systems much more convenient to use. And these are only a few of the developments already adopted in existing systems.

A cellular system offering portable telephone service, enabling a subscriber to have an instrument with him at all times if desired, is clearly possible. Control capabilities necessary for cellular system operations have been made economically feasible by large-scale integrated circuit developments. Unit costs, performance, and convenience capabilities might be further improved by developments in UHF power-transistor technology or by advances in battery technology. Nevertheless, no new "breakthroughs" are required to build a portable telephone system at a cost that could result in widespread acceptance in the long term.

Although our discussion of technologies for land mobile communication has focused primarily upon verbal or printed message capabilities, some other possibilities for cellular systems should be noted. If in the distant future locational ability within cellular systems is greatly improved, either through the use of very small cells or the adoption of sophisticated locational methods (perhaps involving satellites), several interesting functions might be added. Such supplementary functions might include vehicle monitoring and/or routing systems, emergency beaconing service, or automatic telemetry systems. Conversely, the ways these functions are performed now might be simplified if they could be integrated into a cellular system. Some additions to the basic capabilities of widespread wireless telephone systems may be relatively easy and cheap to incorporate, since they would involve only changes in the switching software at base stations. Moreover, portable units will contain microprocessor-like custom LSI chips, to which simple functions can be added cheaply, if incorporated into the basic chip design. Thus, further developments in electronics are likely to make land mobile communication technologies more economical and easier to use, and will make feasible the incorporation of functions that have not previously been available.

NOTES

1. R. T. Buesing, "Modulation Methods and Channel Separation in the Land Mobile Service," *IEEE Transactions on Vehicular Technology* VT-19 (1970): 187.

2. U.S. Department of Justice, *Investigation of Digital Mobile Radio Communications,* prepared for the National Institute of Law Enforcement and Criminal Justice by Thomas C. Kelly and John E. Ward (Washington, D.C.: U.S. Government Printing Office, October 1973), p. 2.

3. Dale N. Hatfield, "Measures of Spectral Efficiency in Land Mobile Radio," Selected Papers on Land Mobile Radio, Technical Memorandum, Office of Telecommunications, Boulder, Colo., November 19, 1973.

4. Named after Agner Erlang (1878-1929), a pioneer in the probabilistic analysis of telephone systems.

5. Hatfield, "Measures of Spectral Efficiency."

6. D. C. Cox and D. O. Reudink, "Layout and Control of High-Capacity Systems," in *Microwave Mobile Communications,* ed. W. Jakes, Jr. (New York: John Wiley and Sons, 1974).

7. Examples of common expressions that are used for relating ρ to W and τ include:

$$\rho = \frac{2W/\tau}{1+2W/\tau} \text{ and } \rho = \frac{(W/\tau\text{-}1)}{W/\tau}.$$

8. V. Agy, "A Review of Land Mobile Radio," Technical Memorandum 75-200, Office of Telecommunications, Boulder, Colo., 1975.

9. R. M. Brown, "Business Radio Communications," *Electronics World* 77 (March 1967):29.

10. V. Agy, "A Review of Land Mobile."

11. "Nature of Inquiry," Docket 18262, *Federal Communications Commission Reports,* 2nd series, vol. 14, p. 311.

12. Ibid.

13. Engineering Director, Transmission and Radio Systems, "Domestic Public Land Mobile Telephone Services Customer Provided Dial Station," Technical Reference PUB 43301 (New York: American Telephone and Telegraph Co., March 1973).

14. Estimates of channel load capabilities have been provided by Bell Laboratories staff.

15. Motorola, Inc., "Technical and Marketing Data on System Design and User Needs at 900 MHz," Formal submission to FCC Docket 18262, Motorola, Inc., Schaumburg, Ill., December 20, 1971, Appendix, pp. 1-4.

16. Because signaling requires only a few tens of microseconds of transmission time, a single channel can page many mobiles without exceeding a maximum allowable delay of one-half second. However, since all cell stations transmit the signal simultaneously, several paging channels may be needed for this purpose in large systems.

17. The FCC has limited channel width to 40 kHz, making the maximum duplex channel width 80 kHz. We shall continue to use the 60 kHz figure to determine a high estimate for subscriber capacity. Wider channels would, of course, decrease the potential subscriber capacity. Use of the 60 kHz figure will also make spectral efficiency comparisons easier between trunked and cellular systems offering mobile telephone service, because the channel bandwidths will be similar.

18. R. A. Kuehn, *Cost-Effective Telecommunications* (New York: American Management Associations, 1975), p. 20.

Chapter 3
LAND MOBILE
COMMUNICATIONS
AND THE
REGULATORY
PROCESS

Mark V. Nadel
Robert E. Glanville
Philip L. Bereano

As the result of the rapid growth rate of systems mentioned in Chapter 1, regulating land mobile communications steadily became a more serious problem for the Federal Communications Commission. In this chapter, we shall review the response of the FCC to this situation. Our review will not be restricted to the content and basis of those decisions, however. Considerable attention will also be devoted to the process involved and the nature of the participants. Because technology assessment is concerned with the inclusion of broader interests in technological decision-making, our assessment of the implications of a new communication system must include a wide-ranging evaluation of the regulatory process that is central to the evolution of the technology.

Until very recently, administrative law was generally considered to pertain only to *formal* agency procedures. The Administrative Procedure Act concentrates on the formal activities of "rule making" and "adjudication" as the two paradigm agency procedures (corresponding respectively to quasi-legislative and quasi-judicial actions). The FCC Dockets discussed in this chapter are formal rule-making proceedings, and thus involve procedural due process requirements, such as the maintenance of a formal record, adequate notice to interested parties, opportunity for these parties to be heard, etc.

Two other aspects of agency activity that will also deserve our attention are the exercise of *discretion* and the existence of *informal agency procedures*. Discretion refers to the legal ability of an administrator to choose from a range of decision options and to exercise judgment in determining the outcomes of proceedings. In reviewing the regulatory activity here, we will see that the FCC enjoyed considerable discretion in making its decisions. Although evaluation of a formal agency procedure is facilitated by the detailed procedural rules, especially the required record of the decision-making process, a large part of the agency's activity involves informal modes. This poses a serious problem because it is difficult to ascertain the extent to which informal activity is involved in agency decision making.

Agencies like the FCC do far more than issue rules and conduct adjudication. Typically they are involved in such informal procedures as: conducting investigations; issuing advice; supervising the activities of actors within their regulatory sphere; conducting public education and public relations programs; engaging in arbitration as an informal mode of adjudication; and promoting various public policies. Frequently these informal procedures are more political and policy-oriented in character than technical, thereby requiring the exercise of a considerable amount of discretion.

Discretion itself is often a necessary aspect of administrative activity. The legislative and judicial branches of government frequently find it hard to handle new problems (particularly those with a major technological component), because they lack the expertise and specialization and the ability to provide a continuity of attention and responsibility. As a result, discretion permeates ordinary law produced by legislatures and courts.

It has been estimated that informal procedures represent 80 to 90 percent of all agency activity,[1] but we do not know whether the FCC is typical in this regard. Furthermore, it is impossible to determine the specific amounts and forms of informal activity involved in the decision making we are assessing. Informal agency activity is not limited to technical and administrative matters, largely involving factual determinations, for instance; important policy determinations, with political ramifications, can also be made in such modes. While the courts have mainly concerned themselves with creating elaborate procedural requirements for formal agency activity (the better to be able to review it), it appears that an increasing volume of agency work has been largely nonreviewable primarily because it is being conducted ex parte and off the record.

DOCKET 18262: BACKGROUND AND OUTCOME

By the early fifties, mobile systems were playing a vital role in police, fire, emergency, public utility, forest, business, industry, and local government communications. Yet the basic frequency allocation to these systems had been established in the 1940s and remained at approximately 40 MHz until Dockets 18261 and 18262 were initiated in 1968.[2] A concatenation of social and economic factors contributed to the enormous and unforeseen explosion in demand for land mobile services. The number of transmitters in operation grew from slightly over 500,000 in 1955 to over 2,500,000 by 1967. Despite a quadrupling of spectrum utilization through channel-splitting efforts, the growth in demand continued to outstrip the technological ability to cope with increasing congestion.[3] Consequently, the FCC was under mounting pressure from a large and vocal group of users to increase the allocation of spectrum available.

In 1957 the commission initiated an inquiry concerning the allocation of nongovernmental frequencies in the spectrum between 25 and 890 MHz.[4] In the Report and Order issued in 1964,[5] the FCC acknowledged the important contribution of land mobile services to the American economy but declared its intention to resist efforts to allocate additional spectrum in this range for land mobile systems, because such a reallocation would encroach upon the 70 channels already committed to UHF television.[6] These 70 channels were perceived as essential to the commission's objective of establishing a "system of competitive nationwide service reaching all parts of the country with the largest possible number of program choices and providing as many outlets of local expression as possible. . . ."[7] The shorter range of UHF signals would permit the operation of many low-power, community-oriented stations.

UHF developed far more slowly than anticipated by the commission. Compared with VHF, it was seriously disadvantaged because of the inferior propagation characteristics of UHF radio waves and the fact that few sets were equipped with UHF reception capability.[8] To enhance the competitive potential of UHF, in 1962 the commission secured enactment of the All Channel Receiver Act (ACRA).[9] which required that all television sets marketed in interstate commerce be equipped with receivers for both UHF and VHF signals. The commission was optimistic that by 1964 UHF could become commercially viable and rejected any thought of reducing its spectrum allocation.[10]

Although the FCC acknowledged the need for relief in the land mobile frequencies, it was determined to achieve that relief without a major spectrum reallocation.[11] In 1964, the commission established the Advisory Committee for Land Mobile Radio Services (AC/LMRS) and directed it to make a thorough study of possible solutions to the frequency problems affecting land mobile services.[12] During its three-and-one-half-year study, the Advisory Committee investigated refinements of technology and improved regulatory approaches that might bring about more intensive use of spectrum already allocated. The committee concluded that, while the adoption of its recommendations would result in a small improvement, adequate relief from the congestion could not be provided without allocation of additional spectrum.[13]

Unfortunately the policies of the commission to stimulate UHF-TV were insufficient to solve the problems of small audiences, low revenues, and inferior programming. Confronted on the one hand, then, with the seemingly nonviable UHF system and on the other with the insatiable demand for land mobile services, the commission announced in April 1967 that it would study the feasibility of reassigning certain UHF channels to land mobile services.[14]

The report of the committee charged with this study was released in March 1968. It stated that the reassignment of four to seven lower channels of the UHF frequency (channels 14–20) would be feasible but would dislocate a substantial number of UHF stations already authorized or established, and the cost of relocation was estimated at approximately $100,000 per station. The committee believed that the utility of a sharing option (geographic assignment of UHF channels not in use) would depend on the number of channels shared and the level of mutual interference that would be tolerated. Such congestion relief would not be uniform and would require that many land mobile operations be seriously limited in congested urban areas—precisely those areas most needing relief. Reallocation of UHF channels 70–83 (the 900-MHz band) was found to cause the least dislocation of operating facilities, since only two assignments had been made in those frequencies. However, this option would not provide immediate relief to land mobile services, because the necessary equipment and systems would require a developmental period of five to eight years.[15]

This report played a major role in persuading the commission to open two new dockets in 1968: Docket 18261 proposed a sharing of UHF channels; Docket 18262 proposed an outright reallocation of UHF channels 70–83.[16] Although we are primarily concerned with Docket 18262, a brief discussion of 18261 is appropriate, since in the early stages it was closely intertwined with the overall problem. Its sharing option was viewed by the commission as offering the greatest potential for *immediate* relief of land mobile congestion with a minimum impact on television reception; its adoption would provide time for a thorough consideration of the Land Mobile Relief Committee's report. Responding to the commission's Notice of Proposed Rule Making, comments and replies were filed by more than 110 parties, mostly representing land mobile and broadcast interests. More than 40 parties participated in two days of oral argument in 1970. In general, land mobile interests argued that sharing was inadequate and that only a major reallocation of spectrum would solve their problems. Broadcasters, on the other hand, asserted that even the limited sharing proposal was unnecessary.[17]

On the basis of this conflicting record, the commission, in June of 1970, reached the decision that "land mobile communications play a vital role and have become indispensable in public safety, as well as industrial, transportation and commercial activities of the Nation,"[18] and that such services "are being hampered because of inadequate radio communications. . . ."[19] Noting that technical progress had enabled land mobile to intensify greatly its use of allocated spectrum, the commission observed that "further reduction of channel width and further improvements along these lines are not practical at this stage of the art."[20] While acknowledging its commitment to further efficiency gains, the FCC found that substantial additional spectrum was needed to achieve long-term relief. The broadcasting groups persuaded the commission that adequate relief could be obtained if the sharing proposal were limited to 10 of the top 25 urban areas and were confined within those areas to one or two UHF frequencies. The measures adopted in Docket 18261 were to provide interim relief pending development of plans for an outright allocation to land mobile services of the 806 to 890-MHz frequencies.[21]

Docket 18262, intimately related to Docket 18261 and initiated simultaneously, was directed at facilitating the expansion of land mobile into Band 9. For a period of nearly 20 years AT&T had been seeking the commission's authorization of sufficient spectrum for the development of a high-capacity land mobile communication system. In Docket 8976 in 1949, Bell Labs had actually submitted a proposal for such a system and had requested spectrum in the 470 to 500-MHz band that was ultimately allocated to television.[22] Finally, in Docket

18262's First Report and Order and Second Notice of Inquiry, the FCC announced a tentative allocation of 40 MHz of spectrum for private land mobile systems and 75 additional MHz for common carrier mobile systems.[23]

Besides signaling a retreat from its general policy of promoting UHF as a viable commercial alternative to VHF, the commission's action was particularly disappointing to supporters of educational television who had long been seeking sizable additional reservations for low-power, economical educational stations in smaller communities.[24] Secondary impacts of the reallocation included the reduction of the Industrial, Scientific, and Medical (ISM) bandwidth from 915 ± 25 MHz to 915 ± 13 MHz,[25] an action resisted by ISM interests on the ground that it would increase product costs and frustrate several years of developmental research on the broader band.[26] Also implicit in the reallocation was a decision not to allocate 26 additional MHz between 890 and 942 MHz so as to provide harmonic overlap with government radio-location systems operating at 420 to 450 MHz; the commission found that land mobile needs were greater than those of aeronautical mobile services or the broadcasting auxiliary service.[27]

Both Commissioners R. E. Lee and H. Rex Lee dissented from the FCC's action, perceiving the decision as a mere palliative that would shortchange educational television. Commissioner Johnson, in a reluctant concurring opinion, was highly critical of the commission for its chronic failure to develop any "consistent, rational policy of spectrum management."[28]

The Second Report and Order on Docket 18261, released on July 22, 1971, addressed technical considerations leading to an implementation of the UHF sharing decisions—including, among other things, frequency assignments, channel spacing, and loading requirements.[29] Despite numerous subsequent modifications, Docket 18261 generally followed the broad outline in the Report and Order of June 1970.

Docket 18262 received considerable additional attention from the commission before it was finally concluded in July of 1975.[30] In 1971, for instance, the FCC reaffirmed its commitment to a 75-MHz allocation to common carriers but retreated from its initial decision to limit the development of common carrier systems to wireline companies only. The commission decided that the exclusion of other land mobile interests from the 806 to 881-MHz frequencies might impede the development of a high-capacity system and that competition offered the greatest promise of rapid and efficient use of the spectrum.[31]

A Second Report and Order issued in 1974 divided the reallocated 900-MHz band and set the terms of service.[32] In an important departure from past practice, undoubtedly in response to general criticism of the block allocation system, the FCC chose to allocate the new frequencies on the basis of the type of system employed rather than the type of service provided. The latter system of allocation had "led to parochialism among the users and inequitable situations where spectrum shortage and abundance exist[ed] side by side in the same cities."[33] This new approach signaled a determination to permit market forces to structure the distribution of spectrum among competing uses.

The FCC concluded that its earlier tentative allocation of 75 MHz to the development of a high capacity cellular system was excessive, since "64 MHz of spectrum would handle the foreseeable demand for mobile telephone and a substantial portion of the dispatch market to the end of this century in our biggest cities."[34] As a result, the allocation was reduced to 40 MHz, with a commitment to release additional spectrum upon a demonstration of need. The capacity of this system was estimated to be about 105,000 telephone subscribers and an equal number of dispatch users per market—sufficient to accommodate anticipated demand through the year 1990 in the largest cities. The commission did note the possibility that the mobile telephone could develop into a consumer item but indicated that available marketing studies did not foresee this.[35]

The tentative allocation of 40 MHz to trunked and conventional systems made in the First Report and Order was also reduced to 30 MHz, with spectrum reserves so located that the services could easily be expanded. The Second Report and Order established numerous reserve bands totaling 45 MHz and provided the necessary flexibility to permit market forces to influence the final configuration of the land mobile services. Because the commission felt that only the Wireline Common Carriers (WCCs) possessed the necessary technical and financial resources to develop the costly and complex cellular system, they alone were permitted access to the 40 MHz allocated to the development of an interconnected mobile telephone service.[36]

To promote competition and prevent manufacturers or equipment companies from monopolizing the market, the Second Report limited such concerns to a single trunked system in any market and to no more than five such systems nationwide, except upon a showing that the limitation should not apply in a particular case.[37] This policy represented a compromise between total exclusion and unlimited entry and was designed to utilize the companies' technical expertise in developing trunked and conventional systems, while at the same time promoting long-term competition.

In order to minimize the danger of cross-subsidization of land mobile operations by the wireline carriers, the commission required that wireline telephone companies establish a separate operating company before they could be authorized to operate a cellular system. Moreover, wireline carriers were forbidden to manufacture, provide, or maintain mobile equipment and were required to offer interconnection services just as they did to their own affiliates or subsidiaries.

Despite objections that permitting cellular operators to provide dispatch services would increase the danger of cross-subsidization, the commission authorized cellular systems to provide all but fleet dispatch services ("all-call" services), since, in the initial stages of system development, dispatch and related services would constitute a substantial share of the cellular market. The commission felt that any other course of action would significantly delay the development of cellular capability.[38]

One of the FCC's most controversial decisions in Docket 18262 involved the creation of a new class of services: the Special Mobile Radio (SMR) Services. This action allowed licensing of SMR trunked and conventional systems and would permit entrepreneurs to make facilities available to those eligible as defined in Parts 89, 91, and 93 of the commission's regulations (essentially everyone but noncommercial individual users) under an open-entry/uncontrolled-price regulatory framework.[39] Hence, eligible users would be able to choose between the trunked or conventional technology, share facilities with other users, or establish nonprofit corporations or associations to manage shared systems.

In March 1975, the FCC adopted a Memorandum, Opinion and Order disposing of petitions directed at action taken by the commission in its Second Report and Order of May 1974.[40] This document reversed the commission's earlier decision to extend the legal monopoly of the wireline carriers into the cellular mobile communications market[41] and concluded that "any qualified entity, in addition to wireline carriers, will be permitted to apply for authorization to develop and eventually operate cellular radio systems. . . ."[42]

Reconsidering the limits that it had placed on equipment manufacturers concerning development of trunked systems, the FCC concluded that permitting manufacturers to operate as many as five trunked systems might suppress competition to an undesirable extent. The commission decided, instead, to allow land mobile radio equipment manufacturers to establish a single 20-channel model trunked system for the purpose of making engineering and financial feasibility assessments. If this limitation were to adversely affect the development of such facilities, the commission expressed a willingness to consider changes. Seeing no present danger of the 900-MHz band being dominated by other entrepreneurs, the FCC rejected a proposal to place similar limits on the number of trunked systems licensed to other developers.

The commission also modified its position concerning frequency tolerance of ISM devices operating in the 915-ISM band, relaxing some of its earlier restrictions. But it reaffirmed its previous decisions concerning the amount of spectrum allocated to cellular systems and the requirement that wireline carriers must establish separate subsidiaries to operate cellular systems, except during developmental stages.

Some wireline carriers did convince the commission that its earlier policy forbidding supply and maintenance of cellular systems by wirecarriers might impede the early development of cellular capability. The FCC decided to permit carriers to supply and maintain mobile equipment manufactured by others; it would not automatically bar the continued use by carriers or users of carrier manufactured equipment used in the developmental systems. However, it affirmed its decision to prohibit manufacture of mobile equipment by cellular operators.[43]

Finally, in July 1975, the commission denied a number of petitions for reconsideration contending that a number of the arguments were substantially the same as those already considered. Subsequently, the decisions in Docket 18262 were appealed to the Court of Appeals. In a decision rendered in January 1976, the court upheld the 1974 order, as modified by the 1975 order.[44] It affirmed the allocation of 40 MHz for development of the cellular system and 30 MHz for use by private mobile service, including a new class of entrepreneurial operators known as Special Mobile Radio Services.

The court devoted considerable attention to the possible antitrust (or anticompetitive) impacts of the FCC's allocation, since competitive factors are properly a component of the public convenience, interest, or necessity standard established by statute to govern the FCC's operation.[45] In its decision, at several places the court notes that the effect of the FCC action will be to increase the power of AT&T, since that corporation will operate most, if not all, of the cellular systems eventually put in operation, will dominate the field of radiotelephone service, and will also become a significant force in the private dispatch services of the market. In addition, the 1975 order eliminated the requirement that wireline operators offer to interconnect radio common carriers into their system, additionally increasing the power of AT&T.[46] The court is obviously concerned that requirements in the 1975 order, which are "designed to prevent cross-subsidization of the activities of the cellular system," probably "will prove largely cosmetic."[47] Although the court concludes "that significant anti-competitive effects may well result . . . ,"[48] it is nonetheless willing to uphold the commission's plan, because the FCC "has stated its clear intention to authorize only a developmental system in the Chicago area, which will utilize only 12.5 MHz of the 40 MHz allocation. The commission retains a duty of continual supervision of the development of the system as a whole, and this includes being on the look-out for possible anti-competitive effects. The serious anti-competitive effects, if they arise at all, will do so only after full implementation begins."[49] In reaching this decision, the court states that it was strongly influenced by the position of the Department of Justice "that the Order as a whole poses no immediate and substantial competitive problems." [50]

The second major issue in the case involved allowing entrepreneurial operators (SMRs) access to spectrum that had previously been used by private dispatch licensees (for their own use), nonprofit cooperative uses, and community repeaters. Competition was an underlying policy concern here also, because the FCC decided to treat the SMRs as noncommon carriers and to federally preempt the process of entry certification in order to stimulate competition that, it is believed, would further hasten technological development. At no point does the court's opinion provide justification (or indicate whether the FCC has provided a justification) for why competition would hasten development of the technology under this allocation but would retard it in regard to the cellular system. The National Association of Radiotelephone Systems petitioned the U.S. Supreme Court for review of the appeals court decision. On May 24, 1976, the Court denied certiorari (review) without giving any reasons.[51] Thus, the Court of Appeals decision stands as the final word to date on this regulatory activity.

DOCKET 18262: INPUT AND CONSIDERATIONS

Following the First Report and Order and the Second Notice of Inquiry in 1970, the terms of the controversy changed and the FCC was subsequently besieged with petitions advancing economic and technical contentions that would ultimately determine the terms of competition and regulation set forth in the Second Report and Order of 1974. These issues will be analyzed in a later section. The present section reviews the considerations that went into the 1970 Order as reflected by the detailed responses in the First Report and Order to submissions made by the interested parties.

The central issue faced in Docket 18262 was whether UHF channels 70-83 should be reallocated to the land mobile services. Major broadcast interests, as well as a number of smaller independent television stations, assembled an impressive array of technical studies purporting to demonstrate that the perceived frequency shortage in land mobile was an artificial one brought about by FCC mismanagement and wasteful block allocation policies. The broadcasters asserted that existing data on spectrum utilization were inadequate to justify the conclusion that the land mobile services required additional allocations and that predictions of land mobile growth in the coming decades were seriously overstated. They suggested that improved coordination among private users, greater reliance upon the more efficient common carriers, and increased employment of the more efficient one-way signaling services would diminish the pressure for expanded spectrum allocations. Broadcasters pointed to the commission's and the Congress' historical commitment to an 82-channel nationwide television network and stressed the importance of channels 70–83 to the establishment of a localized television network, as well as to the full development of educational broadcasting. Finally, they argued that broadcasting was the most efficient and productive element in the national economy and served a much larger segment of the population than the land mobile services.

AT&T, GTE, Motorola, the National Association of Manufacturers, the American Petroleum Institute, the American Automobile Association, and a host of lesser-known organizations and industrial interests fought with equal vigor and resources in support of the proposed reallocation. Their basic position was that the extreme congestion of land mobile frequencies could only be alleviated through the allocation of additional spectrum. By contrast, the anticipated growth of UHF following the ACRA had not materialized and spectrum utilization in the UHF frequencies appeared highly inefficient. It was argued that greater reliance upon common carriers would not provide a solution to the small user problem.

The issue that generated the greatest number of objections from broadcasters, particularly the smaller independent operators, was the proposal in Docket 18262 to reduce by one-half the frequencies allocated to the use of Studio-Transmitter Links (STL). Opponents of this reallocation asserted that the benefits accruing to land mobile from use of an additional five MHz of STL spectrum did not outweigh the burden imposed on the broadcast users of those frequencies, claiming that STL was the only practical means of linking individual studios to transmitters when landline linkages were not possible. Mobile interests, however, argued that these frequencies were required to relieve the congestion in those services and efficiencies achieved by using STL frequencies would counterbalance the loss of spectrum. Despite their conviction, broadcasters were not successful in persuading the commission either to abandon the proposed reallocation or to minimize its impact on STL users by taking the upper five MHz or confining the transfer to urban centers.

A second proposal in Docket 18262 that generated significant controversy involved land mobile sharing with broadcast translators on a coequal basis in the top 25 urban areas. The mobile interests and broadcasters had different objections to this particular provision, but again the mobile interests prevailed. Agreeing that retention of the original cochannel protection proposal would seriously limit the relief offered land mobile by the reallocation, the commission decided to accommodate translators in the UHF channels below 806 MHz (channel 69). Particularly disturbing to the dissenters of this measure was the decision to exclude low-power community educational stations from operating in the 900-MHz band as proposed in Docket 14229. Educational interests, including the Corporation for Public Broadcasting, the Joint Council on Educational Communications, and the National Association of Educational Broadcasters, argued vigorously for preservation of land community outlets by retention of the full 70-channel capacity, although they expressed a willingness to accept interim and secondary sharing of underutilized UHF channels, pending full ETV development.

Rather than attempting to disparage the value or the needs of educational broadcasters, proponents of the reallocation emphasized the importance of land mobile communications and their urgent need for additional spectrum. The commission, persuaded by land mobile interests and satisfied that adequate provision for the needs of educational broadcasters could be made elsewhere in the spectrum, firmly reiterated its belief that exclusive land mobile use of the 900-MHz band was essential.

The preceding considerations are detailed in the submissions of the interested parties and reflected in the commission's own Report and Order. Unlike judicial proceedings, however, the formal record of the agency's decision does not embody the full explanation of the administrative decision, and a panoply of congressional, executive, and private pressures may have been focused upon the agency that played a major role in the formulation of its policy.

BEYOND THE RECORD: INFLUENCES IN THE POLICY SYSTEM

Although nominally independent, federal regulatory agencies such as the FCC are responsive in varying degrees to pressures from Congress, the executive branch, and the private sector. These pressures can be brought to bear on the agency by diverse means.

Given this context, we will look at some of the influences on the FCC in their policy determinations by noting activities of the Congress, input from the executive branch, professional-technical groups and consulting organizations, and actual parties involved in the proceeding of Docket 18262. Congress, of course, is in a particularly strong position to influence the direction of administrative action through its participation in the appointment process, its appropriations policies, and its ability to initiate hearings and investigations.[52] Individual congressmen (especially high-ranking members of influential committees) and their staffs may also wield considerable behind-the-scenes power over administrative personnel. [53]

The House of Representatives, and to a lesser extent the Senate, persistently pressed the FCC for expeditious action to reduce what was perceived as a crisis in land mobile communications. During 1968, the House Select Committee on Small Business, Subcommittee on Activities of Regulatory Agencies, held a series of hearings to dramatize the seriousness of the problem.[54] The committee noted that while broadcast interests had been awarded 87 percent of the nongovernmental spectrum below 960 MHz, land mobile services had been given a mere 4.4 percent—despite their great importance to the economy in general and to small business in particular.[55] The committee therefore advocated "ample, additional, usable frequency spectrum to be allocated without delay for this means of communication," and also suggested that the FCC might consider undertaking a general reappraisal of its position on television broadcasting in an effort to upgrade spectrum efficiency. [56]

In 1969, the committee again held hearings to assess the progress made during the previous year toward easing problems in land mobile. FCC commissioners appeared as witnesses and agreed to hear the testimony of committee witnesses on the plight of land mobile users but no significant action was taken.[57] Federal agencies are vulnerable to influence through congressional and executive control over their purse strings, and in 1967, 1968, 1969, and 1970, the appropriations subcommittees of both the House and the Senate emphasized their impatience with the commission's apparent inability to act effectively on the land mobile problem. [58] By the time appropriations hearings were held for Fiscal Year 1971, rule-making proceedings on Dockets 18261 and 18262 were well under way. FCC Chairman Burch provided the Senate committee with a detailed explanation of the commission's proposed actions, and proclaimed that "this reallocation will provide long-term relief for the land mobile systems. . . ." [59]

It is not easy to assess the precise impact of such congressional activity on the progress achieved by the FCC in its disposition of the land mobile dockets. However, in view of the fact that Docket 18262 was before the commission for eight years, it is difficult to conclude that Congress significantly influenced either the rate or direction of the docket's progress through the regulatory process.

Input from the executive branch seemed to have greater substantive importance. Executive influence may assume a variety of forms: appointment of the FCC commissioners and designation of their chairman, influence in the selection of key staff members, budgetary oversight through the Office of Management and Budget, maintenance of close relationships with FCC personnel through the Office of Telecommunications Policy, and direct participation in administrative action through submissions, both formal and informal, on important matters pending before the commission.[60] In descending order of importance, some of the major executive actions and an assessment of their significance in the land mobile decision are outlined below.

The Office of Telecommunications Policy (OTP), established in 1970, completed in 1973 a report evaluating its own activities over a three-year period and making some general assessments of spectrum management policy.[61] The report acknowledged the critical situation in land mobile and observed that the major barrier to finding adequate frequencies for those services, and for other developing systems of communications, was the dominance of television in the sub-1000-MHz spectrum, much of which was only lightly used.[62] Echoing many of the sentiments of earlier congressional committees, the OTP suggested that the FCC seriously consider a thorough reassessment of its position on television frequency allocations. The commission's actions to alleviate the land mobile problem were catalogued in the report, and the OTP seemed generally to approve the measures being taken. More significantly, OTP's submission to Docket 18262 stated as a major issue the decision as to whether the "increased availability of mobile communications services is best achieved by a regulatory technology or by the creation of a diverse competitive environment."[63] OTP largely succeeded in its attempt to persuade the FCC to encourage the maximum competition consistent with efficient service.[64] In addition, the pattern of spectrum allocation endorsed by the OTP was also highly influential in the FCC's final decision, particularly its concern for flexibility through the creation of substantial spectrum reserves. Moreover, many of the specific suggestions of OTP directed toward stimulating competition also seem to have been influential.[65]

The Antitrust Division of the Department of Justice was also actively involved in Docket 18262. In August 1970, the Justice Department urged the commission to reconsider its tentative allocation of 75 MHz for exclusive use by wireline carriers to develop a cellular system and warned of the undesirability of foreclosing Band 9 to the Radio Common Carriers.[66] Again, in August 1973, the department observed that allocation of 75 MHz to the WCCs would "seriously damage the ability of the RCCs to participate in a meaningful fashion in the growth of the land mobile system envisioned by the Commission."[67] The Department of Justice viewed the docket as providing the commission with a unique opportunity to expand the role of competition in the communications industry.[68] Feeling that competition in the dispatch market could best be encouraged if WCCs were prohibited from offering such services,[69] the Justice Department also urged the FCC to adopt a policy that would stimulate competition in the mobile equipment manufacturing market by forbidding WCCs from "owning, manufacturing, supplying or maintaining user mobile equipment,"[70] and suggested that similar proscriptions might be imposed in the base station market.

The President's Commission on Law Enforcement and the Administration of Justice made an early contribution to the debate on the land mobile problem.[71] Citing the heavy dependence of law enforcement agencies on mobile radio and the deteriorating quality and efficiency of those communications as evidence, the

Task Force Report on Science and Technology urged the FCC to consider allocating additional frequencies and to accord police and emergency services priority in those frequencies, as well as proposing more efficient methods of using spectrum already allocated. [72]

In December 1968, a Presidential Task Force Report on Communications Policy attributed the growing congestion and consequent degradation of land mobile services to spectrum mismanagement by the FCC. [73] The Task Force proposed (1) that land mobile be authorized to use certain UHF channels; (2) that systems be developed that would permit shorter base station separation distances; (3) use of common carrier mobile radio services by smaller volume users; (4) adequate channel-loading criteria; (5) substantial discontinuation of suballocation of land mobile frequencies by user class; and (6) incorporation of public safety radio services into government frequency allocations. [74]

The most politically disinterested input the FCC received during its consideration of the land mobile problem—and possibly the least influential—was provided in studies performed by the Stanford Research Institute (SRI) [75] and the Joint Technical Advisory Committee (JTAC). [76] The Stanford group, working under contract to the commission, labeled their study "the first comprehensive data base and data analysis of the occupancy of the land mobile spectrum." [77] By monitoring radio traffic in New York, Los Angeles, and Detroit, the institute reached the conclusion that serious maldistribution, fostered by the FCC's block allocation practices, existed throughout the land mobile spectrum. [78] It suggested that implementation of more effective management techniques based on regional flexibility and intraservice sharing would relieve much congestion,[79] although observing that the measures it was proposing would offer little relief from the "near-saturation" situation in New York and would accommodate only limited growth in Detroit.[80]

The JTAC, in response to a White House request, consumed nearly four years and two million dollars in its own investigation of the state of land mobile services. This committee found that the FCC's spectrum management techniques tended to be too short-term and that data on spectrum utilization were insufficient to provide the basis for rational allocations,[81] but noted that the FCC possessed neither the resources nor the personnel to formulate and implement long-range planning techniques. Criticizing the current block allocation of spectrum,[82] the report proposed means for more systematic analysis of spectrum needs.[83] Like the SRI report, the JTAC document might be viewed as supporting the broadcasters' positions; nevertheless the commission specifically refused any proposal to consider or appraise the plight of individual services.

Certain parties aggrieved by the regulatory approach adopted in Docket 18262 have suggested that more sinister forces were operating in the background of the administrative process. Jeremiah Courtney, a Washington lawyer associated with the radio common carrier industry, observed that in Docket 18262 "private, off-the-record conferences were . . . the rule, not the exception."[84] Citing an item from *ACTION,* a trade publication of the National Association of Business and Educational Radio, Inc., Courtney reported that NABER representatives had appeared in private session before the five sitting commissioners in

January of 1974–after the time for submission of record comments had expired. He asserted that this constituted a "bold *en masse* attempt *dehors* the record to influence the outcome of Docket 18262 in all of its controversial aspects . . . [t]he harmfulness [of which] . . . cannot be doubted."[85] Courtney also cited an item from *Broadcasting* magazine that reported on the aggressive lobbying tactics of Motorola representatives and commented that "according to FCC insiders, [Motorola Lobbyists were] making periodic visits not only at commissioners' offices but also in technical areas where ultimate recommendations will be made."[86] Furthermore, Motorola president Robert Galvin's service as finance chairman to the 1968 Republican campaign and the favorable treatment accorded Motorola at the hands of the FCC during a Republican administration were, in Courtney's words, "thoroughly disquieting."[87]

It is impossible, in the absence of additional information, to assess the extent to which these diverse inputs are reflected in the FCC's decisions in Dockets 18261 and 18262. What is clear, however, is that the commission's proclivity to adopt ad hoc solutions to frequency management problems and the absence of any comprehensive, long-range policy of spectrum allocation render the agency more vulnerable to external pressures and tend to make such influences less easily detectable. This condition, always undesirable in ostensibly independent administrative agencies, is particularly so in the FCC given its history of responsiveness to ex parte presentations and political influence.[88] That political conditions were not totally without influence in the decision to allocate additional spectrum to land mobile services is indicated in Commissioner Johnson's concurring opinion to Dockets 18261 and 18262:

> Finally, I believe that this scheme is simply a response to a political situation. Land mobile operators have become increasingly upset with the regulation dispensed by this Commission. Our response is not a well-thought out plan; it is not the result of planning; rather it is an immediate response to an immediate need.[89]

REGULATORY ASSESSMENT: THE SUBSTANTIVE ISSUES

We now undertake an analysis of the specific factors considered by the FCC in reaching its decisions in terms of broader themes that relate to the mandate of the FCC.

The Responsibilities of the FCC

In assessing the decisions that were actually made, one must examine the options and requirements set forth for the FCC in its legal mandate. The

Communications Act of 1934 is the primary legal charter of the FCC. Under Title II, the FCC regulates interstate communications common carriers, and under Title III it is responsible for allocating frequencies of the electromagnetic spectrum and for licensing authorized users of spectrum. The general powers of the FCC are outlined in the mandatory provisions of Section 303. In reaching its licensing and allocative decisions, the commission is to be guided by considerations of "public convenience, interest, and necessity." The standard is necessarily broad and flexible because it must be applied in both quasi-adjudicative and quasi-legislative contexts.[90] For a quasi-adjudicative proceeding such as passing upon a license renewal application, where the applicant has a substantial "property" interest at stake and vigorous judicial review is probable if an adverse decision is reached, the standard has been considerably refined and objectified. Within a quasi-legislative context such as spectrum allocation, where the commission is implementing its perception of the relative social utility of alternative uses of the radio spectrum, the standard is far more amorphous and the intensity of judicial review is considerably diminished.

Court rulings commonly emphasize the expertise of the commission and the need to retain flexibility so that rapid changes in communication technology can be accommodated without the necessity for congressional intervention. Although courts are generally less inclined to leave undisturbed administrative interpretations of law (as distinguished from findings of fact), the *Philadelphia T.V. Broadcasting Company*[91] case illustrates the unusual deference the FCC has been accorded when the commission is attempting to incorporate a new technological development into the existing regulatory pattern. The court concluded that the FCC was entitled to considerable leeway in choosing "which jurisdictional basis and which regulatory tools will be most effective in ordering the congressional objective."[92] The decision of the Court of Appeals concerning Docket 18262 also reflects this view.

It is reasonable to assert that in some areas of its jurisdiction, the public interest is indeed what the FCC says it is. In these areas reviewing courts have tended to require only that the commission's decision be rational and have some warrant in the record. Thus, in general, there is a lack of statutory and judicial guidelines, and the FCC, in determining whether particular allocations of spectrum are in the public interest, is not circumscribed by external limits. There is, however, one factor that the commission must consider: the preservation of competition. Statutory language in sections 313 and 314 of the Communications Act has been supplemented by two important U.S. Supreme Court decisions in *FCC v. Sanders Bros. Radio Station*[93] and *National Broadcasting Company v. U.S.*,[94] making clear Congress' intent and the FCC's authority to preserve competition in the noncommon carrier uses of radio.

Competition and Regulation

After 1970 the main focus of the policy deliberation concerning Docket 18262 shifted from whether there would be a frequency allocation to how the reallocated frequency would be divided and regulated. During that second phase of the deliberations, issues of competition in a regulated market were particularly controversial. In order to gain a better understanding of the nature of that controversy, we shall examine some broader questions of regulation and competition.

Beginning with the establishment of the Interstate Commerce Commission in 1867, the last century has witnessed a proliferation of federal and state regulatory bodies impinging upon nearly every aspect of the nation's business activity. The dominant characteristics of regulation by independent commissions are rate control and restrictions on entry. The need for profit control is premised on the notion that since competitive forces cannot or do not operate to equate price with marginal cost, government should intervene to ensure that regulated enterprises receive a "reasonable rate of return," but nothing more. Limitations on entry have been justified by the presence of natural monopolies; the social unacceptability of economic failure and of destructive competition in regulated activities; and the need to guarantee service where it might not otherwise develop at acceptable price levels.

Current sentiment is strong that the administrative and competitive costs of governmental regulation outweigh its public benefits. A Nader report has estimated the annual costs of regulation-generated economic waste at $16–24 billion, including $8 billion in the communications sector alone.[95] Part of the problem is procedural: as economic activity has become more dynamic and complex, limited technical and financial resources and bureaucratic ossification have prevented the commissions "from responding effectively to economic trends and changes in technology, industry structure, and public needs."[96] Economists agree that an administrative body cannot practically—and maybe not even theoretically—regulate prices in a manner that simulates the pattern exhibited in a competitive market. Ineffective price control is believed to conceal monopolistic profits, increase prices, reduce quality, discourage innovation, and encourage inefficient use of resources. These problems result largely, it is argued, from delay, inefficiency, lack of expertise, inadequate information, and undesirable politicization of the regulatory process. While such conditions may be ameliorated by administrative reform, other more formidable objections to regulation remain.

Critics of regulation argue persuasively that much of what is presently regarded as "natural monopoly" is, in fact, unnatural monopoly that could not exist in the absence of government support. They do not challenge the notion that regulation is indicated where the market has failed; rather, they contest popular conceptions of the dimensions of market failure.[97] It is suggested that the removal of regulatory restraints—popularly known as "deregulation"—in areas of the economy where competition can flourish would increase industrial efficiency and reduce governmental expenditures. It has also been suggested that the kind of cross-subsidization presently sanctioned by administrative agencies is inequitable and regressive. Under this view, if it is determined that maintenance of an economically nonviable service is socially desirable, general tax revenues—

not a system of financing that burdens commercially successful systems—should be the source of support.[98]

Advocates of deregulation disagree on the proper scope of that undertaking. Others assert that the market has not failed, indeed that it accurately reflects consumer preferences on these issues.[99] Moreover, deregulation on the scale contemplated by aggressive proponents of the concept would constitute a major departure from the national historical trend toward greater governmental regulation of private economic activity. A comprehensive program of deregulation, although economically judicious, may be politically unfeasible.

A much more modest proposal for introducing increased competition into regulated industries would involve a redefinition of the public interest mandate by the regulatory commissions themselves; the commissions could legally establish a presumption that competition would advance the public interest in many areas of regulated activity. In addition, the usual judicial deference to the decisions of regulatory bodies could be modified to favor competition (within the statutory limits) and the adoption of a more liberal attitude toward applying antitrust laws to regulated industries.

Competition and Innovation

Intimately related to competition in the regulatory process is innovation. The importance of technological innovation in the communications industry cannot be overstated, and a brief review of the major regulatory issues concerning innovation will provide the context for a better understanding of the FCC decision in Docket 18262.

Advances in technology make possible more intensive and extensive use of the radio spectrum. As a result, the FCC's ability to accommodate growing demands for competing uses of the spectrum depends greatly upon the rate of technological progress within the communications industry. Through its absolute control over whether and under what circumstances new uses will be granted entry into the spectrum, the FCC has become a major force shaping the development of communication technology.

Many economists and other students of the regulatory process agree that the beneficial aspects of regulation, such as preventing monopolistic abuses of power, may be outweighed by the deleterious effect of regulatory policies on the rate of technological innovation.[100] Thus, in its 1971 study of regulatory agencies, the President's Advisory Council on Executive Organization (Ash Council) recognized that regulation has tended to inhibit technological progress in regulated industries.[101] Since innovation is frequently a response to the incentive for more efficient performance in a competitive environment, limitations on competition in a regulatory framework—especially via restrictions on entry—are directly related to a poor record of innovation. William Shepard has criticized the FCC for stifling innovation by failing to encourage competition in the communications sector.[102] Furthermore, long delays in the introduction of new services are caused by the protracted character of the administrative process itself.[103]

Although the Ash Council suggested that regulatory inhibition of technological change is largely an unintended product of administrative inefficiency, Roger Noll observes: "Another view is that regulatory agencies intentionally delay or prevent many beneficial technological changes,"[104] citing the ICC and the FCC as two prominent examples. He regards the FCC's prohibition of foreign attachments of the switched communications network and restrictions on developing pay television as impediments to technological progress.[105]

While the administrative process is frequently unreasonably slow, and it is widely believed that the FCC "is close to being overwhelmed by the controversies created by each new application of technology,"[106] the true explanation for some lags in the introduction of new technology may be more complex. Noll asserts that "regulatory agencies have delayed or prevented a number of technological changes that threatened either to shift substantial business from one regulated firm or industry to another or to result in substantial business from one regulated firm or industry to another or to result in substantially less profit for regulated firms generally."[107] He attributes these policies to controversial methods of calculating the costs of abandoning old technology and to a distaste for the uncertainty inherent in any innovation. To this a third and possibly fundamental factor should be added: the regulatory body's perception of the public interest. Few today would maintain that technological change per se should be automatically welcomed. One may argue that the FCC's calculation of the social and economic costs implicit in the introduction of new technology is often more intuitive than systematically mathematical, and one may legitimately disagree with its conclusions; but that the public interest cannot be ascertained in the absence of such considerations is indisputable.

Telecommunications Regulation: Issues of Competition and Innovation

In discharging its common carrier regulatory responsibilities in the area of telecommunications, the FCC is inevitably and continually faced with the looming presence of American Telephone and Telegraph, the largest corporation in the world. Critics have alleged that controversial accounting and rate-base calculation techniques, wasteful rate-base expansion, conservative debt-equity ratio, inefficient capital investment, generous self-dealing, and a host of other abuses deprive consumers of economies that could be made possible by introducing greater competition into the telecommunications industry.

The commission's 1968 *Carterfone* decision permitting interconnection of independently manufactured devices to the Bell system illustrates both the feasibility and the desirability of increased competition.[108] Yet AT&T grudgingly yielded to expanded competition only when it became inevitable. When competition has forced a technological advance, some argue that AT&T has then made a determined effort to dominate the new technology and eliminate potential rivals.[109] Nevertheless, in almost every instance where the FCC consented to expand competition, prices declined and services were diversified.

Contemporary critics of regulatory policy assert that AT&T's "natural" monopoly is much less natural than is widely believed, suggesting that in certain areas greater competition is both feasible and in the public interest.[110] Leonard Waverman argues that regulatory policies often perpetuate monopolies, even after they cease to be economically justified, because of the entrenched monopolies' political strength and the relative weakness of procompetitive interests.[111] But AT&T vigorously rebuts its critics. Arguing that the industry's economies of scale are such that the Bell system as a whole should be considered a natural monopoly, AT&T maintains that introducing greater competition would not be economically beneficial to consumers.[112] Concerning innovation, the corporation asserts that, regardless of the situation in other regulated industries, the Bell system has an enviable record. Moreover, AT&T claims the new entrants into the telecommunications market, allowed by liberalized FCC common carrier policy, have not fulfilled the commission's expectations for new innovative services.[113]

Competition and Innovation in Land Mobile Communications

The commission's historical attitude toward and impact upon technological development in the land mobile radio services has been schizophrenic. As early as 1949 Bell Labs began seeking spectrum allocation for development of a high-capacity interconnected mobile telephone service, yet it was nearly 20 years before the FCC began to respond positively to these demands. At the same time, the commission encouraged technical advances that permitted more intensive use of allocated spectrum by denying additional spectrum to land mobile services. Of course, such advances carry with them higher equipment costs and possible diminutions in quality; in every proceeding designed to reduce bandwidths the FCC must weigh the costs against the disutility of spectrum reallocation.

The commission's decision to open Band 9 to land mobile users represents a decision to encourage the development of both new equipment and new systems of mobile communication. Its decision to allocate 40 MHz immediately and to reserve an additional 30 MHz for use by the interconnected cellular system can also be viewed as a major incentive to technological advance.

Since the 1968 *Carterfone* decision, the FCC's policy on competition as a mechanism for encouraging efficient spectrum use and rapid technical innovation has consistently been positive, and Docket 18262 is one of the most recent manifestations of this healthy development. To be sure, competition is only one factor in the choice of which regulatory scheme will most effectively serve the public interest. The Supreme Court has clearly indicated that there are limits beyond which the promotion of competition is antithetical to the FCC's statutory obligations.[114] In fact, as the Court of Appeals recognizes, the FCC's reliance on a relationship between competition and technological development is inconsistent: the cellular allocation is justified as stimulating technology in a framework that is apt to be noncompetitive, and yet the allocation to the SMRs is claimed to rely on competition to stimulate the technology.

One of the commission's most significant departures from established policy was its abandonment, in Docket 18262, of the controversial system of block allocation. In allocating spectrum instead by system type and service, the commission hoped to "allow the market to determine ultimately how much spectrum is utilized by the various types of users."

We have noted earlier that the FCC has authorized any party evidencing the necessary resource base to proceed to develop an experimental cellular system. Although superficial examination might suggest that a more rapid and efficient exploitation of allotted frequencies will result from this policy, it seems likely that wireline carrier hegemony will be quickly established in the cellular service.

In a decision vigorously contested by the wireline carriers—and heartily endorsed by equipment manufacturers, the OTP, and the Department of Justice—the FCC prohibited the wireline carriers from manufacturing (except in the developmental stages) equipment to be used in the cellular systems. It was apparently believed that measures short of a total prohibition of this nature would enable the wireline carriers ultimately to dominate the equipment market. Although this is consistent with the *Carterfone* thesis that competition breeds innovation, it may unfortunately deprive the industry of the vast research and development resources of the wireline carriers. It is not unrealistic to conclude that even with these limitations, however, the economic incentives are sufficiently strong to precipitate a major commitment of resources by the wireline carriers to developing equipment and systems capable of operating in Band 9.[115] (Another powerful incentive toward the expeditious exploitation of the 900-MHz band by American manufacturers will be Japanese competition. This matter will be discussed in detail in Chapter 13.)

By requiring the establishment of independent operating companies and the filing of all contracts between affiliates, the FCC will attempt to prevent cross-subsidization of wireline carriers operating cellular systems. Aside from their general unfairness, cross-subsidies tend to distort the true costs of a product and may lead to the misdirection of productive resources into economically undesirable areas. The commission, although committed to the development of a nationwide interconnected mobile telephone system, is determined to avoid repeating the UHF debacle. If demand does not materialize or if the technology proves too expensive, the commission appears prepared to reallocate the reserve frequencies tentatively assigned to cellular operators.

The commission has also sought to promote competition in the trunked and conventional systems. To discourage downstream vertical integration and domination of these markets by the manufacturers of mobile communications equipment, the FCC has limited such enterprises to the development of one trunked system for use in assessing the technical and commerical feasibility of such systems. Furthermore, manufacturers cannot establish more than one five-channel conventional system or five one-channel systems in any single market.

It should be noted that the explicit concern with competitive factors discussed here was part of a broader attack on the regulation of rates and entry that has been manifested in recent years. This attack stems not only from concern for the inefficiencies of regulated industries, but also from larger questions of political philosophy: the introduction of market forces reduces the need for administrative intervention. The relationship between competition and regulation is an inverse one; to the extent that competition is present, administrative control need not be. The desire to reduce the dimensions of the federal bureaucracy remains a recurring theme of recent administrations.

THE PROCEDURAL ISSUES: THE SCOPE OF PUBLIC PARTICIPATION

The increased emphasis on competition reflects a decision by the FCC to permit the public, through the mechanism of the market, to express its will. Despite wide endorsement of this policy, the need for more direct and immediate popular participation in administrative decision making is also clear. Public participation is important both as a normative value to maximize political accountability and, in a more pragmatic sense, for its informational value. Often the method by which a political decision is reached will be as important as its substantive content. Broad and effective public participation tends to serve a legitimizing function that renders even an adverse decision acceptable to the contending parties. At the same time, the representation of more diverse interests in the regulatory process will provide the administrative body with more balanced information and an expanded range of alternatives. Thus, as one step in evaluating the decision-making process in Docket 18262, the scope of participation will be reviewed.

As we noted earlier, serious allegations about the integrity of the procedures have been raised. In a 1975 speech before the Georgia Association of Radio Utilities, attorney Jeremiah Courtney lamented the tendency of portions of the FCC staff to "synchronize their thinking and recommendations with the objectives of the tremendous lobbying efforts that Motorola has brought to bear on the staff of the Commission, and on the Commissioners themselves.[116] Although it may be natural for the losing party in a regulatory action to blame the result on lobbying, it should be noted that both on a social and business level the commission heard a great deal more from Motorola and AT&T than it did from other interests. While it is difficult to know whether the ultimate decision was in any way shaped by such contacts, the attendance of FCC commissioners and staff at Motorola-sponsored social events at least gives the appearance of impropriety. Perhaps it is not too much to demand that, like Caesar's wife, the agencies must be beyond reproach in this regard.

Beyond the question of the relative weight given to the interested parties is the larger issue of the general public's role in Docket 18262 and other decisions.

While all immediately affected interests were able to make their views known, there was practically no input from the broader public. In the seven-year rule-making proceeding, hundreds of submissions were filed and all the major affected industries were heard. Nonetheless, apparently no one in the FCC seriously addressed the question of whether, at this stage, the American public would want to underwrite a technology that ultimately promises to allow everyone to carry around his own telephone. There was no public participation and no input from public interest groups.

This is not to suggest that there was anything devious or surreptitious about the decision. But that is precisely the point. This was a decision potentially involving billions of dollars and a major addition to our nation's telecommunication services, and still the public did not participate. Even if there was no obvious detrimental effect to the public, a wider range of public attitudes should have been solicited. And so it is to this problem of public participation that we now turn.

The legal parameters of citizen participation in FCC rule-making proceedings are defined in Section 4 of the Administrative Procedure Act (APA) and in Title 47 of the Code of Federal Regulations at Sections 1.40–427 containing the FCC's own rules of procedure. Rule making is regarded as a quasi-legislative function; public participation enables the regulatory body to inform itself so that it can effectively discharge its statutory responsibilities.[117]

After notice of the proposed rule making is issued, a notice ordinarily appears in the *Federal Register*, and sometimes in the *Consumer News*.[118] The FCC's regulations provide that *any interested person* will be afforded an opportunity to "participate in the rule-making proceeding through submission of written data, views, or arguments. . . ."[119] In theory, the system seems designed to elicit broad popular participation so that the commission has available information and opinions representing a whole range of affected parties. The reality, however, sometimes falls far short of this ideal.[120] In the usual rule-making proceeding, the agency's scope of vision extends no further than the information and considerations brought to its attention by parties seeking to advance their own economic self-interest. The problem is compounded by the FCC's own paltry resources and lack of analytical capabilities.

This phenomenon of the so-called "captive" regulatory agency is neither unique to the FCC nor easily explained. It is not the product of corrupt officials or sinister regulatees but more the natural consequence of the intimate working relationship that develops between business and government. The near-monopoly on information possessed by the regulatees, the agencies' limited ability to engage in independent research and analysis, and the uninhibited talent flow between the regulator and the regulated inevitably impair the objectivity of the regulatory body.

Yet it would be a mistake to conceive of the "industry perspective" as some unchanging and monolithic entity. While this may sometimes be the case, there are frequent occasions when competing views and antagonistic interests contend for the favor of the commission.[121] This competition between regulated industries does not, however, alter the fundamental problem of agency passivity. Thus the issue of participation in decision making is intimately related to the informational needs of regulatory agencies.

Two changes in current regulatory patterns are immediately suggested by a description of the problem: enhancement of the commission's information-gathering and analytical capabilities, and expansion of the scope of popular participation in regulatory proceedings.

More inclusive representation of the public will inevitably mean that agency proceedings will be prolonged and the costs of regulation will be increased. Even though these additional costs may at some point yield a negative cost-benefit calculation, the agency cannot effectively discharge its obligation to act in the public interest without some conception of what constitutes that interest. Of course, if one looks at costs and benefits from a larger social perspective, public participation has frequently forced agencies to include items that would otherwise have been treated as externalities, and has pointed out otherwise unforeseen consequences that have then been taken into account and ameliorated in advance. Therefore, one should not be quick to conclude that the additional, and perhaps substantial, immediate cost of greater public participation is unjustified.

Certain procedural changes in FCC rule-making proceedings could facilitate greater public participation. It is widely believed, for example, that more effective methods of notifying interested groups of pending proceedings could be developed.[122] Deadlines for submissions are often too short—some believe by design—to enable many nonindustry groups to prepare an effective presentation before the commission. The commission requires multiple copies (14) of all comments, replies, and other documents filed in rule-making proceedings,[123] a practice that enormously increases the costs of making a presentation and far exceeds reasonable administrative needs. It has also been suggested that holding public hearings on a regional basis would enable the FCC to gauge more sensitively public sentiment on important issues before the commission.[124]

Such minor procedural modifications are mere palliatives and would not be likely to effect any major changes in the character of FCC rule-making procedures; a significant increase in public participation will certainly require more fundamental reforms. The issue that must ultimately be addressed is the manner in which public interest groups can be assisted in meeting the financial burden incident to participation in the decision-making process.[125] Estimates of attorney's fees range from $4,000 for representation in ICC tariff proceedings to $100,000 or more for participation in major rule-making proceedings before the FTC.[126] Those who believe that the public interest must be represented before regulatory bodies thus focus on the need for providing such groups with effective legal assistance.

The administrative law bar is highly specialized and is occupied almost exclusively with representing regulated interests. These interests have the greatest stake and can underwrite the most extensive legal work. As a result, the law firms that provide legal services to public interest groups seeking to appear before regulatory agencies can be counted on the fingers of one hand.[127]

Proposals for the provision of legal counsel to enable public interest groups to participate more effectively in rule-making proceedings may be categorized as follows: (1) compensation of private counsel for representing participating groups; (2) establishment of intraagency legal staffs to assist groups without resources to retain their own counsel; and (3) creation of an independent, government-wide body to represent the interests of citizen groups before all federal regulatory agencies.[128]

It is reasonable to believe that the most independent, diversified, and aggressive representation of consumer interests could be provided by the legal services model. Aside from the problem of political salability, the difficulty lies in determining who is to be represented by whom and on what basis fees will be awarded. The suggestion that each federal agency might establish an office of public counsel to provide legal assistance to groups wishing to appear before it has also been made. In 1971 an FCC Policy Review Committee report recommended the establishment of such an office, but the proposal was never acted upon.[129] Although consumer groups are concerned that in-house counsel would become dominated by the agency, proponents of the intraagency counsel model observe that what the attorneys lack in independence might be compensated for by their greater effectiveness. Clearly, safeguards would be needed to insulate the public counsel from agency pressures.

Another suggested alternative is the establishment of an independent agency or bureau to provide consumer groups or their interests with legal counsel in appearances before federal agencies. This body is conceived as a roving public interest advocate similar to the proposed agency for consumer advocacy.[130] The political viability of such a proposal is problematical, since past efforts have been unable to mobilize the requisite congressional support.[131] Supporters of the independent agency model argue that it incorporates the best features of both of the other models. It avoids the obvious conflict of the agency's own staff representing outside interests and could be set up in a way that would maximize its independence; yet it would have sufficient continuity, resources, and political clout to compete effectively with industry counsel for the agency's attention.

One additional problem seldom addressed in proposals aimed at providing legal services to public interest groups is the fact that attorneys cannot be expected to function effectively before regulatory agencies without access to a considerable body of expertise. Somehow the resources of the academic community and/or the government must be mobilized to provide public interest groups with the kind of expertise available to private interests. Here the potential strength of the independent consumer agency seems greatest: it could presumably develop a body of expertise separate from that of the regulatory agency, but it would have access to all the resources of the federal government.

The Limits of Public Participation

Facilitating the presentation of diverse views before regulatory agencies will undoubtedly assist these agencies in reaching results more consonant with the public interest than the present regulatee-dominated system. However, it is conceivable—even probable—that broadened public participation in agency rulemaking proceedings will fail to illuminate all issues relevant to rational decision making. Interest groups, public or private, are rarely actuated without a perception that the agency's contemplated course of action impinges upon their interests in a direct or substantial way. Pathbreaking regulatory activity, including, for example, the introduction of an innovative technology or a novel regulatory approach, may generate unanticipated secondary consequences that fundamentally alter the initial cost-benefit calculations. Such potential dysfunctions are likely to remain unrecognized, so long as the preponderance of the information available to the regulatory agency continues to be supplied by parties motivated by self-interest.

The addition of an unstructured public interest component to the present regulatory process constitutes a step in the direction of redressing the current imbalance in favor of regulated industries. It may achieve a state of myopic equilibrium, but only a dispassionate, systematic analysis of the long-term implications of agency decisions can generate conditions favorable to maximizing the public interest. Basically, the problem is that the FCC is ill-equipped to undertake a long-range comprehensive analysis in terms of budget and personnel and hence is forced to be a reactive body—even if its inclinations were otherwise. The FCC is burdened with too heavy an administrative load for its limited resources; much of the commission's personnel and financial resources are involved in routine and administrative matters, with relatively little attention being devoted to central policy planning.

The absence of a long-range planning and analysis capability within the FCC is evident in the proceedings associated with Docket 18262. Our review has shown that very little consideration was given to the social impact of the new systems under examination. Such problems as privacy rights, effects on transportation and traffic safety, and radiation hazards were not considered. In short, despite the large potential impact of these decisions, no broadly based assessment was performed. This is not to say that only negative effects should have been reviewed. As will be discussed later in this volume, the cellular system surely will have many positive effects, particularly in such areas as public safety and productivity, and these effects should be maximized. Unfortunately, however, the broad range of possible impacts (other than economic factors) was simply not dealt with. Nor is Docket 18262 an isolated example. Within the whole range of issues under the general rubric "telecommunications policy," problems of social impact are rarely addressed.

In subsequent sections of this volume, we shall return frequently to the issues raised in this chapter. In particular, an economic analysis of probable consequences of these decisions for the structure of the industry will be presented in Chapter 14. Issues associated with improved planning, mentioned just above, will be taken up again in Chapter 15.

NOTES

Citations of standard legal references are given in the conventional legal format. A list of common abbreviations used is given below. Further elaboration of this format can be found in *A Uniform System of Citation,* 12th ed. (Cambridge, Mass.: The Harvard Law Review Association, 1976).

Legal Reference	*Abbreviation*
Code of Federal Regulations	C.F.R.
Federal Reporter	F.
Federal Reporter, Second Series	F. 2d
United States Code	U.S.C.
United States Supreme Court Report	U.S.

1. Kenneth C. Davis, *Administrative Law,* 6th ed. (St. Paul, Minn.: West Publishing Co., 1977), Chapter 4.

2. U.S. Advisory Committee for the Land Mobile Radio Services, *Report* (Washington, D.C.: Federal Communications Commission, 1967).

3. Ibid.

4. "Order of Inquiry," Docket 11997, *Federal Register,* vol. 22, April 1957, p. 2684. 2684.

5. "Report and Order," Docket 11997, *Federal Communications Commission Reports,* vol. 39, March 13, 1939-June 30, 1965, pp. 567-98 (hereafter cited as *FCC Reports*).

6. Ibid., p. 594.

7. "Fourth Report and Order," Docket 14229, *FCC Reports,* vol. 41, September 7, 1950-June 30, 1965, pp. 1083-96; and *FCC Reports,* vol. 39, pp. 594-96. See also Roger Noll, Morton Peck, and John McGowan, *Economic Aspects of Television Regulation* (Washington, D.C.: Brookings Institution, 1973), pp. 99-104; Mark Green, ed., *The Monopoly Makers* (New York: Grossman Publishing, Inc., 1973), pp. 47-48.

8. See Green, *Monopoly,* pp. 47-48; Erwin Krasnow and Lawrence Longley, *The Politics of Broadcast Regulation* (New York: St. Martin's Press, 1973), pp. 96-102; and Noll et al., *Economic Aspects,* pp. 101-4.

9. 47 U.S.C. 303(s).

10. In the March 26, 1964, Report and Order, the commission observed that "the reallocation of UHF ... was at least worthy of consideration in 1957 (but certain measures taken since then give us) reason to expect that these developments ... will provide the impetus for expanded use of the frequencies allocated to UHF...." *FCC reports,* vol. 39 p. 595.

11. Ibid., p. 597.

12. U.S. Advisory Committee, *Report,* p. 2.

13. Ibid., p. 9.

14. See "Notice of Proposed Rule Making," Docket 18261, *FCC Reports,* 2nd series, vol. 14, August 2, 1968-November 8, 1968, pp. 297-303.

15. Ibid., pp. 298-99.

16. "Notice of Proposed Rule Making," Docket 18261, *FCC Reports,* 2nd series, vol. 14, p. 311.

17. See Henry G. Fischer and John W. Willis, eds., *Pike and Fischer Radio Regulations,* 2nd series, vol. 19, pp. 1681-1728, for a summary of submissions received by the commission on Dockets 18261 and 18262.

18. *Radio Regulations,* 2nd series, vol. 19, p. 1590.

19. Ibid., p. 1591.

20. Ibid., p. 1592.

21. Ibid., p. 1614.

22. "Nature of Inquiry," Docket 18262, *FCC Reports,* 2nd series, vol. 14, p. 311.

23. *Radio Regulations,* 2nd series, vol. 19, pp. 1663-64, released on May 21, 1970.

24. See, for example, *Radio Regulations,* 2nd series, vol. 19, p. 1681 et seq. Summaries of submissions by the Corporation for Public Broadcasting, National Association of Educational Broadcasters, and the Monterey County California Office of Education.

25. Ibid., p. 1664.

26. Ibid., pp. 1669-70.

27. Ibid., pp. 1670-71.

28. Ibid., p. 1615.

29. *Radio Regulations,* 2nd series, vol. 22, p. 1691.

30. *Radio Regulations,* 2nd series, vol. 33, p. 457.

31. *Radio Regulations,* 2nd series, vol. 22, pp. 1727-31.

32. *Radio Regulations,* 2nd series, vol. 30, p. 75.

33. Ibid., pp. 78-79.

34. Ibid., p. 79.

35. Ibid., pp. 79-80.

36. Ibid., pp. 81-83.

37. Ibid., p. 108.

38. Ibid., pp. 83-84.

39. Ibid., pp. 85-90.

40. *Radio Regulations*, 2nd series, vol. 33, p. 457.

41. Ibid., pp. 466-68.

42. Ibid., p. 467.

43. Ibid., pp. 460-63.

44. 525 F. 2d 630 (D.C. Cir., 1976).

45. Ibid., p. 636.

46. Ibid., pp. 637, 639.

47. Ibid., p. 637.

48. Ibid., p. 638.

49. Ibid.

50. Ibid., p. 639.

51. 425 U.S. 992 (1976).

52. See Walter Gellhorn and Clark Byse, *Administrative Law,* 6th ed. (Mineola, N.Y.: Foundation Press, 1974), pp. 109-27; Krasnow and Longley, *Politics of Broadcast,* p. 84: "The power of Congress over the Commission is pervasive and multifaceted. Since the FCC has neither the political protection of the President or a cabinet official, nor an effective means of appealing for popular support, Congressmen have little fear of political reprisal when dealing with the Commission"; also, Erwin G. Krasnow and Harry M. Shooshan, III, "Congressional Oversight: The Ninety-Second Congress and the Federal Communications Commission," *Federal Communications Bar Journal* 26, no. 2 (1973):81-82.

53. Krasnow and Longley, *Politics of Broadcast,* pp. 53-54.

54. U.S. Congress, House, Committee on Interstate and Foreign Commerce, Select Committee on Small Business, *Hearings Before the Subcommittee on Activities of Regulatory Agencies,* 90th Cong., 2d sess., 1968; Ibid., 91st Cong., 1st sess., 1969.

55. U.S. Congress, House, *The Allocation of Radio Frequency and its Effect on Small Business: Report No. 978,* 90th Cong., 2d sess., p. 4.

56. Ibid., pp. 32-35.

57. "We regard it as unfortunate that in the intervening twelve months (since the Committee's earlier report) we have seen little indication that any substantial progress has been made in correcting this situation by affording critically needed relief to land mobile users—and this relief is now much longer overdue." U.S. Congress, House, *The Allocation of Radio Frequency Spectrum and Impact on Small Business: Report No. 982,* 91st Cong., 1st sess., 1970, p. 28.

58. See U.S. Congress, Senate, Appropriations Committee, *Hearings on Independent Offices and Department of Housing and Urban Development Appropriations for Fiscal Year 1969*, 91st Cong., 1st sess., 1968, pp. 73, 17.

59. Senate, Appropriations Committee, *Hearings on Independent*, Fiscal Year 1971, p. 986.

60. Gellhorn and Byse, *Administrative Law*, pp. 128-42; Krasnow and Longley, *Politics of Broadcast*, pp. 44-50.

61. "The Radio Frequency Spectrum," 1973.

62. Ibid., pp. E-2 - E-3.

63. Executive Office of the President, Office of Telecommunications Policy, "Conclusions and Recommendations of the Office of Telecommunications Policy Regarding Land Mobile Radio Service in the 900 MHz Band," Formal submission to FCC Docket 18262, Executive Office of the President, Washington, D.C., 1973, pp. 14-15.

64. Ibid., p. 2.

65. OTP recommended that 40 MHz be allocated to "any mobile service to be offered on a non rate-regulated competitive basis. . . ." It contended that any suballocations within particular user categories (aside from the allocation to cellular systems) would be antithetical to the FCC's objective of encouraging technological innovation and economical service. It did not, however, propose the exclusion of the WCCs from the dispatch market. Instead, OTP supported WCC provision of dispatch services at regulated rates, but only as an adjunct to their mobile telephone operations.

OTP suggested authorizing the radio common carriers to provide "licensed but otherwise non-regulated mobile service . . . on a competitive basis," and noted that federal preemption might be necessary to ensure that competition not be impeded by state rate regulation.

To minimize the occurrence of interservice cross-subsidy, OTP would have prohibited telephone carriers from participating in "the non-regulated portion of the mobile communications market in their own telephone service area." OTP expressed no opinion on the issue of whether the WCCs should be permitted to manufacture and service equipment necessary to the operation of the cellular system, but it did express the opinion that mobile radio equipment manufacturers should not be prohibited from operating systems, as long as interoperability was required.

66. See Clay T. Whitehead to Hon. Dean Burch, August 17, 1973, Memorandum, Office of Telecommunications, Executive Office of the President, Washington, D.C., p. 2.

67. Ibid., p. 3.

68. Ibid., p. 2.

69. Ibid., p. 6.

70. Ibid., p. 8.

71. President's Commission on Law Enforcement and the Administration of Justice, *Task Force Report: Science and Technology*, prepared by the Institute for Defense Analyses (Washington, D.C.: U.S. Government Printing Office, 1967).

72. Ibid., p. 29.

73. President's Task Force on Communications Policy, *Final Report* (Washington, D.C.: U.S. Government Printing Office, 1968), p. 2.

74. Ibid., pp. 50-68.

75. T. I. Dayharsh and W. R. Vincent, *A Study of Land Mobile Spectrum Utilization* (Menlo Park, Calif.: Stanford Research Institute, 1969).

76. Joint Technical Advisory Committee of the IEEE and EIA, *Spectrum Engineering--The Key to Progress* (New York: Institute of Electrical and Electronics Engineers, 1968).

77. Dayharsh and Vincent, *Study of Land Mobile*, part A, p. iii.

78. Ibid., p. 74.

79. Ibid., pp. 153-55; part B, pp. 40-42.

80. Ibid., part B, pp. 41-42.

81. IEEE and EIA, *Spectrum*, p. 17.

82. "At present, substantial numbers of frequencies though idle, are rarely considered for possible selection and assignment because they have been allocated to users at classifications different from those of most applicants." Ibid., p. 77.

83. Ibid., p. 76.

84. Jeremiah Courtney, "FCC Docket 18262—Due Process of Law," *Communications,* September 1974, p. 6.

85. Ibid., p. 16.

86. Ibid., p. 17.

87. Ibid.

88. "The available evidence indicates that it (the FCC), more than any other agency, has been susceptible to ex parte presentations, and it has been subservient, far too subservient, to the subcommittees on communications of the Congress and their members." James M. Landis, "Report on Regulatory Agencies to the President-elect" (Washington, D.C.: U.S. Government Printing Office, 1960), p. 53.

89. *Radio Regulations,* 2nd series, vol. 19, p. 1622.

90. See Louis Caldwell, "The Standard of Public Interest, Convenience or Necessity as Used in the Radio Act of 1927," *Air Law Review* 1 (1930): 295-96.

91. 359 F. 2d 282 (D.C. Cir., 1966).

92. Ibid., p. 284.

93. 390 U.S. 470 (1940).

94. 319 U.S. 190 (1943).

95. Green, *Monopoly Makers,* p. 24.

96. President's Advisory Council on Executive Organization, *A New Regulatory Framework* (Washington, D.C.: U.S. Government Printing Office, 1971), p. 18.

97. See, for example, ibid., pp. 10, 90-94; Mark Green and Ralph Nader, "Economic Regulation v. Competition: Uncle Sam the Monopoly Man," *Yale Law Journal* 82 (1973):871-89; W. Shepard, "The Competitive Margin in Communications," in *Technological Change in Regulated Industries,* ed. William M. Capron et al. (Washington, D.C.: Brookings Institution, 1971); Leonard Waverman, "The Regulation of Intercity Telecommunications," in *Promoting Competition in Regulated Markets,* ed. Almarin Phillips (Washington, D.C.: Brookings Institution, 1975), pp. 207-39.

98. For example, rural electrification and the extension of telephone lines into rural areas at rates comparable to those obtaining in sectors of peak demand would be directly subsidized by tax revenues rather than cross-subsidized by users in low cost areas.

99. Ralph Winter, "Economic Regulation vs. Competition: Ralph Nader and Creeping Capitalism," *Yale Law Journal* 82 (1973):890.

100. Capron, *Technological Change,* p. viii.

101. President's Advisory Council, *A New Regulatory Framework.*

102. Shepard, "The Competitive Margin," p. 95.

103. Kenneth Goodwin, "Another View of the Federal Communications Commission," in *The Role of Analysis in Regulatory Decision Making,* ed. Rolla Park (Lexington, Mass.: Lexington Books, D.C. Heath and Co., 1973), p. 43.

104. Roger Noll, *Reforming Regulation: An Evaluation of the Ash Council Proposals* (Washington, D.C.: Brookings Institution, 1971), p. 24.

105. Ibid.

106. *Electronics,* July 25, 1974, p. 107.

107. Noll, *Reforming,* p. 24.

108. "In the Matter of Use of the Carterfone Device," Docket 16942, *FCC Reports,* 2nd series, vol. 13, pp. 420-27.

109. Beverly C. Moore, "AT&T: The Phony Monopoly," in Green, *Monopoly Makers,* pp. 84-90; Shepard, "The Competitive Margin," pp. 113-15.

110. Moore, "AT&T," pp. 93-94.

111. Waverman, "The Regulation of Intercity," p. 202.

112. Among other studies, AT&T cites as support for its position the following reports: Ira Horowitz, "Natural Monopoly and the Bell System," Formal submission to FCC Docket 20003, Exhibit 27, American Telephone and Telegraph Company, New York, N.Y., April

21, 1975; Stanford Research Institute, "Analysis of Issues and Findings in FCC Docket 20003," Formal submission of FCC Docket 20003, Exhibit 65a, American Telephone and Telegraph Company, New York, N.Y., April 1977; Systems Applications, Inc., "An Evaluation of AT&T's Multiple Supplier Network Study," Formal submission of FCC Docket 20003, Exhibit 66, American Telephone and Telegraph Company, New York, N.Y., April 29, 1977.

113. Arthur D. Little, Inc., "The Relationship between Market Structure and the Innovation Process," Formal submission to FCC Docket 20003, Exhibit 52, American Telephone and Telegraph Company, New York, N.Y., January 1976. See also Richard E. Wiley, "Address" (Remarks at the 87th Annual Convention of the National Association of Regulatory Utility Commissioners, Boston, Mass., November 3, 1975).

114. See *FCC* v. *RCA Communications, Inc.,* 346 U.S. 86, 93 (1953), where the Supreme Court commented "encouragement of competition as such has not been considered the single or controlling reliance for safeguarding public interest."

115. In fact, the Bell system has acknowledged already spending $20,000,000 on systems development for Band 9 operations. J. A. Baird, "Affidavit in Support of AT&T Motion to Modify Stay," Formal submission to FCC Docket 18262, May 15, 1975, p. 2.

116. Jeremiah Courtney, "FCC's New Common Carrier Mobile Radio Game Plan—Dealer's Choice, with Deuces, One-Eyed Jacks and Hole Cards Wild" (Remarks at the Annual Meeting of the Georgia Association of Radio Utilities, April 26, 1975).

117. Kenneth C. Davis, *Administrative Law Treatise,* 3rd ed. (Brooklyn, N.Y.: Foundation Press, 1972), pp. 139-40.

118. See James R. Michael, *Working on the System—A Comprehensive Manual for Citizen Access to Federal Agencies* (New York: Basic Books, 1974), pp. 265-68, for a description of information services available on FCC activities, their cost, and how to obtain them.

119. 47 C.F.R. 1.419 (1974).

120. See comments of Rosel Hyde in U.S. Congress, Senate, Committee on the Judiciary, Subcommittee on Administrative Practice and Procedure, *Responses to the Questionnaire on Citizen Involvement and Responsive Agency Decision-Making,* 91st Cong., 1st sess., 1969, p. 19.

121. Typical of this type of conflict has been the battle between broadcast and land mobile interests in Docket 18262 over the disposition of the UHF spectrum and the confrontation between AT&T and Motorola on the one hand and the radio common carriers on the other concerning the distribution and terms of use of the spectrum reallocated to the land mobile services.

122. See, for example, comments of Commissioner Johnson, U.S. Congress, Senate Committee on the Judiciary, *Responses to Questionnaire,* p. 67; Michael, *Working on the System,* pp. xiii-xiv; Ernest Gellhorn, "Public Participation in Administrative Proceedings," in U.S. Administrative Conference, *Recommendations and Reports* (Washington, D.C.: U.S. Government Printing Office, 1972), pp. 415-20.

123. 4o C.F.R. 1.419 (1974).

124. U.S. Congress, Senate, Committee on the Judiciary, *Responses to Questionnaire,* p. 63.

125. See Gellhorn, "Public Participation," pp. 405-6; Roger C. Cramton, "The Why, Where, and How of Broadened Public Participation in the Administrative Process," in U.S. Administrative Conference, *Recommendations and Reports,* p. 435.

126. Cramton, "The Why," p. 435.

127. Michael, *Working on the System,* pp. 278-88.

128. See Gellhorn, "Public Participation"; Cramton, "The Why."

129. Michael, *Working on the System,* pp. 289-90.

130. See "Introductory Remarks of Senator Kennedy," in U.S. Congress, Senate, Committee on the Judiciary, *Hearings on S. 3434 and S. 2544 to Provide for the Establishment of a Public Counsel Corporation,* 91st Cong., 2d sess., 1971, pp. 1-3.

131. U.S. Administrative Conference, *Recommendations and Reports,* pp. 13, 71. (Recommendation 5: Representation of the Poor in Agency Rulemaking of Direct Consequence to them.)

Chapter 4
THE COST OF
SYSTEMS AT 900 MHz

Erwin A. Blackstone
Harold Ware

A s the result of the regulatory decisions described in the previous chapter, most of the growth in land mobile communication systems will occur within the new allocation of spectrum near 900 MHz. In this frequency range, as at lower frequencies, costs will be determined by a number of factors, including service parameters such as message length, wait time, and signal-to-noise ratio, as well as range and capability for random access. The cost to the user depends greatly upon the number of users that can be accommodated, and this number is affected by the nature of service provided. For example, because the requirements for dispatch service are less stringent than those for telephone service, costs for the latter are higher. Thus, the nature of the service, the degree of technical complexity, the efficiency of spectrum use, and economies of scale all influence costs.

The factors determining costs are interrelated—service parameters will determine possible applications, and specification of applications will determine required service parameters. Similarly, technological complexity is directly related to spectrum efficiency and service parameters. The ultimate cost of the system, for instance, depends on its signal range. Signals at 900 MHz have a much shorter range than those at 150 and 450 MHz; in order to achieve the same range at higher frequencies, transmitters of higher power and receivers of higher sensitivity are required and are inherently more costly.[1] However, this explicit cost disadvantage may be partially offset by the fact that propagation characteristics at 900 MHz facilitate the use of small cell systems that are spectrally efficient. Yet, although individual base transceiver costs may not be higher than those at lower frequencies, more of them will be required and wirelines will be needed to link the cells. Also the mobile unit itself may be somewhat more costly than lower frequency units. Overall, then, one would expect the 900-MHz systems to cost more than those operating at lower frequencies, if they are produced at the same scale and use the same techniques. Nevertheless, costs projections to be discussed below suggest that some future 900-MHz systems may provide some services at lower costs than present lower frequency systems, because they are likely to be produced in larger quantities using more advanced methods. Were there sufficient spectrum available at lower frequencies, this cost advantage would not occur.

In this section the potential direct costs to the user of various systems are analyzed. Indirect costs, spectrum costs, and externalities will be treated later. Most of the data presented are from sources written between 1969 and 1972. While a summary table is presented at the conclusion of the chapter in constant dollars, the individual estimates have not been adjusted; disparities between the various quoted figures caused by inflation are not substantial, because inflation rates were moderate during the relevant period. It should also be noted that seemingly precise figures are estimates for developing systems, and the precision of the figures is, in fact, limited by a number of uncertainties associated with the projections.

ONE-WAY PAGING

One-way paging systems require the least investment per user of all mobile communication technologies (with the possible exception of Citizens Band). The cost depends, of course, on the type of paging system. For example, a tone-voice system accommodates fewer mobiles per channel or per system than tone only, requiring a higher investment per user.[2] In addition, the cost of the actual paging units is higher in a tone-voice system. A study by Arthur D. Little, Inc., reports that current pagers sell for $200 to $300 in small quantities—presumably the lower figure refers to tone-only pagers, while the higher figure corresponds to tone-voice devices.[3] Though estimates from various sources differ, the average total investment per mobile is approximately $300. To gain more insight into the factors determining this figure, we shall examine some of the individual cost estimates in more detail.

Richard Lane points out that because a large number of users can be accommodated on a given system, investment per user for fixed equipment (i.e., the base station and a manual or automatic control system) is no more than $100 for systems filled almost to capacity. He attributes roughly $200 to the mobile unit itself, and thereby obtains a range of $200 to $325 for the total investment per mobile.[4] More detailed information has been presented by Ax and Hatfield. They state that the radio equipment costs roughly $4,000 for a 300-watt base station, the antenna about $2,000, and the control equipment ranges from as little as $1,500 for a simple 400-subscriber system to more than $150,000 for a computer system capable of accommodating 10,000 mobiles. Table 4.1 indicates the range of costs for the various size systems.

Table 4.1: *Investment for One-Way Paging*

	100-mobile system	400-mobile system	10,000-mobile system
Total cost of base station equipment	$6,000	$7,500	$156,000
Base station equipment cost per mobile	60	19	16
Mobile unit cost	240	200	200
Total investment per mobile	$ 300	$ 219	$ 216

SOURCE: Figures are from U.S. Department of Commerce, "One Way Paging," Working Paper prepared for the Office of Telecommunications by Gene Ax and Dale N. Hatfield, Washington, D.C., May 20, 1974, pp. 22-23.

It is clear from Table 4.1 that there are economies of scale. With larger systems, the cost of the base equipment per user becomes a small part of the total investment per user. This is probably the reason most paging services are currently public systems supplied by the radio common carriers and the wireline or telephone common carriers.

The monthly charges for paging systems are based on the costs of maintenance, depreciation, labor, interest, and a profit margin. Typical monthly charges in the United States range from $15 to $25. Our estimates, using a five-year depreciation period, suggest that these prices could be lowered by about $5 per month and still provide profit margins that are similar to those in the rest of the communications industry; the difference may be caused by the demand presently exceeding the supply of these systems.

Monthly prices for more complex systems can be inferred from current costs in Europe (where several countries are planning nationwide and even Europe-wide paging systems). For example, in West Germany the caller telephones a code to a radio center, then the center transmits it to a receiver, triggering a beeper and one of up to four signal lights, each of which signifies a particular message. The German system involves a $20 per month charge for a single-code receiver plus a charge of $20 for each additional code. Thus a single-code pager in West Germany would cost the user less than 20 percent of the price of a conventional automobile telephone ($109 per month), with the initial cost providing similar savings ($800 versus $4,000). An extensive paging system is being planned for Sweden and will involve subscriber charges of $50 per year plus $0.05 per call. These paging units may be sold for under $250.[5] The cost of Japanese systems is discussed in Chapter 13.

At the present time, it appears that the demand for one-way paging systems can be met within the frequency ranges at which they presently operate. It is useful to consider the impact on cost, though, if demand rises sufficiently to necessitate the use of the 900-MHz band. We noted earlier that, other things being equal, operating at higher frequencies should tend to raise system costs. However, if demand becomes large enough to warrant using the 900-MHz band, it is possible that this disadvantage will be mitigated by economies of scale. A study by Arthur D. Little, Inc., states that "a personal paging receiver in the $50-70 cost range is possible utilizing modern MOS/LSI techniques with large production runs to meet the anticipated demand."[6] Similar savings resulting from economies of scale might also be achieved at lower frequencies. We do not have adequate data to give reliable estimates of the relative costs of paging systems operating in the high- and low-frequency bands, but it is reasonable to assume that use of the 900-MHz band for paging is not advantageous with respect to costs.

Our main conclusion concerning paging is that it provides a low-cost technology for a highly restricted range of mobile communications; it will be very competitive with other technologies for certain specific areas.

CONVENTIONAL SYSTEMS

Conventional systems are the least complex of two-way systems and are among the least costly (though not in terms of spectrum). This discussion will emphasize conventional systems used only for dispatch service, since these systems are not well suited for the provision of telephone service.

For a 150-MHz single-channel private system, the investment per mobile is roughly $1,000.[7] At 900-MHz, comparable systems of similar range will be more costly. Lane has given a figure of $1,090 for the investment per user at 900 MHz. While this implies only a 9 percent difference in costs,[8] others have suggested that a 30 percent cost differential is more likely.[9] Lane also suggests that the investment per user can be reduced to $700 through the use of shared repeaters.

Let us consider the cost of 900-MHz conventional dispatch systems in more detail. We shall use 1969 estimates made by Staras and Schiff[10]—$3,000 for the base station and $1,000 per mobile. (Note that this 1969 figure is larger than 1972 estimates for more complex cellular and trunked units to be discussed later. Perhaps the discrepancy can be explained by the fact that the complex units are intended to be purchased in large numbers for common carrier systems,

Table 4.2: *Initial and Monthly Costs for Conventional Systems*

Number of Mobiles	Investment/Mobile	Monthly Charge/Mobile
2	$2,500	$58
5	1,600	44
10	1,300	39
20	1,150	36
50	1,060	35

while the conventional units are to be used in small, private systems. In addition, the assumption of rapid production advances may be implicit in the estimates for cellular costs.) We shall also assume a depreciation period of five years and a monthly charge of $17 for other costs, including costs of the required lines from office to base station, to compute the investment and the monthly charge per mobile. This is displayed in Table 4.2.

Obviously, the rate of depreciation affects the monthly charge. If we use seven years instead of five years as the period of depreciation, the charge for ten mobiles falls from $39 to $32 per month.

The figures indicate that a conventional system with a large number of users would be a comparatively low-cost technology even at 900 MHz. However, because relatively large investments will be required to establish repeater stations, small firms with few mobiles will have a strong incentive to share the cost of base equipment (including repeater or relay stations) in order to maintain the cost per user at 150 or 450-MHz levels. Estimates of the monthly charge for these low-frequency systems, based on the above method of analysis, would be about $25 to $30 per mobile for relatively large-scale operations used to capacity. Typical common carrier tariffs for such service are considerably higher.

The table also illustrates the effect of the system's size (i.e., number of mobiles) on the cost; increasing the number of mobiles from two to ten decreases the investment per user by a factor of 1.9. Range is another important determinant of cost, since achieving greater range requires more powerful receivers and transmitters and/or repeater stations, especially at 900 MHz.

Our analysis of conventional systems has been limited to their use in dispatch service, because the economics of use for mobile telephone service is so unfavorable. Fewer mobiles can be accommodated with a given amount of spectrum and base station equipment for mobile telephone service (MTS) than for dispatch, since the quality requirements specified for 900-MHz MTS are far more stringent. It has been estimated that only 20 mobiles could be accommodated per MHz—assuming the same quality as the wired telephone system—compared with almost 2,000 mobiles per MHz for dispatch service. Consequently, even with shared repeaters, a conventional system would involve an investment per mobile telephone user of $5,741, compared to $669 for ordinary dispatch service.[11] It is clear why private conventional systems are not perceived to be a practical means of providing mobile telephone service.

TRUNKED SYSTEMS

Trunked systems are more complex than conventional ones. Additional costs are required for computers, and the mobile units are more expensive because they must operate on a greater number of channels. On the other hand, more users can be accommodated per MHz and per system, so that the additional costs may be offset by a greater sharing of the fixed equipment costs.

We can gain some insight into the costs associated with trunked systems by examining the "3-C" system proposed by Motorola. Designed for dispatch use and operating at 900 MHz, this system would require a total base equipment investment of $600,000 for base station transmitters, antenna, and control computers. Each mobile receiver would be priced at $750, and thus the total investment per user would be $1,050 when the system is filled with 2,000 units. Since this system is designed for dispatch purposes, the mobile unit will neither be duplex nor require a dialing device.[12] The expected monthly charge is computed in Table 4.3.

In the provision of dispatch service, the economies of scale inherent in trunked systems are evidently limited. Lane estimates that investment per dispatch user falls from $822 on a 4-channel system capable of accommodating 800 mobiles to $790 on a 16-channel system with roughly 4,000 mobiles. This relatively small difference may not be significant in view of the uncertainties in the analyzed data. In dispatch systems, the use of trunking should not result in major increases in efficiency or number of users, because dispatch channels tend to be intensively utilized. The potential for substantial savings from multiple channels is much greater for mobile telephone service, as will be discussed below. The FCC has restricted trunked systems to a maximum of 20 channels, perhaps because of the limited economies of scale beyond 20 channels.[13]

As with other systems, investment per user and monthly cost depend upon range and use of the system, that is, dispatch or mobile telephone. The cost of the mobile unit will also depend upon the number of frequencies on which it can operate.

We now turn to the cost of mobile telephone service provided by multichannel trunked systems. Lane estimates that the investment per user for MTS should be $2,539 on a 4-channel fully loaded system capable of accommodating 44 users, and $1,166 on a 16-channel system with 330 mobiles; thus the advantage of a larger number of channels is quite substantial. As expected, MTS is significantly more expensive than dispatch service, primarily because the fixed costs of the control and base station are spread over a smaller number of users. It

Table 4.3: *Expected Monthly Charge: Trunked Dispatch Service*

Base equipment: $600,000 for 2,000 units or $300 per unit
Mobile cost: $750

	7-Year Depreciation	5-Year Depreciation
Depreciation per month	$13	$18
All other costs, including* *maintenance and landline rental*	17	17
Total monthly charge	$30	$35

*We are assuming such costs to be the same for trunked as for conventional systems.

should be noted that for mobile telephone service, the number of mobiles per MHz approximately doubles when going from 4 to 16 channels (220 to 413), whereas for dispatch service there is only a 20 percent increase (4,065 to 4,827).[14]

We can obtain some independent estimates of the cost of mobile telephone service from a reexamination of the Motorola 3-C system. While it is designed for dispatch service, it could be modified to provide telephone service, and the cost can be extrapolated from the published figures for dispatch use. The 3-C system can accommodate 2,000 mobiles for dispatch purposes; if used for MTS, it could accommodate roughly 500 users.[15] The fixed investment per mobile would increase from $300 to $1,200 ($600,000/2,000 and $600,000/500). The mobile units would also have to be duplex, raising the cost of the mobile from $750 to perhaps $1,000. As a result, the investment per user could become $2,000 or $2,500, figures only slightly above the estimates of current cost for mobile telephone service. An annual revenue requirement of one-third of the capital investment (which is the capital turnover ratio for the telephone industry)[16] would result in a monthly charge of about $70. Actual charges could be somewhat lower than this figure, because the increased competition expected in the 900-MHz band may make it difficult for mobile service suppliers to maintain the same ratio of revenue-to-investment that has characterized the monopolistic fixed telephone industry.

Another estimate of the cost of MTS at 900 MHz using trunked systems can be obtained by considering current costs of systems operating at 450 MHz. AT&T's "Improved Mobile Telephone Service," which employs a four-channel trunked system at 450 MHz, has a per user investment of $2,000.[17] About 85 percent of that investment is attributable to the mobile unit. If we again assume that the annual revenues in the mobile sector are roughly a third of the total investment, as they are in the rest of the telephone industry,[18] we can calculate the annual revenues per mobile telephone to be $667 with a monthly charge of $55. For a 900-MHz system, one should add 15 percent to account for higher frequency operation, bringing the monthly charge to $63, which would cover all MTS costs, including maintenance.

It is worth noting that equipment for 450-MHz systems is probably not being produced in quantities large enough to take advantage of economies of scale. Consequently, future costs at 900 MHz could be considerably lower than current costs at lower frequencies. More widespread use could lead to economies of scale that would compensate for the increased costs associated with higher frequency operations. While using a greater number of channels may raise equipment costs, again the resultant trunking efficiency could lead to lower costs per user by accommodating more units.

On the basis of various estimates given above, we conclude that the monthly charge for 900-MHz trunked telephone service would range from $50 to $70.

CELLULAR SYSTEMS

The cellular system can be used to provide only mobile telephone service, only dispatch service, or a combination of the two. The combination of services is the most likely, followed by a system used exclusively for dispatch. Because the initial market for MTS is expected to be far smaller than that for dispatch service, a system providing only mobile telephone service is the least probable. It is important to emphasize that the initial investment and monthly charge per user depends significantly upon service parameters; systems with higher blocking probabilities can accommodate more users and hence reduce the shared equipment investment and cost per user. Therefore, we shall discuss the impact of modifying some of the service parameters.

AT&T has presented some estimates for the cost of a combined dispatch-MTS cellular system,[19] using the Philadelphia area as a case for analysis. The example included 25,000 automatic dispatch users and 25,000 MTS users in a 1,300-square-mile area. The resulting figures for full capacity use of a 64-MHz allocation indicate a common user investment of $410 per automatic dispatch user and $560 per mobile telephone user.[20] AT&T also presented an estimate of $950 for the mobile unit itself.[21] Since the mobile unit accounts for roughly two-thirds of the cost to the user, competition in its provision or greater economies of scale through increased production could be quite important in reducing investment per user. The cost of the mobile unit does not vary substantially with the service parameters; rather, it depends to a large extent on the required range (one to four miles for the cellular system) and the number of frequencies on which it can operate (which could be standardized).

Motorola has presented a report that questions the validity of AT&T's production assumptions; this report suggests a price above $950 for the mobile unit. If Motorola is correct, economies of scale in producing the unit would be of increased significance. The study commissioned by Motorola reported that common user investment will include electronic switching equipment, radio equipment, exchange circuits, outside plant exchange, buildings, and land.[22] The total investment (shown in Table 4.4) was subdivided into these categories on the basis of information from previous state rate cases, earlier FCC dockets, and other analysis of telephone costs.

The figures from Table 4.4 would imply a monthly charge of $40 for MTS users, assuming revenues equal one-third of the investment. This figure includes all mobile unit and base station costs. Adding a reasonable charge for line costs not already included would increase the estimate of the monthly charge to approximately $50. Motorola presents a market report that suggests the monthly charge will be $59:[23]

Common equipment depreciation	$18
Mobile unit depreciation	25
Maintenance	10
Exchange charge	6
Total	$59

Table 4.4: *Investment Per Mobile Telephone Service User*

Common Equipment	Percentage of Total	Amount
Electronic switching equipment	53.6	$300
Radio equipment	33.1	185
Exchange circuits	0.5	3
Outside plant-exchange	4.6	26
Buildings	6.4	36
Land	1.8	10
Total common equipment	100.0	$560
Mobile units		$950

SOURCE: Motorola, Inc., "Reply Comments," Formal Submission to FCC Docket 18262, Motorola, Inc., Schaumburg, Ill., July 20, 1972, p. A-7.

Thus, the monthly charge for mobile telephone service on a fully utilized cellular system should fall within the range of $40 to $60.

Another important element in determining costs is the presence or absence of economies of scale; costs will be higher if firms are too small to realize such economies. Data presented by AT&T indicate that the shared investment per user, or average fixed cost, is constant once a subscriber level of 20,000 users is attained.[24] In this case, average fixed costs of $560 for MTS and $410 for dispatch can be expected. Costs for smaller systems can be much higher; for example, if the number of users is only 2,500, the costs are roughly double the above figures.

Lane's figures for a cellular system used only for MTS also suggest substantial economies of scale. He presents the following data for such a system: 25 MHz could accommodate 1,604 mobiles/MHz at an average cost of $1,795 per subscriber; 40 MHz could accommodate 1,973 mobiles/MHz at $1,607 per subscriber; and a 60-MHz system could accommodate 1,985 mobiles/MHz at $1,521 each. Computation of monthly charges on the same basis we used earlier (annual revenue equal to one-third of total investment) yields figures of $50, $45, and $42 for the three cases.

These figures seem low; we would have expected the costs of a system dedicated to MTS alone to be greater than the costs for providing a combination of services. While the reasons for the discrepancy are not immediately identifiable, some possibilities do present themselves. Lane's figures may exclude landline charges. Also, all cellular systems are not identical; for example, his 40-MHz cellular MTS system can accommodate roughly 40 percent more users than the combined system described above, and more than three times as many subscribers as the dispatch only system to be discussed below. Thus, in the systems envisioned by Lane, there are more users to share the fixed cost. Finally, the relatively small monthly charge is at least partially attributable to the use of different estimating techniques.

We shall now examine the costs for a cellular system providing only dispatch service. As is true for all systems, the quality of service is considerably lower for dispatch and more users can be accommodated per channel; hence, investment per user and the monthly charge will be substantially lower than for mobile telephone service. The investment for this system, as proposed by RCA is shown in Table 4.5.[25]

Staras and Schiff suggest that the expected monthly charge for this cellular system when filled with 25,000 users would be about $22. This figure results from adding $150,000 (the monthly cost of depreciation of the $9 million equipment, maintenance, and a reasonable profit) to $87,500 (monthly charges for landline telephone), dividing by 25,000 to obtain average fixed cost, and adding depreciation on each $700 mobile unit. If one uses a revenue requirement of one-third of the total investment to compute a monthly charge, the result is about $30. Thus, a cellular dispatch system could probably entail monthly charges of roughly $22 to $30 (1969 figures). Even adding 15 percent for inflation to make the figures comparable with our previous analysis, which is based on 1972 figures, the monthly charge would be substantially below that for the combined cellular-dispatch mobile telephone service.

Another factor not yet mentioned that will affect the costs of cellular systems is the number of cells used to cover a given service area. As the number of cells increases, the amount of complexity of the stationary equipment increases, and fixed costs will rise. However, reducing the cell size can partially offset this tendency in two ways. First, as the number of cells increases in a given area, cell size is reduced and the required range decreases, hence costs associated with transmitting and receiving will decline. This effect will probably be small. Second, since greater spectral efficiency can be achieved with small cells, the system will be able to accommodate more subscribers, among whom the fixed costs are shared.

AN OVERVIEW OF THE COST DETERMINANTS

It is clear from the preceding discussion that several key factors affect costs. These include the following interrelated factors: (1) the quality of service or

Table 4.5: *Investment for Cellular Dispatch System (1969 Estimates)*

Central processing unit, dispatcher panels, cell base stations	$8 million
Vehicle location systems	1 million
Total	$9 million
Each mobile unit	$ 700
Shared investment per mobile (for 25,000 units)	360
Total investment per mobile	$1,060

values of the principal service parameters, such as message length, range, blocking probability, and waiting time; (2) the type of service provided by the system; (3) the level of technological complexity; (4) spectral efficiency; and (5) economies of scale. In addition, depreciation rates and accounting practices will affect the monthly charges.

In the preceding analysis of land mobile communication systems, we have encountered several cases where the values of the service parameters have an important influence on costs. Generally, as service parameters become less stringent and more users can be accommodated, the cost per user declines. Not only will the cost for shared equipment be less, but also the cost of the mobile unit itself, if economies of scale are significant. This situation is crucial in the case of MTS, where the specified values of the service parameters—comparable to those in standard telephone service—may be excessive for mobile use. For example, if average wait time is increased from 3 to 15 seconds on a 16-channel trunked system, investment per user declines from $1,166 to $789.[26] This reflects an increase of spectral efficiency from 413 to 915 mobiles per MHz; gains achieved by longer waiting times are substantial. These figures contrast markedly with the trivial change in investment per user for ordinary dispatch service ($790 to $785) that can be achieved by the same proportional variation in waiting time (20 to 100 seconds). Only a small increase in spectral efficiency, from 4,827 to 5,083 mobiles per MHz, is achieved by this fivefold increase in waiting time.[27]

The relationship between spectral efficiency and cost per user is further demonstrated by the fact that trunked systems have approximately the same cost per user as conventional systems. If it were not for the advantage in spectral efficiency of the former, the investment per user would be much higher. Similarly, the highly complex cellular system would cost far more per mobile were it not for its high degree of spectral efficiency.

Finally, economies of scale can be realized in three ways: increasing the frequency allocation, increasing spectral efficiency, and weakening the quality parameters. Obviously, larger systems that serve more users become feasible when the allocation of spectrum is increased; less obvious is the fact that trunking efficiency is also increased. The result is likely to be lower cost per user.

SUMMARY AND UPDATE

There are several major assertions that can be supported based on these cost estimates and analysis. In the provision of dispatch service there are significant economies of scale manifested by cellular and, to a lesser extent, by trunked systems. However, neither means of providing dispatch service is less expensive than conventional systems. Indeed, the relatively low cost of conventional dispatch service will help small-scale operations to compete with larger trunked

Table 4.6: *Summary of 900-MHz Cost Estimates (In 1976 Dollars)*

System and Service	Number of Users or Subscribers	Total Base Station Investment	Base Station Investment Per User	Mobile Unit Cost	Total[a] Investment Per User	Source— See Note #:
Cellular MTS (as part of a combined system)	25,000	$22,400,000	$ 900	$1,520	$2,420	19
Trunked MTS (20 channel)	500	924,000	1,850	1,540	3,390	Calculations based on n. 12 see text
Conventional MTS	5	---	---	---	7,800	4
Cellular dispatch (as part of a combined system)	25,000	16,400,000	660	1,520 (1,300)[b]	2,180 (1,960)[b]	19
Trunked dispatch (20 channel)	2,000	924,000	460	1,160	1,620	12
Conventional dispatch	2	5,160	2,580	1,720 (1,200)[c]	4,300 (3,780)[c]	10
	50	5,160	100	1,720 (1,200)[c]	1,820 (1,300)[c]	10

NOTE: All dollar figures are rounded to the nearest $10. Costs for 900-MHz paging systems are not given because demand for these systems can be met at lower frequencies. See "One-Way Paging" section for cost details.

a. Total investment may not equal mobile unit cost plus base station investment per user due to rounding.

b. The figure in parentheses takes account of the likelihood that a cellular dispatch unit would be less costly than a mobile unit used for MTS.

c. The figures in parentheses reflect our belief that conventional mobile units should be less costly than either trunked or cellular units used for the same purpose. The figures above the parentheses are the original estimates corrected for inflation.

systems on almost equal footing, while the cellular system, which is somewhat more costly, will have to rely on its superior capabilities to compensate for its higher cost. On the other hand, in providing MTS, large cellular systems used to capacity will be less costly than smaller cellular systems; they would also be less costly than trunked and conventional MTS, if the FCC had allowed noncellular systems. Lastly, paging service has been seen to provide a low-cost alternative to the two-way services offered by the above systems.

In Table 4.6 we have collected data based on the preceding analysis in order to facilitate comparisons. The estimates presented have been adjusted for inflation using the wholesale price index. Such a compensation for inflation is not entirely reliable, however, because these estimates were made for developing systems and the estimates may have attempted, in varying degrees, to correct for this factor already. The monthly charges mentioned in the text are not summarized in the table, since we do not know if current estimation techniques will be used to determine rates for these future systems. Simple "cost plus" techniques can be used as a very rough guide to likely charges. The reader can, for example, multiply total cost per unit by 1/3 or 1/36 to determine the yearly or monthly revenue "requirement," but these methods may result in misleading estimates because competition in the provision of dispatch services may cause a change in the revenue-investment ratio. It is impossible at this time to determine how MTS rates will be regulated or how much the additional charges for wireline costs will affect these rates.

It is noteworthy that, in a 1977 submission to the FCC, American Radio Telephone reported that Motorola would supply mobile telephone equipment at costs that are extremely close to corresponding estimates presented above.[28] In particular, the contract between the two firms stipulates that the mobile units will be supplied for $1,800 each and that the base station equipment will be provided at a total cost of $2,500,000, or $2,500 for each of the 1,000 units that the system is initially intended to serve.[29] These figures correspond to 1971 estimates which indicated that mobile units would cost roughly $1,000 and base station equipment about $1,120 per MTS subscriber on a 2,500-user combined MTS-dispatch system; translation of these figures into 1977 dollars yields approximately $1,700 and $2,000, respectively. Considering that the figures for a larger, combined system should be lower, the compatibility between the two sets of figures is encouraging. Although this measure of agreement does not guarantee the precision of all the figures we have presented, it gives us some confidence that the cost estimates presented are sufficiently accurate for the purpose of our assessment.

CAPITAL REQUIREMENTS

Thus far we have examined costs on the micro level, which is a suitable basis for dealing with the expected prices for service and a number of issues related to regulation. We turn now to an assessment of capital requirements that is needed to deal with macro level issues. The analysis that follows will help us to put the data discussed above in a broader perspective.

The technologically complex systems under consideration will require large investments for their deployment, and the magnitude of the capital requirements will determine the impact and feasibility of their deployment in an economy with numerous potential investments and scarce capital. The magnitude depends on many factors. Obviously, the overall level of penetration (i.e., the total number of mobile units deployed), the market shares of each system, and the cost per unit are key factors. Geographical and temporal patterns of deployment will influence capital cost, as will the use for which subscribers employ the units.[30]

We shall base our evaluation of capital requirements on three plausible scenario states that we characterize as low, medium, and high deployment. These scenarios will be discussed in detail in Chapter 15. At this point it will be sufficient to know that the "low" scenario describes a state in which the adoption of 900-MHz devices is very limited—250,000 cellular devices in 5 markets and 875,000 noncellular dispatch units. The "high" scenario describes major diffusion of the devices, though on a far smaller scale than the "universal" deployment of the standard telephone; the parameters assumed are 17,000,000 cellular devices in 20 markets and 13 million noncellular dispatch units. The "medium" state describes an intermediate level of deployment. Figures for costs of these systems have been assigned to each level of deployment. In order to avoid understating capital requirements and consequently underestimating potential problems of capital scarcity, we shall adopt the upper bound when a range of values is associated with a particular scenario.

The capital requirements for cellular systems are shown in Table 4.7. The capital cost for each service (MTS and dispatch) is obtained by multiplying the following figures: number of users per market, number of markets, and cost per user. Total capital cost is then computed by adding the capital costs of the two services.

Since the scenarios incorporate the concept that competing (noncellular) systems will share a significant fraction of the major markets, and may in fact provide the only mobile service in markets unable to support cellular systems, we will now examine the capital requirements of noncellular systems.

Trunked and conventional systems can provide dispatch service for about the same cost per subscriber. According to our earlier estimates, these noncellular dispatch systems may be introduced at a cost of approximately $1,000 per subscriber, a figure comparable to the cost of cellular dispatch in the low scenario state. If we assume that the costs of these systems are subject to the same influences that are hypothesized in the cellular scenarios, average costs for the medium and high states would be $850 and $550, respectively. The number of subscribers for noncellular dispatch can be derived by subtracting the number of cellular dispatch users from total demand for all types of dispatch service. Multiplying the difference by the cost per subscriber gives us the total capital

Table 4.7: *Capital Costs for Cellular Systems Under Various Scenarios*

Scenario Titles	Type of Service	Scenario Parameters			Capital Cost		
		No. of Users Per Market (× 1,000)	No. of Markets	No. of Mobile Units (× 1,000)	Cost Per Unit	Capital Cost Per Service (× 1,000,000)	Total Capital Cost (× 1,000,000)
Low	Dispatch	25	5	125	$1,300	$ 162.5	$ 362.5
	MTS	25	5	125	1,600	200	
Medium	Dispatch	250	10	2,500	1,100	2,750	4,050
	MTS	100	10	1,000	1,300	1,300	
High	Dispatch	350	20	7,000	700	4,900	12,900
	MTS	500	20	10,000	800	8,000	

Table 4.8: *Capital Costs for Noncellular Dispatch at 900 MHz*

	Scenario Parameters				
Scenario Titles	Total No. of Dispatch Units (× 1,000)	No. of Cellular Dispatch Units (× 1,000)	No. of Noncellular Dispatch Units (× 1,000)	Cost Per Unit	Capital Cost for Noncellular Dispatch (× 1,000,000)
Low	1,000	125	875	$1,000	$ 875
Medium	10,000	2,500	7,500	850	6,375
High	20,000	7,000	13,000	550	7,150

cost for noncellular dispatch systems. Table 4.8 displays the parameters used in the computation and the resulting estimates of capital costs.

The capital requirements for paging must also be considered. In 1975 there were almost one-half million pagers in use; an Arthur D. Little report predicts that by 1985 the total will grow to 2.9 million.[31] Based on these and similar estimates, 1 million, 2 million, and 4 million are reasonable figures for the number of units to be assumed in the low, medium, and high scenario states. Arthur D. Little also predicts the price per pager will fall to $125 by 1985. As a result of this prediction, the level of current cost, and some other analysis contained in Chapter 14, we shall use figures of $220, $160, and $120 as the cost per user in the low, medium, and high states. Paging capital requirements based on these parameters are shown in Table 4.9.

Since published data on capital markets are based on investment per annum, we must convert our total capital stock requirements into yearly investment flows. In order to do so, we must make some assumptions about the length of time it will take for the systems to be deployed. Taking a conservative stance on the issue of capital scarcity, we shall assume that the time span is ten years for the low and medium states and 15 years for the high state. After totaling the capital requirements for each system, the average yearly net capital requirements can be computed; these are shown in Table 4.10.

Now we can evaluate the relative size of the investments under consideration. But the question arises: What should we use as a basis for comparison? Examining totally new capital sources external to corporations would overstate potential difficulties of raising capital, because much corporate investment is financed with retained earnings. Using figures for *gross investment* is relevant to the extent that new systems will replace older low-frequency equipment; however, the extent to which this will occur is quite uncertain and using gross investment figures would unnecessarily complicate the analysis. *Net investment* provides a more desirable basis for comparison, because it is a measure of *new* additions to the stock of capital.

In 1972, the base year for our cost calculations, aggregate net investment was almost $80 billion in the United States. Assuming the economy continues its recent trend rate of real growth (about 3 percent) and that net investment as a

Table 4.9: *Capital Requirements for Paging Service*

Scenario State	Scenario Parameters		Capital Cost (× 1,000,000)
	No. of Users (× 1,000,000)	Cost Per User	
Low	1	$220	$220
Medium	2	160	320
High	4	120	480

Table 4.10: *Capital Requirements and Net Investment Per Year for Mobile Communications (In Millions of Dollars)*

	Capital Requirements				Total Net
Scenario Titles	Cellular	Trunked and Conventional	Paging	Total	Investment Per Year*
Low	362.5	875	220	1,820	182
Medium	4,050	6,375	320	10,745	1,074.5
High	12,900	7,150	480	20,530	1,369

*Extended over a ten-year period in low and medium scenarios; over a fifteen-year period in high scenario.

fraction of the GNP continues along its rather steady path,[32] an investment figure of approximately $100 billion (in 1972 dollars) can be expected by 1980. This means that the net yearly investment for new mobile communications will represent the following percentages of aggregate net yearly investment: roughly 0.18 percent in the low state, 1.07 percent in the medium state, and 1.37 percent in the high state. These percentages are small enough to indicate that mobile communications can probably meet its capital needs without serious problems.

To the extent that mobile communications replaces other inputs and increases efficiency, as well as reducing crime, accident, and injury costs, it releases other resources, so that competing demands for capital will be somewhat reduced. Moreover, if people become more mobile and use mobile communications in place of fixed telephones, the net addition to capital needs will also be reduced. Therefore, the above estimates exaggerate overall capital demands. On the other hand, with regard to cellular systems and especially in the provision of MTS, our figures may be somewhat low, because the cost data on which they are based take into account only part of the wireline facilities that will be utilized.

Since the operators of cellular systems will not have to build all of their own "plant" but can pay for part of the landlines as an operating expense, capital entry barriers will be somewhat reduced. The impact of this consideration on competition, however, is unclear because of the potential problem of cross-subsidization. Existing companies can use their control over landline systems to inhibit entry in several ways: in addition to subsidization and price discrimination, they can withhold information from a potential entrant that is necessary for the design of an efficient cellular system.

The capital demands of mobile communications must also be evaluated in the light of projections that foresee a "capital shortage" appearing in the next ten years. After noting the limitations of even the most advanced econometric models, James O'Brien of the Philadelphia Federal Reserve Bank states:

> In addition, long-term forecasts, particularly the capital needs forecasts, face several major obstacles in making accurate predictions. One, of course, is knowing the future underlying forces that will be affecting our economy. For example, the earlier capital needs studies (made during 1973 and 1974) assumed that the mid-seventies would be a period when the economy was operating at or near full employment rather than being in the doldrums of a deep recession. But even if we guess correctly on the basic forces, economic behavior will remain hard to predict. For example, it is suggested that households' savings may be shrinking (relative to GNP) because of rising tax rates, a shift in the population mix toward young families, and inflation. However, the speculativeness of this suggestion is indicated by the fact that each of these forces has been operating since the mid-sixties, yet personal savings rates have been rising.[33]

He also notes that the capital needs forecasts in question do not take sufficient account of the ability of technological change to mitigate the need for capital generated by situations such as the energy crisis. Thus the extent (and even the existence) of the capital shortage is far from clear.

Whether or not a capital shortage exists, scarcity is always a fundamental element in the allocation of resources between competing demands. If the private sector is going to develop mobile communications, it will make the required investments assuming their rates of return compare favorably with alternatives. If, on the other hand, these private returns do not adequately reflect the social returns or costs, then public action will be needed to obtain the necessary resource allocation. Thus, the issue should not be narrowly framed in terms of a capital shortage; it should be approached more generally. We must ascertain whether the net benefits of expanded mobile communications are sufficient to warrant the use of the requisite private, and possibly public, resources.

An evaluation of the relevant costs and benefits should be based on an examination of present and possible future uses of mobile communications. This examination will be the central theme of the next five chapters.

NOTES

1. Motorola, Inc., "Reply Comments," Formal submission to FCC Docket 18262, Motorola, Inc., Schaumburg, Ill., July 20, 1972, p. 100.

2. Tone-voice requires an average of 12 seconds per message and has a capacity of 400 users per channel. Tone only requires air time of 2 or 3 seconds per page and can handle up to 2,000 mobiles per channel. U.S. Department of Commerce, "One Way Paging," Working Paper prepared for the Office of Telecommunications by Gene Ax and Dale N. Hatfield, Washington, D.C., May 20, 1974, pp. 18-19.

3. See U.S. Department of Commerce, "A Study of the U.S. Radio Paging Market," prepared for the Office of Telecommunications by Arthur D. Little, Inc., Contract no. OT-117, Washington, D.C., April 30, 1975, pp. 4-5.

4. Richard N. Lane, "Spectral and Economic Efficiencies of Land Mobile Radio Systems," *IEEE Transactions on Vehicular Technology* VT-22 (November 1973):98.

5. John Gosch, "Germans to Start Nationwide Paging," *Electronics,* July 11, 1974, pp. 69-70.

6. U.S. Department of Commerce, "Study of the U.S. Radio Paging Market," pp. 4-5.

7. U.S. Department of Commerce, "Land Mobile Radio Cost Analysis," Working Paper prepared for the Office of Telecommunications by Dale N. Hatfield and Wesley Harding, Washington, D.C., p. 6-6. The investment per mobile is highly dependent upon size of the system, i.e., the number of mobiles that can be served.

8. Lane, "Spectral," p. 98.

9. U.S. Department of Commerce, "Land Mobile," p. 6-23.

10. H. Staras and L. Schiff, "A Dispersed-array Mobile Radio System" (Princeton, N.J.: RCA Laboratories, 1969). The authors assume a seven-year depreciation period. Moreover, since the study was published in 1969, current costs will be higher because of inflation.

11. Lane, "Spectral," p. 98.

12. Motorola, Inc., "Reply Comments," p. 91.

13. "Memorandum, Opinion and Order," Docket 18262, *Federal Communications Commission Reports,* 2nd series, vol. 51, p. 972, par. 79.

14. Lane, "Spectral," p. 98.

15. A 16-channel system can accommodate 400 telephone users. See Lane, "Spectral," p. 98. Since we are considering a 20-channel system, we assume a capacity of 500.

16. This means that three dollars in new investment will generate an additional one dollar revenue. See U.S. Congress, Joint Economic Committee, *Economic Report of the President: Hearings Before the Joint Committee on the Economic Report* (Washington, D.C.: U.S. Government Printing Office, 1975), p. 177.

17. U.S. Department of Commerce, "Land Mobile," p. 6-14.

18. Lane uses roughly the same ratio. See "Spectral," p. 98.

19. Bell Laboratories, "High Capacity Mobile Telephone Service Technical Report," Unpublished document, Bell Laboratories, Holmdel, N.J., December 1971.

20. Ibid., p. 5.

21. Ibid., p. 10.

22. Motorola, Inc., "Reply Comments," Appendix A, pp. 5-6.

23. Ibid., p. A-8. This estimate uses a relatively short period of depreciation, approximately 3 years.

24. Bell Laboratories, "High Capacity," pp. 4-5.

25. Staras and Schiff, "Dispersed-array Mobile."

26. Lane, "Spectral," p. 101.

27. Ibid.

28. American Radio Telephone Service, Inc., "An Application for a Developmental Cellular Mobile and Portable Radio Telephone System in the Washington-Baltimore-Northern Virginia Area," American Radio Telephone Service, Inc., Baltimore, Md., February 1977.

29. Ibid., exhibit 7.1. The costs in question are specified in "1977 dollars adjusted by the lower of the Consumer Price Index or the Communications and Equipment Expenditure Index."

30. Aside from differences in cost for providing various types of service, if mobile units employed for personal services are purchased by subscribers, capital requirements are reduced since the expenditures involved would be acts of consumption. This factor will not be included in our analysis because capital will still be required to produce these units and because most subscribers may prefer rental to purchase.

31. U.S. Department of Commerce, "A Study of the U.S. Radio Paging Market," pp. 1-2.

32. James O'Brien, "Capital Needs Projections: A Need for Perspective," *Business Review,* May-June 1976, p. 9.

33. Ibid., p. 7.

Part II
USES OF
MOBILE
COMMUNICATIONS

Chapter 5
ORGANIZATIONAL USES OF LAND MOBILE COMMUNICATIONS

Cary Hershey
Eric Shott

INTRODUCTION

Land mobile devices are now used for many purposes in a wide range of private and public organizations. In a highly industrialized country such as the United States, these organizations frequently operate within an environment characterized by complexity, uncertainty, and change; their ability to anticipate and adapt has become a condition of survival.[1] Technology plays a major role in both causing organizational change and assisting organizations in responding to change.[2] Telecommunication technologies—the telephone network, satellite communications, data transmission systems, mobile communications—are used extensively by businesses and other organizations as a means of coping with an increasingly changeable world. They allow personnel to send and receive voice and digital information virtually instantaneously. Telecommunication technologies permit managers to increase their monitoring, feedback, and control capacity over dispersed personnel; to coordinate the activities of widely separated operations; to organize suppliers and customers into coherent networks; and to provide information for operational and strategic decision making.

American society is highly mobile,[3] characterized by long commutations to and from work within expanding metropolitan areas; a large amount of intercity travel using an advanced highway system; a high degree of dependence on trucks to transport goods; a well-developed public service delivery network that makes extensive use of cars and other vehicles; and an elaborate system for providing commercial services to homes and offices. Mobile communications can be viewed as a technological response to this service-oriented, mobile society. They facilitate the exchange of information between people who work out of vehicles and their coworkers, supervisors, managers, or customers who are either based at fixed sites or are themselves on the move.

In this chapter we shall lay the groundwork for anticipating some of the potential organizational usage patterns and impacts of the 900-MHz cellular system. Since this system is not yet deployed, it cannot be studied directly. Therefore, it is necessary to examine technological proxies for the cellular system.[4] Fortunately, mobile communications already deployed by organizations—paging systems, dispatch radio, and the conventional mobile car telephone—can be used to obtain baseline data on how mobile communication technologies are currently utilized and what their impacts have been. We have collected both qualitative and quantitative primary data from public and private organizations currently using mobile communication equipment.

The data collection and subsequent analyses have been motivated by a need to address the following questions:

What types of organizations use mobile communications?

Why do business organizations use mobile communications? What specific organizational tasks are accomplished by their use and what functions do they serve?

What factors might influence the attractiveness of the cellular system to organizational users of mobile communication technologies?

Consideration of these questions will begin with a description of the methods employed in gathering relevant data. The subsequent three sections present data and analyses on currently available paging, dispatch radio, and conventional mobile telephone systems. An examination of current user perspectives concerning the cellular system is followed by a brief concluding discussion of the role played by mobile communications in the management of organizations.

Methods for Acquiring Baseline Data

There is no standard methodology for analyzing usage patterns or impacts of communication technologies. A review of the telecommunication research literature indicates that a wide variety of methods are employed in this diverse field, including highly quantified cost benefit studies of spectrum congestion, carefully controlled laboratory experiments on the impact of new communication technologies on behavior, and speculative qualitative analyses of the telephone's impact on American society.[5] In carrying out our work, we have attempted to follow an intermediate course between developing a qualitative understanding of how mobile communication technologies are deployed, and undertaking a more rigorous, quantitative analysis of usage patterns and impacts in specific organizations. We have employed a number of widely accepted social research methods, including field research, mail and telephone surveys, and content analysis, to obtain baseline data.

None of these methods, either individually or collectively, constitutes a comprehensive approach to data acquisition for technology assessment; the time and resources needed for more comprehensive data acquisition were not available. The fact that we were studying a subset of telecommunication technologies that had not received systematic attention from social scientists led us to an exploratory approach involving a series of relatively small-scale studies. In this way, we were able to alter our research design as our understanding of the subject increased and as significant changes occurred in the field during the research, like the upsurge in Class D Citizens Radio Service usage after 1973.

Webb et al. suggest that any technique suffers from inherent weaknesses that can be overcome only by cross-checking with other techniques.[6] In designing our research strategy we have been sensitive to the strengths and weaknesses of the methodologies employed. Case studies allowed us to examine the more subtle aspects of social processes and individual attitudes, to develop hypotheses about relationships between variables, to improve the questionnaires, and to refine and interpret out statistical results. The mail and telephone surveys permitted us to distinguish between those relationships that are unique to a particular case and those that are generalizable. Both fieldwork and survey research, however, must confront the problem of respondent bias, interviewer effect, and other interaction contingencies that can produce spurious results.[7] The content analysis approach, used in the study of Citizens Band radio to be discussed in Chapter 9, enabled us to gather relatively uncontaminated data on usage patterns, but shed little light on users' attitudes and perceptions toward the technology.

The first stage of the research involved open-ended field interviews with seventy employees in twenty four public and private organizations, of various sizes and functions, where mobile communications were deployed. The organizations studied included a large medical school, a television station, an electric utility, a multinational oil company, a city police department, and a small plumbing contractor. All of these organizations employed one or more types of mobile communication systems. The information obtained from this fieldwork stage was used to develop more systematic survey instruments.

The next stage consisted of a national survey of 1,000 randomly selected organizations using the "business radio" class of dispatch radio service, involving such technologies as two-way dispatch service, portables, and paging systems. The sample was drawn from the 12,000 members of the National Association of Business and Educational Radio (NABER).[8] The instrument was pretested and modified, and covered four basic areas: demographic characteristics of the organizations and individual respondents; current uses of business radio; impacts of business radio on the organization; and interest in the 900-MHz cellular system. The results of this analysis appear in the section entitled "Business Radio Use," and the tables are subtitled "National Survey."

Concurrent with the NABER survey, an intensive survey of organizations in a large metropolitan area was undertaken in cooperation with a communication common carrier to obtain information on users of a mobile communication technology—mobile telephone service—not covered in the national survey, as well as additional information on dispatch radio and paging system users. A telephone survey was utilized during this phase of the research. A total of 899 interviews were completed. The respondents were selected at random from two preselected groups: (1) organizations that are currently using mobile telephone service, or are on an active waiting list to obtain the service, or formerly subscribed to the service (599 respondents), and (2) organizations that do not use mobile telephone equipment (300 respondents).

The instrument, which took between 30 and 45 minutes to administer, was designed to elicit the following types of information: demographic characteristics of the organization and individual respondents; information about use of mobile telephone service and other mobile communication technologies; attitudes about the importance of mobile communications; interest in the 900-MHz cellular system.

While the design of the survey was a joint venture, the selection of the sample and the actual interviewing were undertaken entirely by the common carrier. Two separate efforts were undertaken to analyze the data, one by the common carrier for their own purposes, and the other by the Cornell team for the purpose of developing a data base on present users, uses, and impacts of mobile communications. Tables based on data from this source are labeled "Metropolitan Survey."

Data from both surveys are presented in the form of percentages in order to minimize confusion that might result from trying to compare tables containing raw numbers. By using derived frequencies, it becomes possible to directly compare information based on different sample sizes.

Each of the following sections on paging systems, dispatch radio service, and mobile telephone will be divided into three subsections—the first providing information on the technical and service configurations of the respective systems, the second presenting information on users and uses of the systems within their organizational contexts, and the third analyzing the impacts of the system on the user organizations.

PAGING SYSTEMS

Technological and Service Configurations

Paging systems permit one-way communication between a person activating the device and the possessor of a small portable receiver. The messages transmitted are simple and brief; paging systems are not designed or employed to deliver a wide range of complete messages. If a two-way dialogue is necessary, the recipient of the page must locate a fixed telephone or another type of mobile communication technology.

A large proportion of the pagers in use today are owned either by a telephone company or by independent radio common carriers (RCCs). (There are also several large systems owned and operated by private firms such as IBM and Xerox, the operators of which have made substantial contributions to the development of more sophisticated paging devices.) The user often leases the equipment and service from the owner; sometimes the user buys the equipment and leases the service. In many cases the systems are directly linked, or interconnected, with the telephone system to permit direct dial paging.

From the user's perspective, pagers are the easiest of all mobile communication devices to operate. They require no technical knowledge and only minimal instructions to ensure competent use. Little more is required of the user than making sure that pagers have fully charged batteries.

Users and Uses

Pagers are used by people in a wide range of professions and occupations who may need to be contacted at any time while they are on the move. Page users include doctors, lawyers, clergymen, contractors, real estate salespersons, and servicemen.[9]

Table 5.1 provides information on the variety of organizations using paging systems. The table illustrates that the use of pagers is widely diffused throughout the economy. There seem to be several rather well-defined groups of people who actually use the devices. In our field interviews, it was found that the users were primarily professionals (doctors, lawyers, etc.), mid-level managers (supervisors, technical "experts," etc.), small business owners or top management, and "pale blue collar" service workers (office and home service repairmen, elevator repairmen, etc.).

The respondents to our metropolitan survey identified three of these four groups—professionals, middle managers, and small businessmen—as the primary users of pagers within their organizations. This is shown in Table 5.2. The fact that the fourth category of primary user—service workers—was not identified by the respondents probably results from the identity of the respondents, most of whom were executives who may have been more familiar with the managerial and professional uses than with the service or dispatch type use in their organizations.

In terms of organizational functions and usage patterns, paging systems can be classified into two major categories: managerial use and service dispatch use. The first category involves users such as professionals, managers, and small business owners who can be termed "decision makers." (It should be noted that the head of a small business and the mid-level manager from a larger organization often perform similar functions, since the owner in a small firm may well be the primary supervisor as well as the technical expert.) Paging systems permit members of this group to be "on call" when they are away from the office. A doctor might be paged when a patient suddenly becomes ill, or a field supervisor contacted if there is a problem with one of his work crews. For these alerting purposes, the pager is adequate. However, because of the service characteristics inherent in their technical specifications, especially the stringent limits on message length and their one-way nature, pagers are not useful for extended dealing with substantive matters. The user must find a telephone to engage in a two-way communication that permits questioning and feedback.[10]

The second category of paging system use—the service dispatch type—serves a different organizational function. Pagers are used to contact lower level organiza-

Table 5.1: *Percentage of Organizations Employing Paging Systems by Sector (Metropolitan Survey)*

Sector	Percentage
Medical	19
Transportation	15
Construction	10
Trade (wholesale/retail)	10
Manufacturing	9
Government and nonprofit	8
Repair/service	6
Financial	5
Business services	5
Communications	4
Police	4
Real estate	2
Personal services	2
Utilities	1

NOTE: This table includes only those organizations that reported organizational use of paging systems (n = 221).

Table 5.2: *Primary Paging System User in Organizations by Sector (Metropolitan Survey)*

Type of Organization	Primary User			
	Professional	Top Management	Supervisory	Service Worker
Medical	x			
Transportation		x		
Construction		x		
Trade (wholesale/retail)		x		
Manufacturing			x	
Government		x		
Repair/service		x		
Financial		x		
Business services		x		
Communication		x		
Police		x		
Real estate			x	
Personal services		x		

tional employees engaged in field service and maintenance activities primarily to reorder their tasks, change priorities within their schedules, or inform them of a situation requiring immediate attention. Again, however, the pager can only alert the individual; a more complete, two-way communication may be required for the individual to respond adequately.

These two major organizational communication functions accomplished by pagers are most easily differentiated by analyzing the direction of the message and the action that is taken as a result of the message. In the first category— managerial use—the individual contacted has to be alerted about a nonroutine situation or impending decision requiring immediate attention. The page is frequently initiated by a lower level employee, operating out of a fixed site. The page holder then must contact an individual within or outside of his organization to obtain further information or clarification, and/or instructions for necessary action. The second type of page communication—service dispatch—is directed to a person who implements *decisions made by others.* Here the page is initiated by an organizational decision maker or supervisor and transmitted to the field personnel.

Data obtained in our metropolitan survey show that there is a remarkably even distribution of use in organizations of all sizes, presumably because of their ease of use and low cost. However, one can discern in Table 5.3 some important differences with respect to those who use pagers within an individual organization. These differences appear to be related to the size of the organization. Table 5.3 shows a ratio of actual/expected pager users in small and large organizations. One-third of all organizations reporting use of paging systems have less than 50 employees. If size had no influence in determining the primary user in the organization, we would expect that the 33.5 percent of the organizations citing top management as the primary users would be small firms (less than 50 employees). However, almost 50 percent of those specifying top management as primary user were small organizations (actual ratio: 46.5/33.5 or 1.39). It is clear that in smaller organizations members of the top management are the

Table 5.3: *Ratio of Actual/Expected Frequency of Each Class of User in Small and Large Organizations (Metropolitan Survey)*

Class of User	Small Organizations <50 employees	Large Organizations >50 employees
Top management	1.39	.80
Professional	.92	1.04
Supervisory and middle level management	.48	1.26
Emergency personnel	.30	1.35
Other field personnel	.57	1.22

NOTE: This table includes only those organizations that reported organizational use of paging systems (n = 221).

individuals most commonly furnished with pagers. In large organizations, supervisors and middle level management personnel are more frequently equipped with pagers. There are, comparatively, fewer lower level operatives equipped with pagers in small organizations than in large ones. Thus, it appears that the service dispatch use of paging systems is at this time primarily a large organizational phenomenon.

Impacts

As with all mobile communications, the stated purpose for having a paging system is to increase the productivity and efficiency of organizational personnel, through programming and scheduling of field service activities, and securing the attention of higher level decision makers and professionals. It is obvious that pagers can decrease response time in many situations. Although we know of no studies to determine if the use of pagers has actually increased the economic efficiency of an organization, almost all of the users interviewed indicated that this technology had made their operations more efficient.

The results of our survey suggest that the users of pagers feel a high level of satisfaction with these systems. More than 88 percent of the respondents were at least relatively satisfied with their paging service, and 51 percent expressed extreme satisfaction. This figure is higher than that found for other mobile communication services, as we shall see.

When pagers are utilized as a management tool, there is little *obvious* impact of the paging system on the decision-making process. One exception occurs in the case of field personnel in small organizations: when supervisors or top level management are pager-equipped, there seems to be a tendency for field workers to rely more heavily on superiors for decisions. It is interesting that this pattern did not seem to occur in the larger organizations.

The use of pagers may cause a slight change in the level of centralization of an organization. Where the pager is used by decision makers, it is likely that the organization will become somewhat more centralized after their introduction. This can result from the users being increasingly accessible, and therefore able to participate directly in more decisions. On the other hand, where the pager is used in the dispatch mode, the organization is likely to become somewhat less centralized; because he can be paged when needed, each of the workers has the ability to follow his own initiative and need not constantly report back to the home base with an updated status report.

It appears from our interviews that there was very little resistance to the introduction of pagers among employees. Where workers were required to use the pagers, they generally seemed to feel that it was a part of their job and, in fact, that it helped them perform the job more effectively. Additionally, in a number of cases, having the pager enabled the worker to develop considerably more autonomy.

In addition to the more utilitarian reasons for acquiring a paging unit, we have noted in our field interviews that there appears to be a growing status phenomenon associated with having a pager. People seem to feel important, perhaps indispensable, when they are so equipped. This may be related to the fact that many current users of pagers are "professional" people in high-level positions. Consequently, a pager is associated with an important job. Individuals who initially acquire a pager for status reasons may not conform to the usage patterns described earlier, since for status reasons may not conform to the usage "functional communication need." In several cases examined during the field stage of the research, it was found that pagers acquired apparently for status aggrandizement were often utilized for very routine messages.

Overall, the pager seems to be an effective mobile communication technology, and there are several advantages to using pagers rather than other forms of mobile communication. The cost is relatively low. The easily portable unit is much smaller and more reliable than other available mobile communication devices, and radio interference is infrequent.

Pagers can be of great value for key organizational personnel called upon to make decisions about nonroutine situations, as well as service personnel performing functions that might need to be rescheduled. The pager's primary utility is in instances that require the immediate attention of the paged party. Within the organizations themselves, the impact of pagers seems to be limited mostly to the individual users—the structure of the organization being only minimally affected.

Nevertheless, there are important limitations on the communication ability of pagers due to message length restrictions and their one-way design. The pager, by itself, is not a very effective managerial tool for coordination and control, since there is no capacity for two-way feedback. While their value should not be minimized, especially in nonroutine and emergency situations, it should be recognized that pagers are mobile communication devices that can only serve special purposes.

DISPATCH RADIO SERVICE

Technological and Service Configurations

Dispatch radio systems are utilized by a wide variety of "radio services" (a term used by the FCC when allocating radio frequencies that is defined as a collection of similar uses of mobile communications having some specific amount of spectrum allocated to its use), including the Police Radio Service, the Special Industrial Radio Service, and the Motion Picture Radio Service. This diversity of uses makes it difficult to discuss all of the various dispatch services in detail. Consequently, we shall begin our analysis with a discussion of those characteristics that appear to be common to all dispatch radio users and uses, followed by a more intensive analysis of one specific service classification, "Business Radio," based on a national survey of members of the National Association of Business and Educational Radio.

Users and Uses

As with pagers, dispatch radio is potentially useful to any organization whose employees are mobile. As a general rule, organizations initially acquire dispatch radio to assign and reassign personnel in the field, to locate and deploy equipment to handle emergencies, and to decrease response time in servicing dispersed customers and clients. Accordingly dispatch services are more commonly utilized by field-oriented service organizations, or service-oriented departments of larger organizations.

Table 5.4 illustrates the broad range of organizations using dispatch radio. Predominant use is concentrated in those organizations frequently involved in emergency situations (medical, government, and police), and organizations that need to respond quickly to customer demands (transportation/distribution, repair/service). Within such organizations, the actual users are generally quite different from the typical users of paging devices. As shown in Table 5.5, use of the radio within emergency service-oriented organizations (medical, police) is primarily by emergency personnel, and secondarily by top officials and their drivers. In the transportation/distribution sector, principal use is by drivers and dispatchers, and in the case of repair/service, by the actual maintenance person-

Table 5.4: *Percentage of Organizations Employing Dispatch Radio by Sector (Metropolitan Survey)*

Sector	Percentage
Transportation/distribution	20
Police	14
Government (nonpolice)	13
Repair/service	10
Medical	9
Construction	8
Business services	6
Trade (wholesale/retail)	4
Utilities	4
Real estate	3
Communication	2
Financial	2
Other	3

NOTE: This table includes only those organizations that reported organizational use of dispatch systems (n = 142).

nel. Thus, the systems are used chiefly by those field personnel who are engaged in nonroutine operations, including higher level officials engaged in field coordination, supervision, and inspection.

It is clear from the data presented in Table 5.4 that dispatch systems are utilized by organizations of all types. We also found that small organizations are as likely to use dispatch systems as large ones, although the patterns of use between small and large organizations do display important differences.

Two-way mobile radios are often diffused at all levels within small, predominantly single function organizations, such as plumbing contractors, electrical contractors, and home service repair firms. In such organizations, the head of the firm is usually involved in many of the operational activities of the organization. He is often the primary user of the system, and the system is shaped by his needs and managerial style; frequently away from the office, he often needs to be contacted. This pattern of use is comparable to the nonroutine, decision-related managerial uses of paging systems. Top personnel in these small organizations also use two-way radios in the traditional dispatch mode to contact their employees in order to monitor, supervise, assist, or redirect their efforts. In smaller organizations, then, the two-way radio is used in ways similar to both the decision-making and more routine "dispatch" modes.

Patterns of use in larger organizations can be quite different, however. The highest level personnel do not directly supervise routine field activities except in rare emergencies. As a result, their communications tend to be less routine, more strategic, and typically shaped by the patterns prevailing in a nonmobile office. If a two-way mobile communication is needed, these executives are likely to require a more private medium that allows them to engage in conversations of longer duration than can be provided by dispatch service. For these reasons, use of this type of mobile communication is not prevalent at the executive level of large organizations.

In our fieldwork we found that the actual diffusion of such systems within larger organizations tends to follow a definite pattern. The first field units to be equipped with a dispatch system are the ones that frequently handle emergen-

Table 5.5: *Dispatch Radio Users by Sector (Metropolitan Survey)*

Sector	Most Frequently Mentioned User	Second Most Frequently Mentioned User
Transportation/distribution	Drivers	Dispatchers
Police	Emergency personnel	Top management
Government	Top management	Emergency operatives
Repair/service	Maintenance/repair	Drivers
Medical	Emergency personnel	Top management and their drivers

NOTE: This table includes only those organizations that reported organizational use of dispatch systems (n = 142).

cies. These are followed by units handling more routine assignments (e.g., construction) and fewer emergencies. Units performing operations such as field inspection, which closely follow a predetermined work schedule and are generally highly programmed, install dispatch systems at a later stage. In general, the less routine a field worker's task, the more likely he is to have access to a radio dispatch system in its first stages of introduction. As noted above, diffusion rarely reaches the executive level.

While dispatch radio is usually acquired for use in emergency situations, it is mainly employed by organizations of all sizes for routine transmissions. In our fieldwork we found that more than 75 percent of dispatch radio communications involve headquarters-to-field transmissions and are predominantly locational in content. Even field-to-headquarters transmissions for assistance and technical advice, though more frequent in the smaller organizations studied, do not occur as often as locational transmissions. Also, especially in larger organizations, these locational transmissions tend to be highly programmed; in many cases, field personnel are instructed to call in when they arrive at a job site, when they complete an assignment, and at specific time intervals.

Impacts

In both the small, single purpose organizations and larger, multipurpose organizations, almost all of the personnel interviewed in our field studies asserted that the use of dispatch radio service resulted in an improved organizational capability for assigning personnel and material, more intensive and effective servicing of their existing clientele, and an expansion of the number of customers they could service within a given geographic area. Smaller firms also used the mobile system to centralize technical advice by keeping specialists at the home base, thus increasing their productivity. This situation was not observed in large organizations where field workers, such as office equipment servicemen, were themselves technically sophisticated and their activities more programmed, so that they had less need for technical advice. Such organizations had a greater tendency to employ paging systems.

The overall response of workers using dispatch radio services appears to be favorable, an attitude that may be explained by the following factors. First, most of the field workers, unlike higher level executives and certain kinds of emergency personnel, only have to use the system during working hours; they are not "on call" at other times. Second, because many of the workers are responsible for handling nonroutine situations, they tend to view the radio technology as helping them to respond to these assignments more effectively. Third, the system reduces frustrating errors and unnecessary trips. Moreover, they know that the system allows rapid contact with their supervisors or technical specialists if assistance is needed. Both field-based workers and supervisors/managers agree that dispatch services provide greater control over and support of field personnel; in fact, many found it difficult to separate these two activities, actually stating that dispatch services integrate support and control.

While workers can accomplish tasks more efficiently using the radio system, job structure itself does not seem to change markedly. The alterations in job structure that have occurred were viewed as beneficial, tending to increase freedom and flexibility, even if higher productivity is expected. Many users are able to maintain less rigid schedules, since they can be contacted when needed, and some organizations even allow employees to conduct personal errands during slack working hours, because they can be in constant communication.

A Case Study: Business Radio Service

The Business Radio Service is the largest component of a group of services designated by the FCC as Industrial Radio Services. With the exception of the Citizens Radio Service, the Business Radio Service has been the fastest growing land mobile radio service during the past ten years. Its primary purpose is to provide two-way land mobile communications for use by businesses and non-profit organizations. It differs from other dispatch services in that service is not restricted to a particular segment of the public or private sector, as is the case with Motion Picture Service, Police Service, etc. Because of the diversity of its users, we have explored this class of mobile communication in greater detail.

SIZE AND SPATIAL CHARACTERISTICS OF ORGANIZATIONS.   Table 5.6 illustrates that business radio is being used in a wide variety of businesses. It is fairly clear, however, that use of Business Band radio is concentrated in those sectors that are either field-oriented (i.e., construction, farming, etc.) or provide service to a customer (communications/electronics, transportation, road service).

Our survey results indicate that while business radio is employed by organizations of all sizes, use of the service is substantially higher among small businesses.

Table 5.6: *Percentage of Organizations Using Business Radio by Sector (National Survey)*

Sector	Percentage
Construction	20
Communications/electronics	20
Personal services	10
Business sales and services	9
Farming and food processing	9
Transportation	7
Road service	7
Retail sales and service	5
Security	3
Medical	2
Manufacturing	2
Other	6

Organizations with less than 50 employees account for 76 percent of the total using business radio. We also found that 53 percent of firms responding to the survey operated in only one location. These figures are especially significant since our other studies have indicated that the usage patterns in small and large organizations are somewhat different.

The firms characterized their areas of operation as follows: urban/suburban— 54 percent; rural—32 percent; both—14 percent. The percentage of rural operations is somewhat higher than expected; it may reflect the fact that rural-based firms tend to service wider areas than their urban counterparts. Also, mobile communication technology is easier to use in rural areas than in congested urban areas, because channels are more accessible.

EQUIPMENT.   The median investment in business radio equipment is approximately $8,000 per firm. Table 5.7 shows that approximately 75 percent of business radio users operate only one base station and from two to nine mobile units. A number of the users, particularly those in the construction field, also make use of additional portable units.

When analyzing the future of mobile communications, information concerning the age of existing equipment is of interest. From the survey, we found that 35 percent of both the base stations and mobile units are more than five years old, and 16 percent of the mobiles and 14 percent of the base stations are more than ten years old.

Another important factor is the "upgrading" phenomenon. More than 70 percent of the current business radio users had prior experience with mobile communications before acquiring their present equipment. Table 5.8 presents information on the type of equipment involved in this prior experience. More than 40 percent of the current users had employed Citizens Band radio before acquiring their business radio. This suggests that business radio may continue to grow as increasing numbers of businessmen using Citizens Band service decide to "upgrade" to a more reliable service requiring more sophisticated equipment.

Table 5.9 shows that there are important differences with respect to prior equipment use between organizations operating in urban/suburban areas and those in rural areas. Thirty-six percent of the urban/suburban organizations with prior experience had used Citizens Band radio, as compared to 81 percent of the rural operators. Urban/suburban organizations used paging systems (22 percent)

Table 5.7: *Percentage of Organizations Using Business Radio Service with Specified Number of Units (National Survey)*

	Number of Units				
Type of Equipment	1	2-4	5-9	10-15	More than 15
Base stations	74	23	2	1	0
Mobile units	5	46	29	8	12

Table 5.8: *Type of Mobile Communication Equipment Used Prior to Introduction of Business Radio Service (National Survey)*

Type of Equipment	Percentage of Organizations Indicating Prior Use of Mobile Communication Equipment
Citizens Band	40
Paging System	15
Mobile telephone	21
Special industrial	1
Other	11
More than one of above	12

and mobile telephone (26 percent) considerably more often than did the rural operators (6 percent and 13 percent, respectively). The frequent use of Citizens Band equipment in rural organizations is understandable; rural firms are generally smaller than their urban counterparts and probably less well financed, and therefore less able to afford the higher expenditure for other services. It should also be noted that mobile telephone and paging services are often not available in rural areas. Moreover, frequency congestion within the Citizens Band service is less serious in rural areas, making Citizens Band both a more useful tool and a viable option for rural firms.

USES. Business radio is employed in much the same way as other types of dispatch radio. As expected, this particular segment of the radio service is used more often for programmed than for nonprogrammed transmissions. As shown in Table 5.10, the most important functions of the communication system, for both emergency and nonemergency purposes, is to send or receive locational and operational information and to deploy personnel in the field. The least mentioned reason for using business radio is that of maintaining contact with clients and conducting business transactions. These functions are usually performed by higher level personnel; they generally involve interorganizational communication with clients, customers, and suppliers, rather than the intraorganizational communication characteristic of business radio usage. The business radio is not well suited to interorganizational communication, since the client (or other party) would have to have a radio operating on the same frequency.

Table 5.9: *Use of Mobile Communications Prior to Deployment of Business Radio by Area of Responsibility (National Survey)*

	Citizens Band	Paging System	Mobile Telephone	Special Industrial	Other
Urban/suburban	36	22	26		16
Rural	81	6	13		
Both	40		20	20	20

Table 5:10: *Functions of Business Radio Communication Ranked by Category of Use (National Survey)*

Purpose	Ranked Frequency of Use Emergency Use	Routine Use
To provide information (including locational and operational)	1	1
To receive information (including locational and operational)	2	2
To deploy personnel	3	3
To maintain contact with office while in the field	4	4
To maintain contact between employees in the field	5	7
To deploy materials and/or equipment	6	5
To maintain contact with other employees in the organization	7	6
To provide technical assistance	8	8
To receive technical assistance	9	9
To conduct business transactions	10	10
To maintain contact with clients while in the field	11	11
Other	12	12

NOTE: Each respondent was asked to indicate the major purposes for which business radio was used by their firm. They could respond positively to more than one of the listed reasons, as well as list their own reasons. The ranking was achieved by summing the positive responses for each use. Thus, for example, more respondents indicated that they used their radios to provide information than to conduct business transactions.

While communications during emergency situations are of utmost importance, emergencies are relatively infrequent occurrences. As a result, the primary function of the mobile system is to aid in the coordination and control of field units. Although providing field personnel with technical assistance may be necessary in some situations, it is not the primary use of the system. The business radio is rarely used by managers for other purposes.

We found that 44 percent of the users were "very satisfied" and 27 percent were "quite satisfied" with the service. Of those who were less satisfied (either "fairly satisfied"—23 percent—or "not very satisfied"—4 percent), the major reason given was mechanical or electrical problems with the equipment. These reasons had little to do with the intrinsic capabilities of this type of mobile communication system or the way in which the systems are integrated into organizations.

IMPACTS.   The primary impacts reported by business radio users were economic in nature, specifically, capital savings in terms of equpment, fuel economies, improved service provision to customers, and increased business. The most striking result reported by many of the business radio users involves tradeoffs between transportation and communication. As can be seen in Table 5.11, more than 70 percent of the organizations reported that the use of business radio had enable them to decrease the number of vehicle trips necessary to accomplish their objectives. Over 14 percent indicated that the business radio had enable them to elimate some trips but had created the need for others. The creation of new trips may have resulted from new business opportunities.

Table 5.12 shows that an even larger portion of business users (77 percent) realized an actual decrease in the number of miles being driven as a direct result of using two-way radio. Most of the organizations reported that there had been a decrease in mileage in the range of 10 to 20 percent, though some claim a reduction of as much as 50 percent (see Table 5.13). In light of the continued energy scarcity and increasing fuel costs, this finding is especially significant. If business or dispatch radios can bring about fuel savings of this magnitude, their importance in the future seems assured.

The most significant effect of business radio for the organizations surveyed is presented in Table 5.14. Almost 60 percent of the organizations reported an increase in the amount of business transacted that could be directly attributed to the use of business radio. When this finding is coupled with the capital and energy cost savings, it becomes quite clear that dispatch radio is a means of substantially increasing efficiency and productivity.

In conclusion, dispatch radio generally, and business radio in particular, are communication media through which directions and instructions are transmitted from higher levels to lower level field personnel. Also, because of the radio's two-way capability, higher level personnel in smaller organizations and mid-level supervisors in large organizations can respond to requests by field-based

Table 5.11: *Effect of Business Radio on Number of Vehicle Trips (National Survey)*

Effect on Number of Trips	Percentage of Total Organizations
Eliminated trips	75
Both eliminated trips and created new trips	14
No effect	7
Other	4

Table 5.12: *Effect of Business Radio on Actual Mileage Driven (National Survey)*

Effect on Mileage Driven	Percentage of Total Organizations
Increased mileage	2
Decreased mileage	77
No effect	21

Table 5.13: *Percent Decrease in Mileage Driven (National Survey)*

Percent Change in Mileage Driven	Percentage of Organizations Reporting Decrease in Mileage Driven
2-5	8
6-10	30
11-15	10
16-20	25
21-25	9
26-30	9
31-40	3
41-50	6

Table 5.14: *Effect of Business Radio on Amount of Business Transacted (National Survey)*

Effect on Business Transacted	Percentage of Total Organizations
Decreased business	0
Increased business	59
No effect	41

employees and modify or change their instructions as the situation warrants.

Dispatch radio services provide a much needed type of communication capacity. A large majority of the present users are satisfied with service they are currently receiving. One should not infer from this, however, that there is little interest on the part of such users in obtaining improved service. On the contrary, it is fairly evident that as improvements are made in dispatch equipment—which increase the efficiency and effectiveness of this type of mobile communication service—users will seriously consider adopting them.

Nevertheless, it is important to note that, with the exception of small businesses, the higher level personnel to whom the field personnel report are generally supervisors, dispatchers, and technical experts, instead of high-level managers and executives. Thus, while dispatch service meets many of the communication needs of personnel at the lower and middle levels of these organizations, it is not especially relevant to the mobile communication needs of higher level managers.

MOBILE TELEPHONE SERVICE

Technological and Service Configurations

Mobile telephones are the most sophisticated mobile communication devices presently available to the general public. They differ from other types of two-way radios by being interconnected with the wireline telephone network, either through an operator or by electronic switching that is computer controlled. (Although other two-way radio systems can, both legally and technically, be interconnected with the wireline telephone system, such interconnection is cumbersome and provides less satisfactory results.) Another characteristic of mobile telephone service that differentiates it from other two-way mobile communications is its relative privacy. When a mobile telephone is in use, a radio channel is dedicated exclusively to that use. The level of privacy thereby permitted is greater than that characterizing dispatch systems, but it still is only partial. Others who locate the channel being used can monitor the conversation.

Mobile telephone service (MTS) is provided by common carriers. In many areas, the telephone company can provide mobile telephone service, and independent radio common carriers provide MTS to a large number of areas as well. In either case, the customer may lease or purchase the mobile telephone equipment. Our metropolitan survey shows that 55 percent of the organizations using any type of mobile telephone service leased their equipment. The percentage of those leasing equipment who utilized service provided by the telephone company is even higher—63 percent in the survey and 78 percent on a national basis.

Users and Uses

In a number of major metropolitan areas there are more potential mobile telephone users than can be accommodated, because of the limited number of

Table 5.15: *Percentage of Organizations Using Mobile Telephone Service by Sector (Metropolitan Survey)*

Sector	Percentage
Manufacturing	12
Construction	11
Trade (wholesale/retail)	11
Transportation/distribution	11
Medical	10
Real estate	8
Financial	7
Business services	7
Repair/service	6
Government	3
Personal services	2
Communications	2
Utilities	2
Legal	2
Other	6

NOTE: This table includes only those organizations that reported current organizational use of mobile telephone systems (n = 394).

channels that are currently available. This excess of demand over supply has made it necessary for the FCC to create a priority system for the allocation of new mobile telephone service as it becomes available. For the past two decades, first priority has been given to those organizations that are involved in the protection or preservation of life and property. Therefore, one might expect that current mobile telephone subscribers in these congested metropolitan areas would be involved principally in life safety activities. However, as Table 5.15 shows, MTS is utilized within almost every sector of the economy in the metropolitan area studied. In fact, it appears that mobile telephone users are more concentrated in the business sector than are the users of other forms of mobile communication, with the exception of those services specifically desig- nated for business use (Business Radio Service and Industrial Radio Service). We also found that mobile telephones are as likely to be utilized in small as in large organizations.

Within these organizations, the personnel actually using mobile telephones are almost always high-level managers or persons performing direct services for them. As Table 5.16 shows, more than 80 percent of all users are top manage- ment personnel. This is a substantially different user group from that found in other mobile services, the reason for which depends primarily upon two factors. It is not practical to use mobile telephone systems for dispatch-like purposes; since each mobile unit would have to be contacted separately, often with operator assistance, it requires more time to make a call than with traditional dispatch service. Also, in present systems, frequent delays in finding a free channel limit the use of mobile telephones in emergency situations.

It is also not economical to utilize the mobile telephone as a replacement for the other mobile communication services discussed earlier. The mobile tele- phone customer is billed for equipment, service, connection to the wireline network, local messages based on charge per minute, and long-distance calls. The average total bill for monthly service in the metropolitan area studied was over $100 per unit. Thus, an organization must highly value the MTS user's ability to communicate for it to bear this expense. Routine field-oriented, supervisory, and support tasks can be handled effectively and more cheaply by dispatch radio.

Table 5.16: *Position of Primary Users of Mobile Telephone Service (Metro- politan Survey)*

Organizational Rank	Percentage
Top management	81
Professional	10
Supervisors	2
Sales	2
Drivers	1
Other	4

NOTE: This table includes only those organizations that reported current organizational use of mobile telephone systems (n = 394).

Furthermore, if a field worker had to wait for a channel, which is likely in many areas, using a mobile telephone could result in a loss of productivity.

While current MTS is not well suited for either routine uses or emergency situations, there is an intermediate range of uses involving important managerial and decision-making functions where it can play a useful role. Higher level managers employ mobile telephones as an extension of the regular wireline telephone network, using the technology to gather information for decisions, manage organizational affairs, maintain contact with other top level management personnel, monitor the implementation of decisions, and maintain contact with customers and key personnel in other organizations, much as they would if they were working in their office.

An important difference between using a mobile and a fixed telephone is that the manager usually has a secretary who can screen calls while in an office. This gatekeeping function is quite important, since an excess of information can be as detrimental to a decision maker as too little information. The mobile telephone user can partially replicate this gatekeeping function, however, by not making his mobile telephone number available; currently, many mobile telephone numbers are unlisted numbers by the request of the user or because of telephone company policy.

Impacts

The explicit purpose of the mobile telephone, when used for business, is to increase the efficiency and effectiveness of executive level communication. This technology extends the range of communication for such personnel from a fixed office site to a mobile state, potentially increasing the usefulness of their traveling time. In short, the mobile telephone allows the executive's car to become an extension of the office. It is difficult, however, to translate these potentialities into quantifiable economic and social impacts, because managerial productivity is exceedingly difficult to measure.

There are long waiting lists for MTS in a number of metropolitan areas. This suggests that certain types of executives perceive that mobile telephones will allow them to perform their jobs more effectively; the potentially positive effect on their respective organizations is recognized, if difficult to quantify. Of the executives using mobile telephones who were surveyed, 64 percent expressed at least moderate levels of satisfaction with the system, while 29 percent expressed various degrees of dissatisfaction. This is the lowest level of satisfaction expressed by any class of mobile communications user. Discontent was largely a result of channels frequently not being available for immediate use and radio interference (see Table 5.17).

There appear to be certain potential costs and disadvantages associated with higher level managerial use of mobile telephones. As noted earlier, "information overload" is a possibility, since there is no secretary/gatekeeper to screen calls. In addition, the busy executive may consider his vehicle a "sanctuary," a place where he can get away from the pressures of the office, particularly those

Table 5.17: *Reasons for Dissatisfaction with Mobile Telephone Service (Metropolitan Survey)*

Reasons	Percentage
Can't get line	21
Poor reception	15
Service errors	7
Too costly	6
Insufficient range*	5
Equipment problems	5
Other reasons	5
No reasons given	36

NOTE: This table includes only those organizations that reported current organizational use of telephone company supplied mobile telephone service (n = 328).

*May also be part of poor reception.

associated with the telephone. This is especially likely if the executive has responsibilities that place him "on call" 24 hours a day. Installation of a mobile telephone in the vehicle may cause an increase in productivity, but it may also result in increased stress on the user.

Another attribute of the mobile telephone is the status it confers on those who possess it. The importance of status aggrandizement should not be underestimated, especially in influencing the scale and pattern of deployment. Individuals have been known to place old telephone sets on the rear dash of a car so that others will believe they have mobile telephone service. One department store in New York City actually sells kits that include an antenna and a dummy telephone set to be mounted in the car. Real or perceived influences on status have been an important factor in the evolution of many business related technologies.

Because of the relatively small number of mobile telephone units within organizations (74 percent of the organizations have only one unit), it is unlikely that MTS would be responsible for any significant decrease in fuel consumption as was observed among dispatch radio users.

In summary, mobile telephone service has characteristics that differ significantly from the other mobile communication services discussed. While it permits easy interconnection with the telephone system, it is not suitable for dispatch services and has limited utility for emergency communications. As a result it is used by an occupational stratum—high-level managers and top executives. Such personnel are infrequent users of other forms of mobile communication technologies. MTS enables these higher level employees to engage in two-way extended conversations with persons inside and outside their organizations while traveling by car, leading to more productive use of their travel time. Yet despite the existence of waiting lists that indicate unmet demand, there is only a medium level of satisfaction with the present form of this service.

POTENTIAL USES OF THE CELLULAR SYSTEM

Service Characteristics of the Cellular System

The cellular system represents a substantial improvement over existing mobile telephone service (both MTS and IMTS) in terms of technical and service characteristics. The service improvements, potential uses, and some possible impacts of the system will be the subject of the following discussion.

The cellular system's primary improvement over existing mobile telephone service is that an adequate number of radio channels will be available so that users will have no more than a three- to ten-second wait for a dial tone, which should substantially decrease the dissatisfaction experienced by current MTS users. As was seen in Table 5.17, 21 percent of current mobile telephone users expressed dissatisfaction because they "couldn't get a line" when they wanted to place an outgoing call. In addition, there may be many persons who are frustrated by attempts to contact mobile telephone subscribers but are unable to do so because channels are not available. Hence, it can be expected that the availability of an adequate channel capacity in the cellular system will make the service much more useful in both routine and emergency situations, and one aspect of user dissatisfaction with mobile telephony should decrease.

A second important feature of the proposed cellular system is improved quality of reception. In fact, proponents of the new system claim that the reception will be comparable to that of the fixed telephone. Again, referring to Table 5.17, this could result in considerably decreasing the discontent expressed by current mobile telephone users, 15 percent of whom mentioned poor quality of reception as a reason for dissatisfaction. Moreover, another 5 percent indicated that existing service had insufficient range, which may also be connected with poor reception.

The effect of improved reception on future mobile communications is difficult to determine with certainty. However, some speculation is possible. Higher quality service will result in more efficient communication—message clarity and reliability will be improved while, at the same time, reducing the need to repeat garbled messages. Clearer reception of at least vocal clues—such as inflections, accents on certain words or passages, changes in pitch—is likely. This should reduce ambiguity and increase understanding.

A third service characteristic of the cellular system that is a direct result of the technology is the vehicle location potential. It will be possible to determine the location within a service zone of a vehicle equipped with cellular equipment. In some circumstances, this could be useful in routing and coordinating field units.

User Perspectives Toward the Cellular System

It is clear, then, that the 900-MHz cellular system represents a significant technological advance in the development of communication capacity for people on the move. This does not necessarily mean that there will be widespread adoption of the system, but the likelihood of adoption will increase if the system is perceived by potential users to address a mobile communication need not currently being met or to outperform existing systems. Of course, the desire to have the most sophisticated system available for status or other reasons may also play a role in the decision to adopt the cellular system.

In order to obtain information on the attractiveness of the proposed system, those interviewed during each phase of the research were asked how important the various features of the cellular system were to them. Respondents to the national survey were told that the system would have all the features of a fixed (wireline) telephone, including direct dial from all telephones and privacy. They were told that the system would have adequate channels; that users would be able to reach the mobile party without knowing his exact location; and that the units could be mounted in a car or used as a portable unit. In addition to these features, respondents to the metropolitan survey were told that several "add on" features would be available, including "hands free" usage, call waiting and forwarding, and a special dispatch feature allowing all units of a particular company to be contacted at one time.

As shown in Table 5.18, we found that 41 percent of the business radio users in our national survey identified the direct dial capacity of the cellular system as its most important feature. This seems to indicate that these business radio users would like to increase the number of people who can be contacted. Both adequate channels and privacy were also viewed as important features by a substantial portion of the respondents (18 percent and 17 percent, respectively).

Table 5.19 presents similar information gathered from the metropolitan survey. Among those responding to the survey, having adequate channels (and thus rapid access or dial tone within three seconds) was considered one of the most important service features; it was mentioned by 36 percent of all the respondents. Thirty-eight percent of the respondents viewed the call waiting/forwarding feature as the most attractive characteristic of the new system, suggesting that the respondents not only want to be able to contact others with greater frequency and speed, but that they also want to be accessible for incoming messages. The unlimited direct dial capability and high quality of reception were substantially less important to the users. Hands free usage and

Table 5.18: *Business Radio User Perceptions of Most Important Features of Cellular Mobile Telephone Service (National Survey)*

Most Important Feature	Percentage
Direct dial from all phones	41
Adequate channels	18
Privacy	17
Vehicular and portable	13
Reach without knowing location	8
NA	3

Table 5.19: *Perceptions of Respondents Regarding the Most Important Features of Cellular MTS (Metrolpolitan Survey)*

Most Important Feature	Percentage of Respondents
Call forwarding/waiting	38
Hands free usage	1
High quality reception	8
Touch-tone dialing	3
Adequate channels—including fast dial tone (3 seconds) and rapid accessibility	36
Direct dial capability	13
	n = 712

touch-tone dialing were of almost no concern to the respondents. Thus, the users placed the highest value on accessibility and their ability to communicate with others. Once that is achieved, they will be more concerned with the actual quality of the transmission and reception.

Data from the national survey, presented in Table 5.20, indicate that the cellular system is perceived as a technology that will be used primarily by executive level personnel. Twenty-six percent of the respondents indicated that they thought the president of the organization would be the primary user; another 26 percent thought that both the president and some other personnel, such as engineers, foremen, or department heads, would be the most likely users.

It can be inferred that current mobile communication users view the cellular system as a means of increasing the effectiveness of executive time. This is consistent with current usage patterns of mobile telephones. In addition, the fact that engineers, department heads, and foremen were frequently mentioned as potential users may indicate that the increased capacity of the cellular system will make the mobile telephone more useful in situations where they are not now widely deployed. Since the cellular mobile telephone service does not

Table 5.20: *Most Likely Users of Cellular System (National Survey)*

Most Likely Users	Percentage of Respondents
President	26
Engineer	9
Foreman	8
Department head	7
Other officer	6
Sales manager	3
Administrative personnel	2
Other	13
President and others	26

appear to possess particular service quality advantages over present systems when used in a dispatch mode, it seems reasonable to infer that the system will be used for executive and management purposes, and possibly by lower level personnel in mainly nonroutine emergency situations. As mentioned earlier, the primary drawback in using paging systems for these purposes is that they are only one-way communication devices. But, while the deficiency of paging systems could be overcome by the cellular system, the cost of the system will be fairly high in comparison to pagers.

It is also important to attempt an assessment, at least in an approximate way, of the *perceived need* for this service at various prices among current users. Consequently, in our national survey we asked the business radio users to state the maximum price they would pay for the specified new service. Several limitations of this approach should be noted, however. In the first place, the price categories do not necessarily represent the amount that an organization would be willing to pay for cellular service. Because most people find it difficult to perceive their need for a technology not yet in existence, respondents are considering a hypothetical question and the answers may be misleading. Furthermore, the information presented in Table 5.21 presents only a partial picture of interest in the cellular system; the information was derived from current business radio users, and not from all current mobile communication users or nonusers of these technologies.

From Table 5.21 we see the level of interest on the part of business radio users is closely related to the service's cost to the user. Only 14 percent of the respondents indicated they would be interested in the service if it were priced at $95 or more. At the same time, a total of 57 percent of the respondents would be interested in the service at $45. This percentage includes all those willing to pay $45 or more for the service.

One must realize that while this 57 percent figure is extremely speculative, it seems to imply that a substantial proportion of the business radio users contacted are willing to consider cellular services if the price is low enough. Since

Table 5.21: *Maximum Amount Respondents are Willing to Pay for Service Per Month (National Survey)*

Maximum	Percentage of Respondents
More than $95	7
$95	7
$70	13
$45	30
$15*	12
Not interested at any price*	31

*The distinction between the categories of "not interested at any price" and "under $15" is somewhat vague. Since the current fixed telephone generally costs businesses at least $15 per month, a respondent specifying this amount may either be reacting negatively to the new technology (that is, not making a decision based on perceived need) or may not understand what the new system will be like.

the business radio users are currently one of the most satisfied mobile communication user groups, it is possible that other user (and nonuser) groups may be even more interested in this form of mobile communication. Although we cannot assess the actual numbers of potential users, the data suggest that there is substantial interest in the cellular system and that there is a sizable potential market for cellular service in the business sector.

It should also be noted that there does exist a substantial negative reaction to the service. Thirty-one percent of the respondents are unwilling to even consider the service, and another 11 percent are willing to consider the system only if it were less costly than existing fixed telephone service. But one must remember that some of the most pervasive technologies currently used in business (e.g., the fixed telephone) were initially received with the attitude of "who needs it?"

Speculations on Where the Cellular Mobile Telephone Service May Be Used

Given the information about service characteristics and the perceptions of potential users presented above, it is possible to speculate on where cellular mobile telephone service may be used. Two areas appear most probable: as a replacement for current mobile telephone service and as a complement to, and perhaps even a substitute for, current mobile communication systems used in nonroutine, nonprogrammed, and emergency situations.

It is likely that, where available, the cellular service will replace existing mobile telephone service. The service quality is designed to be better than and the cost to be at least comparable with, and probably significantly lower than, the current cost for MTS. Moreover, virtually immediate access to a channel would be assured for use in emergency situations. Thus, those individuals or organizations seeking mobile communication service for the purpose of making more efficient use of managerial time will be potential cellular users.

Also, it appears that the nonroutine and emergency situation in the field requiring immediate reliable mobile communications for redirecting and rescheduling operations is another important area where the introduction of the cellular system might prove quite useful. As discussed earlier, mobile telephones as they are currently constituted are too unreliable for these functions. Conventional dispatch service is crowded in many areas and generally not portable; pagers, although adequate for an alert function, do not provide two-way capacity.

In large organizations, the individuals who are most often involved in these situations are supervisory personnel and field workers. In smaller organizations, the owner or manager of the firm is also often involved in these nonroutine activities. We therefore anticipate that cellular service will provide a useful alternative to existing mobile communications employed for such functions. On the other hand, less costly dispatch and paging systems will be preferable for many organizational purposes, especially when the activities are more routine and programmable, and these systems will remain highly competitive with cellular mobile telephone systems.

MOBILE COMMUNICATIONS AND THE MANAGEMENT OF ORGANIZATIONS: A CONCLUDING NOTE

The need for and use of electronic communications have their roots in changing social patterns. They reflect, for example, the fact that American business has become increasingly dominated by large national and multinational corporations, as the populace has moved from predominantly rural and small town environments to large metropolitan centers. Urbanization, industrialization, and dominance of the economy by large organizations are clearly related; they are manifested by increases in the size of societal activity, the centralization of control, the integration of regions, and multiplication of points of concentration.[11] The benefits of the resultingly complex and highly specialized system of production, though unevenly distributed, are obvious to all through the vast quantity of goods, services, and information produced.

The complexity inherent in such systems of production frequently leads to increasing uncertainty about achieving organizational and societal objectives, an uncertainty that grows with the number of system components, their differentiation, and their degree of interdependence.[12] Moreover, as Emery and Trist note, the components themselves are operating in a rapidly changing environment.[13] Consequently, the effectiveness of organizational policies and programs is increasingly influenced by the actions of individuals and entities operating outside of the control of the organization,[14] and the task of managing organizations under such conditions represents a formidable challenge. While the coordination and control of highly specialized operations, complex divisions of labor, and widely dispersed activities have always been extremely difficult, it becomes even more so in an unstable environment. Unpredictability of resource supplies and market demands, growing interdependence of organizational components as well as the public and private economies, and difficulties in obtaining timely and accurate decision-making information become the norm. Problems of strategic planning, operational coordination and control, and outcome evaluation are compounded by these conditions.[15]

The element of *time* is becoming more and more significant. As a result of modern systems of production, distribution, and service—and the associated pressures for increased productivity, efficiency, and profits—employees are assigned additional tasks to accomplish in a fixed amount of time.[16] At the management level, an emphasis on the improvement of executive skills has resulted. At the operational level, more attention is being devoted to efficient scheduling, which is easier for certain types of organizations to achieve than others. Organizations engaged in activities that involve continuous and repetitive tasks (such as those of the assembly line worker or the readers of gas and electric meters) find it relatively easy to schedule activities for maximum efficiency. On the other hand, organizations delivering predominantly nonprogrammable services (such as hospitals, or sales and service organizations) face greater scheduling difficulties.[17] These organizations frequently resort to communication technology in order to transfer information necessary for coordination and efficient scheduling.

Timing is also important to the client or customer, whether he is the victim of a crime or an office manager faced with a malfunctioning duplicating machine. Delayed response to requests for service can generate substantial economic, social, and personal costs.[18] Conversely, missed appointments create expensive idle time for service delivery organizations and individual service providers.

Faced with these conditions, organizations adopt various strategies to cope with time and productivity pressures, and the attendant costs of delay. These include increasing the number and/or training of personnel responsible for monitoring and servicing customers and clients, the development of cooperative relationships with other organizations, and the adoption of a more decentralized organizational structure.[19]

Each of these organizational responses produces a structural need to communicate, and the importance of "open" channels of communication is widely recognized.[20] When employees are operating in the field out of cars, trucks, or other vehicles, communication technologies must be mobile. The "tempo" of events—that is, the number of unique events per unit of time—in an organization's external environment is often greater than that in its internal environment. Because prediction and standardization of response are so difficult, especially in the service sector, organizations often employ mobile communication technologies to adjust the rhythm of their internal production processes to the changing demands of the external environment. These technologies are agents for promoting coordination and control of organizational units, synchronization of the internal and external environments of the organization, and increased productive efficiency. They decrease the uncertainty level associated with decision making in a complex and unpredictable environment. Given these important functions, it is likely that their importance will increase in the future.

In this chapter we have examined some general characteristics of the use of mobile communication devices within organizations. We now turn to uses for specific purposes. The next two chapters analyze use within the trucking industry and a component of the public transportation sector. The subsequent chapter will focus on use within public service delivery organizations.

NOTES

1. See Daniel Bell, *The Coming of the Post-Industrial Society* (New York: Basic Books, 1973); F. Emery and E. Trist, "Causal Texture of Organizational Environments," *Human Relations* 18 (1965):21-32; Todd R. LaPorte, "Organized Social Complexity: Explication of a Concept," in *Organized Social Complexity*, ed. Todd R. LaPorte (Princeton: Princeton University Press, 1975), pp. 1-39; Zbigniew Brezinski, *Between Two Ages* (New York: Viking, 1973); Warren Bennis and Philip Slater, *The Temporary Society* (New York: Harper and Row, 1968); and Alvin Toffler, *Future Shock* (New York: Bantam Books, 1971).

2. There is considerable literature on the subject of technology and organizations. Researchers have been motivated to study the effect of technology on organizational

structure and functioning primarily for two reasons: first, to determine whether modern technology has any deleterious effects on worker satisfaction and, second, to specify variables that need to be taken into account for a comprehensive theory of organizations. Many of these studies are summarized and reviewed in Jack Rothman, *Planning and Organizing for Social Change* (New York: Columbia University Press, 1974), pp. 152-94.

3. Some social implications of the high degree of transcience and mobility in American society are discussed in Bennis and Slater, *The Temporary Society*, pp. 53-83; Philip Slater, *The Pursuit of Loneliness* (Boston: Beacon Press, 1976); Bertram M. Gross, "Space-Time and Post-Industrial Society" (Bloomington, Ind.: Comparative Administrative Group, American Society for Public Administration, 1966), pp. 1-66; Vance Packard, *A Nation of Strangers* (New York: McKay, 1972); Sheila Johnson, *Idle Haven* (Berkeley: University of California Press, 1971); and George W. Pierson, *The Moving American* (New York: Alfred A. Knopf, 1973).

4. For a discussion of this approach to collecting baseline data for a technology assessment, see Cary Hershey and Elizabeth Sachter, "Acquiring Baseline Data on Potential Uses of New Mobile Communication Technologies," *Journal of the International Society for Technology Assessment* 2 (Spring 1976):52-61.

5. Ibid., pp. 53-55.

6. E. Webb et al., *Unobtrusive Measures* (Chicago: Rand McNally, 1966).

7. For a discussion of how fieldwork and survey research methods can be profitably employed together, see Samual Sieber, "The Integration of Fieldwork and Survey Methods," *American Journal of Sociology* 78 (May 1973):1335-57.

8. It must be noted that this sample of "business radio" users does not represent an unbiased response group. Since NABER is a private association, its membership represents a self-selected universe. Furthermore, the sample size of 1,000 was not chosen because it was the optimal size for reliability purposes, since any sample size over about 200 would have given approximately the same level of reliability. Rather, our purpose was to obtain the maximum amount of information possible. More than 400 valid responses were received prior to the start of the analysis.

9. See U.S. Department of Commerce, "One Way Paging," Working Paper prepared for the Office of Telecommunications by Gene Ax and Dale N. Hatfield, May 20, 1974, pp. 18-19.

10. For discussions of the role of two-way feedback in promoting effective communications, see International Joint Seminar, *Attitudes and Methods of Communication and Consultation Between Employees and Workers at Individual Firm Level* (Paris: Organization for Economic Cooperation and Development, 1962); and A. C. Spence, *Management Communication, Its Process and Practice* (London: Macmillan, 1969), p. 129.

11. Philip Hauser, "Urbanization: An Overview," in *The Study of Urbanization,* ed. Philip M. Hauser and Leo F. Schnore (New York: John Wiley and Sons, 1965), p. 9.

12. LaPorte, "Organized Social Complexity," p. 6.

13. Emery and Trist, "Causal Texture of Organizational Environments," pp. 21-32.

14. LaPorte, "Organized Social Complexity," p. 3.

15. See James D. Thompson, *Organizations in Action* (New York: McGraw-Hill, 1967), pp. 159-61; Gary Walmsley and Meyer Zald, *The Political Economy of Public Organizations* (Lexington: D.C. Heath, 1973); and Harold Wilensky, *Organizational Intelligence* (New York: Basic Books, 1967).

16. Barry D. Schwartz, *Queuing and Waiting* (Chicago: University of Chicago Press, 1975), pp. 1-2.

17. See Gerald Hage and Michael Aiken, "Routine Technology, Social Structure, and Organizational Goals," *Administrative Science Quarterly* 14 (September 1969):366-76.

18. Schwartz, *Queuing and Waiting,* pp. 2-3 and 44.

19. Thompson, *Organizations in Action,* pp. 160-61; Michael Aiken and Gerald Hage, "Organizational Interdependence and Interorganizational Structure," *American Sociological Review* 33 (1968):912-30; and Wilensky, *Organizational Intelligence.*

20. See Spence, *Management Communication,* p. 129; and Richard C. Huseman, Carl M. Logue, and Dwight Fleshley, eds., *Readings in Interpersonal and Organizational Communication* (Boston: Hobrook Press, 1960).

Chapter 6
LAND MOBILE COMMUNICATIONS AND THE TRANSPORTATION SECTOR
The Trucking Industry

Arnim Meyburg
Russell Thatcher

INTRODUCTION

We now focus on the role of mobile communications in the transportation sector. Although it is possible to assign a transportation element to almost all uses of mobile communications, the next two chapters concentrate mainly on those aspects that contribute to operational efficiency within the trucking sector and the demand-responsive passenger transportation sector. Highway motorist aid and information will also be considered.

Communication can affect transportation activities and services in three ways. First, where the purpose of the trip is to transfer information, as distinct from goods or persons, electronic communications can offer an alternative to transportation. This type of use has been termed communication substitution. Second, use of communications can result in additional trips, as well as eliminate the need for them. In cases where communications facilitate an increase in human interaction, the effect is referred to as stimulation of transportation activities. Third, the operation, safety, efficiency, and effectiveness of transportation services can be affected by communication use. We shall refer to such uses as complementary.

As might be expected, the difference between substitutive and complementary relationships is not always clear. For example, through the use of mobile communications, a freight vehicle can be sent to make a nearby pickup that was not requested in the original dispatch, making a separate trip unnecessary. Such a change can be considered either a substitute for part of a trip, or as a complement to vehicle efficiency. This potential for confusion is removed by limiting substitutions to those cases in which a trip is entirely replaced by communications.

Our investigation has concentrated on complementary interactions, that is, the use of mobile communications as a means of reducing transporation costs and improving efficiency in goods or passenger movement. The substitution of electronic communications for trip making was discussed briefly in the previous chapter and has been the subject of much research.[1]

Any communication among vehicles or between vehicles and fixed stations by visual, electronic, or other signals can be considered mobile communication; it can be achieved by manual, automated, or semiautomated methods. Within the context of transportation, several categories of communication should be distinguished. Vehicle control from a stationary center includes air traffic control systems, as well as electronic monitoring and control facilities used in rail transit operations. But probably the most common form of this type of control is the routing and dispatching of freight pickup and delivery vehicles, taxis, and other demand-responsive passenger vehicles, usually accomplished with two-way radio equipment. Control systems can also be vehicle activated, however. Public transport vehicles can be equipped with automated or driver actuated devices that give these high-capacity vehicles priority at intersections or in congested areas. Fixed-rail transportation systems also include vehicle activated control mechanisms for switches and signals.

There are other types of communication with moving vehicles that cannot be considered *mobile* communications in the strictest sense. For example, automobile traffic can be electronically monitored from central locations, and traffic flow can be improved by road-surface signal controls that are activated by wheels of vehicles. Telephone or microwave call boxes are located along highways for use in emergencies. Changeable message signs, controlled from a central location, can alert auto drivers about hazardous road, weather, or traffic conditions. While these systems involve communication with moving vehicles and are used to improve transportation, they fall outside of the scope of our analysis, which is principally concerned with two-way systems. Nevertheless, in the future, it is likely that some of these communications will be accomplished by devices located within vehicles. Mobile phones could reduce the use of call boxes; Citizens Band radio has had this effect already. Automatic, in-vehicle transmitters could act as proximity warning devices or could alert traffic signals of an approaching vehicle; the latter device is already used in several mass transit systems. A number of new applications are becoming steadily more feasible as the cost and complexity of transmitting devices are reduced.

MOBILE COMMUNICATIONS AS A SUBSTITUTE FOR TRANSPORTATION

Past and present impacts of mobile communication on transportation have been largely complementary. In 1970, in the cities of Chicago, Los Angeles, New York, and Philadelphia, approximately 98 percent of licensed two-way mobile units were used by dispatch-oriented businesses. Only two percent were described as mobile telephones. These figures suggest that mobile radio is employed largely to coordinate vehicle fleet movement and control—a complementary impact. However, the introduction of a high-capacity mobile telephone system and the widespread acceptance of such a technology by the general public could allow for greater substitution and stimulation of transportation by mobile communication. It is useful to speculate whether the technology will generate additional travel (stimulation) or reduce the need for travel (substitution).

In a recent publication by the Office of Telecommunications of the U.S. Department of Commerce,[2] it was claimed that the ratio of telecommunications to transportation costs was 1:26, in favor of telecommunications. Similarly telecommunication was shown to have a favorable energy consumption ratio of 1:500 when compared to transportation. Such figures suggest that increased use of electronic communication is a desirable social goal; indeed, the OT report specifically refers to the promise of mobile communication for improving energy consumption.

Examination of the use of mobile communications by the general public and in the industrial and business sectors has led us to the conclusion that a widely deployed mobile telephone system might cause automobile travel (both the number and length of trips) to increase. Consequently, energy consumption could rise. For illustrative purposes, one can consider the following differences between standard telecommunications and mobile communications with respect to the potential for travel substitution.

1. Standard telecommunications (specifically teleconferencing and technologies such as the "electrowriter" and "facsimile service") have the potential to allow decentralization of work forces. Transfer of documents, as well as audio information, gives the user access to all the information of a central office while being located at a site closer to their residence—an ability that can reduce the number of miles traveled to work. In the short run, mobile communications will probably not offer such visual and digital capabilities, and consequently reduction of the work trip is not likely.

2. While standard telecommunications can only be used at fixed sites, mobile communications will make it possible to conduct business during travel. Thus, mobile communications could allow more profitable use of travel time and would therefore decrease the disutility of travel to work. Given the tradeoff between the amenities of a residential area and the wasted travel time to work, effectively decreasing travel disutility could lead to further decentralization of residential locations and increased commuting distances.

3. Due to the high cost of telecommunications services, the impact of telecommunications on transportation has been limited to the work and business trip. But inexpensive mobile telephone service is likely to be used by the general public during travel for social, personal business, shopping, and recreational purposes.

In terms of substitution and stimulation, the effects of mobile communications on these more personal types of travel will be threefold. First, by providing continuous accessibility to others (as opposed to the intermittent accessibility provided by landline telephones), a mobile telephone would eliminate the "unnecessary" trip (substitution), examples of which might include the unsuccessful attempt to meet a friend or associate, or travel to a number of business establishments for comparative shopping.

Second, continuous contact with others is likely to increase the use of communication technology for human interaction. Many social and personal business trips might now be accomplished by telephone (substitution). On the other hand, being continually in contact with others electronically could increase the desire for face-to-face contact, resulting in more travel (stimulation).

The final effect would also be one of stimulation. At present, the location of landline telephones defines the places at which a person is accessible to the outside world, and vice versa. Consequently, some persons are reluctant to travel far from such points of contact. Mobile communications would offer areawide accessibility and would, therefore, allow persons to venture farther from home and office. Overall vehicle-miles traveled for social, shopping, personal business, and recreational purposes could increase.

The combined impact of the above effects is a matter of speculation. It is felt that the latter effects are likely to be of a greater magnitude than the elimination of "unnecessary" travel and the increased use of communications for human interaction. Increased energy consumption is implied. Moreover, the ability to conduct work and personal business while traveling could also increase energy consumption, because the desire to converse in private might increase the incentive to take one's own vehicle and decrease the attractiveness of car-pooling.

COMMUNICATIONS AND FREIGHT
TRANSPORTATION—THE TRUCKING INDUSTRY

The Industry

Within the transportation sector, trucking and taxi firms are the largest users of land mobile radio (LMR) services. The trucking industry began using LMR extensively as early as the mid-1950s, making it one of the first large-scale exploiters of mobile communications for commerce.

In analyzing the impact of communications on freight transportation, it is important to recognize that the industry is very diverse and at the same time highly specialized; many types of operations exist, yet a number of firms concentrate on a specific type of operation. We shall briefly review the different types of trucking operations within the "for-hire" industry. (Private truck fleets are not included in this discussion.) For-hire trucks carry about 42 percent of all goods moved by truck.

A trucking company can be considered a local or a long-haul carrier. A combination of these two types permits the development of the "spiderweb" network that facilitates coverage of a region. Local carriers are responsible for the collection and distribution of goods within their allotted region, while long-haul carriers link major service areas, transporting goods between cities for subsequent distribution on a local basis. Terminals are established in each local region and direct service between these terminals becomes the function of these "over-the-road" operations.

Routing arrangement also characterizes the diversity of the industry and affects the communications needs of a carrier. The Interstate Commerce Commission (ICC) recognizes five distinct types of routing: regular route/scheduled service; regular route/nonscheduled service; irregular route/radial service; irregular route/nonradial service; and local cartage service.[3] Local cartage service is usually performed by local truckers in intracity operations, although large intercity motor carriers sometimes supplement their through service by maintaining a fleet of small trucks in the large cities to perform their pickup and delivery services. Local and long-haul operators must correlate their activities to a high degree. As would be expected, local carriers operate for the most part using irregular routes, adjusting them to suit customer demands. The long-haul carrier, however, is restricted to fixed routes by regulation of the ICC.

The distinctions made between different services according to routing arrangement are not clear-cut. For example, an irregular route pickup and delivery (P&D) operation is not necessarily an exclusively local service. In a number of cases, P&D operations have expanded to a regional level with terminals at various locations throughout the service area. Yet, although there is a certain degree of interterminal shipping, the regional pickup and delivery will generally operate on an irregular route basis.

Routing arrangements of the local operator do not provide exact distinctions either. It should not be assumed, for instance, that local operators do not employ regular routes. Many companies, because of the nature of their customers' requests or the road network in their service area, choose to operate fixed routes. Nevertheless, the local operator may adjust that route to serve new demands. In contrast, it is illegal for a long-haul trucker to alter the network of routes specified by regulation.

Carriers can also be classified according to their economic characteristics. The ICC uses the following categories:

Class I Carrier—annual gross revenues in excess of $3,000,000
Class II Carrier—annual gross between $500,000 and $3,000,000
Class III Carrier—annual gross under $500,000

In 1976, the U.S. had 3,488 Class I and II carrier firms and 12,984 Class III carrier firms,[4] with a total of approximately 400,000 over-the-road vehicles. Because the former employ a much larger number of vehicles, however, the difference between the classes with respect to total equipment is much smaller.

This stratification has significant implications for the use of mobile communications. Approximately 50 percent (about 1,750 companies) of the Class I and Class II carriers use radio equipment, whereas only five to eight percent of the Class III firms employ mobile technology (about 700 companies).[5] Two factors seem to be responsible: the capital expenditure necessary to obtain radio equipment, and the reduced problem of communication when only very few trucks are involved.

This taxonomy of the industry can be completed by considering two other classifications: (1) the type of "contract arrangement," which refers to the ownership of the cargo being transported (the important distinction is between "for-hire" carriers that provide freight movement for other businesses and "private" carriers transporting their own cargo); and (2) the type of commodity transported. Of these two additional classifications, the latter is the one most relevant to the use of mobile communications. Moving high value or hazardous cargo might necessitate the use of radio equipment for safety and security reasons. Overall, the characteristics that are most relevant to an analysis of mobile communications are "local" versus "long-haul" operation, routing arrangement, and firm size.

The Use of Communications Equipment

Two basic communication needs can be specified for the trucking industry: to relay new customer requests from terminal dispatcher to truck driver, and to control the fleet of vehicles. The purposes of fleet control include monitoring drivers for performance and trucks for information concerning arrival and departure times, as well as maintaining control of situations such as breakdowns, hijacking, or other emergency situations. Regulation of cargo and business transactions between terminals, that is, data communications, should also be regarded as a part of "control communications."

The relative importance of these two categories depends on the type of trucking operation. Long-haul operators, because they are largely regular-route carriers, do not encounter a great deal of rerouting per se, although additional stops or a slight deviation from the company's allotted network may occur. The most important need of the long-haul trucker is one of control. In companies with many terminals and a large vehicle fleet, monitoring of transferred cargo and knowledge of vehicle arrival times at each terminal (for more efficient loading) are essential. Communications could also be used to aid in emergency situations and to monitor the performance of the driver.

Local service, typically serving irregular routes, relies more on communications for relaying new assignments to drivers so that a truck can accomplish several successive pickups and deliveries before returning to a terminal. Local operators also require a certain amount of control over vehicles, drivers, and cargo; while these requirements are less stringent than those of the larger long-haul truckers, monitoring of vehicles and drivers is useful for dealing with breakdowns and assessing driver performance, etc.

Communication Systems Presently Used

A wide variety of communication systems are used by the trucking industry. These include various conventional telephone arrangements, on-line data transmitting and teletype systems, as well as mobile technologies such as pagers, two-way radios, and digital equipment.

The conventional telephone has been and remains a mainstay in the trucking industry. Before the introduction of the first radio dispatch equipment in 1950, communication was necessarily conducted by telephone. Drivers were often instructed to call the local terminal every few stops to get a "batch" of new assignments; by reducing the need to return to the terminal for orders, more efficient and flexible service were possible. A number of modifications have been made to conventional telephone service to facilitate these communications, such as local service and WATS, Foreign Exchange (FX), and tie lines. The latter (otherwise known as direct lines) are full-period lines that connect terminals to each other or connect terminals to headquarters. They serve many important purposes including interterminal cargo control, coordination of billing, and transmission of other business transactions; they can also be used for conferences between terminal managers.

Larger trucking companies that transmit a great deal of data between terminals have found it useful to change from voice communication to digital transmission, including both printout and display equipment, for interterminal control.[6] The deployment of microwave repeater networks has also occurred,[7] but their use is limited because of the extremely high price (between one dollar and three dollars/mile/month).

Another nonmobile technology in use is intraterminal closed-circuit television. Approximately 80 percent of all losses in the trucking industry due to pilferage or hijacking occur within the terminal in "across the loading dock" operations.[8] Monitoring by television helps to reduce this loss.

Many types of mobile communication technology are employed by truckers. Two-way radios are used in local pickup and delivery operations. Citizens Band radios are often purchased by individual truck drivers and used for personal communication with other drivers, enabling them to avoid police speed traps and to warn one another of adverse traffic and road conditions. Also, pagers are used in local systems to alert the driver to call the terminal for a new assignment; UHF and VHF two-way radio dispatching systems are more frequently used for this purpose, however.

The most recent additions of communication technology to the trucking industry are digital transmitting base stations and mobile printout units.[9] These have several advantages. For example, digital mobile systems provide the driver with written assignments rather than verbal ones, thus eliminating voice inconsistencies and the need to write down the assignment. Information transmitted back to the terminal from the driver can be processed directly. While digital technologies are capable of handling a much greater number of transmissions, again, because of costs, few digital systems are presently in operation. Two of the more extensively used systems are Motorola's MODAT system and the Cadec system developed by General Systems Development Corporation.

Efficiencies and Impacts of Communication Technology

The use of mobile radio has two general kinds of economic benefit: increased revenue and decreased costs. Immediate relay of incoming requests to an available vehicle enables a larger number of customers to be served per day (increased revenue). In addition, fewer trucks are needed to service a particular area if they are radio-equipped. Plotkin reports that mobile equipment is able to increase efficiency by 15 to 25 percent.[10] This means that four radio-equipped trucks can perform the task of five trucks without radio equipment, resulting in savings of driver wages, fuel consumption, etc. Plotkin studied a particular radio-equipped carrier, recorded the kinds of radio transmissions that occurred,

Table 6.1: *Communications Messages and Delay Consequences*

	Message	Delay Consequences*
	Request for directions or address	c
	Business or delivery problem	c
	Vehicle problem	c
Mobile-	Delayed at some location	a
to-	All deliveries or pickups done	a
Base	Traffic jam	a
	Checking in to report status	a
	Arriving at_____	a
	Leaving _____	
Mobile-		
to-	Request direction information	c
Base		
	Assign new pickup or make other changes	c &/or d
	Request for location or status	a
Base-	Given information regarding some other	c
to-	problem	
Mobile	Paging some particular vehicles	a
	Request driver to telephone the base	a

*Delay Consequences: a. decrease of efficiency
                                    b. loss of customer goodwill
                                    c. additional overtime
                                    d. additional mileage for the truck

SOURCE: Adapted from H. A. Plotkin, "Mobile Radio—A Pickup and Delivery Tool," **The Logistics and Transportation Review** 9, no. 4 (1973):361.

and analyzed the consequences of delaying such transmissions. Table 6.1 illustrates the numerous types of messages communicated and the results of postponed response.

Plotkin also introduced the concept of a "delay cost," which refers to the additional mileage and overtime, loss of customers' goodwill, and decrease of efficiency that results from a lack of instantaneous communication. He arrives at these figures by postulating that a delay may prevent a driver from sending a message in time to achieve intended savings. A delay is thus associated with a lack of communications equipment. Figure 6.1 shows these delay costs to be quite substantial.

Based on the estimated 20 percent efficiency, Plotkin develops what he terms a "basic radio value." A graph showing the radio's value as a function of the number of trucks in a fleet, obtained by Plotkin from observing a specific motor carrier, appears in Figure 6.2. According to this analysis, the value of the radio seems to far exceed its costs, since the figure indicates a savings of about $15 per truck per day.

Besides direct economic savings, there are a number of indirect benefits. Many customers prefer to do business with mobile radio users because of the greater speed and quality of service provided. Questions arising during pickup and delivery that cannot be answered by the driver can be clarified immediately. In the event of an emergency, increased safety can be provided through the use of radio equipment. The driver is able to get aid through the local terminal or from other company vehicles in the area, if mobile-to-mobile capabilities exist, and less reliance on outside help means greater security for driver and cargo. Furthermore, communications can reduce the danger of hijacking.

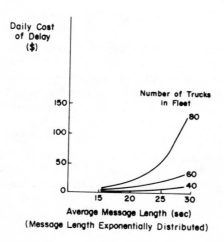

Figure 6.1: *Daily cost of message delay*

SOURCE: H. A. Plotkin, "Mobile Radio—A Pickup and Delivery Tool," **The Logistics and Transportation Review** 9, no. 4 (1973): 364, figure 1.

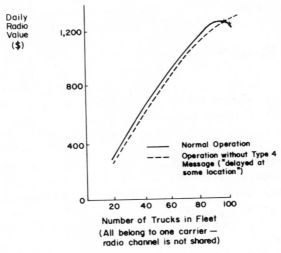

Figure 6.2: *Effect of message elimination on daily radio value for the particular carrier in the analysis*

SOURCE: H. A. Plotkin, "Mobile Radio—A Pickup and Delivery Tool," **The Logistics and Transportation Review** 9, no. 4 (1973):365, figure 2.

Latest figures published by the U.S. Department of Transportation[11] indicate that the total cost of cargo theft and pilferage exceeds $1 billion a year in direct claims (excluding processing costs), with the trucking industry experiencing the largest percentage of that total. Theft occurs: (1) during loading and unloading; (2) in the terminal yard—the DOT study states that "85 percent of stolen cargo goes out the front gates of transportation facilities during normal operating hours and in the possession of persons and in vehicles authorized to be on facility premises for legitimate reasons";[12] (3) while the vehicle is in transit between terminals. Hijacking has recently become more frequent, but the exact value of hijacked cargo is difficult to ascertain. Increased terminal security has cut down on the first two types of loss, so that the problem has moved to the road. Hijacking involves more than the loss of cargo, however, since in many cases the driver is in danger.

In an effort to reduce the increasing number of hijackings, two communication systems have been developed under the authorization of the Department of Transportation.[13] The most promising of these involves equipping a truck with a small transmitter that responds to an ultrahigh-frequency signal sent from an electronic "interrogator" unit. The transmitter can be hidden in the cargo before it is sent out of the terminal. Interrogation can be performed by means of helicopter and other aircraft, or by a ground-based unit, such as an unmarked police car or a company vehicle. The second system, developed by Hoffman Electronics Corporation, utilizes the same principle but employs an electronic license plate instead of a small "bug." With this system, though, if cargo is removed from the truck, the ability to locate the stolen items is lost.

The concept underlying these two systems is basically sound, but two problems are evident. First, it is necessary to interrogate a vehicle that has already been hijacked in order to recover the cargo. Although the increased chance of being caught would no doubt deter hijacking, no sure method for detecting hijacked trucks is now available. Second, the cost of providing an interrogation vehicle, such as a police helicopter, is prohibitive, and police cars and company vehicles do not have sufficient speed or mobility to be fully effective.

Although only a portion of all thefts result from on-the-road hijacking, the vast sums of money involved in cargo loss would seem to justify the cost of modifying a cellular system to locate hijacked trucks. In addition, theft prevention is highly "profitable" for the industry. For each $100 theft claim prevented, as much as $50 can be profit, compared to a return of about $2 for each $100 of new business.[14]

POTENTIAL IMPACTS OF NEW MOBILE COMMUNICATIONS ON THE TRUCKING INDUSTRY

A Brief Review of the Past Impacts of Communications on the Trucking Industry

As was seen in the previous sections, the trucking industry experienced significant operational, financial, and organizational changes through the widespread introduction of mobile radio in the early 1950s. The ability of a growing number of firms to have immediate contact with drivers has resulted in much more flexible and dynamic operating practices, which has increased vehicle efficiency, reduced response times, and decreased operating expenses. Using mobile radio has also allowed pickup and delivery business to expand the geographic areas they serve and induced many local carriers to become regional carriers. Radio has provided vehicle control throughout a much larger area: when combined with the ability to keep vehicles in the field for an entire day, this control enabled local operations to offer fast, one-day response to the requests of customers located hundreds of miles from the central terminal.

In addition to the impact on vehicle control and utilization, the development of telecommunication technology is beginning to change the organizational structure of trucking firms and has increasingly controlled the flow of goods between terminals.[15] Financial and operational practices of widely separated terminals can now be coordinated from central headquarters. This has led to more efficient transfer of goods and streamlined administrative operations.

New landline and radio capabilities have affected the usage of mobile communications, as have price reductions of older devices. A manifestation of the latter is the extraordinary rise in the use of Citizens Band radio systems by truckers. Although the limited range and other technical limitations of this equipment greatly restrict its utility for dispatching and vehicle control, private use by drivers has been responsible for numerous operational impacts. Citizens Band radio has also become a means of personal communication between truckers on the road, a form of entertainment in an otherwise potentially boring job.

Benefits resulting from mobile radio and telecommunications also accrue to persons outside of the trucking industry. Increased efficiency and decreased operating costs can lead to reduced rates and a higher level of service for the shipper. Energy savings are also a direct result of greater vehicle utilization. Another important, though less obvious, benefit involves truckers' extensive use of Citizens Band radio, which enables them to aid stranded motorists. Finally, the fact that drivers can be in radio contact with others may increase their sense of security and may even reduce the risk of hijacking.

The Limitations of Present Technology

Several features of presently available communication systems limit the realization of additional and more widespread benefits. The first of these limitations concerns the high capital and operating costs of two-way radio, teleprinter, and telephone service (see Chapter 4). High capital investment for the communications equipment greatly reduces the total derived benefits and, in some cases, may even preclude the purchase of communications equipment altogether. Small, Class III operators are particularly affected by high costs. For these firms, with gross revenues of less than $500,000, the anticipated increase in profits would probably not be equal to the cost of communications equipment, and at present only five to eight percent of Class III carriers employ mobile radio technology.

A second serious limitation of present technology is its unsuitability for dealing with the communications needs of carriers involved in long-haul operations. Even if ICC route regulations were relaxed, allowing for off-route servicing by over-the-road carriers, few firms would attempt dynamic routing and scheduling simply because radio control of vehicles is not economically feasible. Also, in areas where the radio channels are congested, delay costs (see Figure 6.1) effectively reduce the value of mobile radio equipment and further restrict its implementation.

New Technologies and Their Potential Impacts

The many advantageous characteristics of the multichannel trunked system and the cellular system—such as increased user capacity and increased range— hold promise for new impacts on trucking operation, organization, and finance.

Increased range, for instance, might allow for more flexible and dynamic operation by regular-route, long-haul truckers. Reducing spectrum congestion in metropolitan areas might stimulate investment in and use of mobile communications because of eliminated delay costs. In addition, introduction of these new technologies could lower the investment required for communications equipment, thus making mobile communication more accessible to many firms who presently find it uneconomical. Other new capabilities, such as mobile-to-mobile communications and the ability to monitor vehicles, could improve operating efficiency and safety.

The extent of the impacts will ultimately depend on the new systems' costs and capabilities, since these will determine the scale of adoption. We shall discuss future impacts in terms of the scenarios mentioned briefly in Chapter 4, and which will be elaborated upon in Chapter 15. For this purpose, the potential impacts have been divided into the following categories: operational, financial, and organizational.

OPERATIONAL IMPACTS. Introducing cellular and/or MCTS technologies into the trucking industry is likely to affect the efficiency and safety of trucking operations. Delays and hazards caused by mechanical difficulties are one particularly serious problem of the long-haul trucker to which the new technologies are applicable. Trucks can be stranded many miles from the nearest town, at any hour of the day or night. Equipping the truck with CB radio provides the best available, but inadequate, solution. A cellular system whose network extended over major trucking routes (high deployment scenario) would provide the range necessary to allow the driver to notify the company of his situation. Such a network might also help reduce hijacking, attempts at which have been plagued by the cost inherent in the surveillance of a large number of trucks. A cellular or MCTS system, though, with appropriate modification could solve this problem; the locational ability in the cellular method could be augmented by the incorporation of an antihijacking "bug." Since cellular systems are to be installed initially in major metropolitan areas, it is fortunate for this application that hijacking is almost totally concentrated there.

Another operational impact might be an increase in dynamic routing and scheduling in the industry. Firms that had previously elected to operate on a "call at each stop" or "bunch assignment" basis, because of the high initial investment required to purchase mobile communications, might be able to afford radio equipment at the projected monthly service costs of a widespread cellular system (see Chapters 4 and 10).

Perhaps the most important operational impact might occur in long-haul operations. Without communications equipment capable of keeping the dispatcher in contact with vehicles, the present long-haul trucker operates in a manner that is analogous to the local P&D trucking firm of preradio times. Changes in routes and schedules must take place either when the driver reaches his destination or at a time when he has been instructed to call in. With the introduction of long-range radio capabilities, as envisaged in the high deployment scenario, the long-haul trucker would be able to increase vehicle load factors by informing drivers of additional on-route assignments, or even assignments requiring slight route deviation.

It could be argued that because long-haul trucks typically have a high load factor—excluding the effects of empty backhauls, which are a result of ICC regulation—dynamic routing and scheduling would not greatly influence the overall efficiency of this segment of the industry. Considering the large distances traveled by long-haul trucks, however (the average length of haul is approximately 280 miles),[16] any elimination of duplicative routing could increase the potential for energy savings and have a significant effect on profit margins. But the realization of improved efficiencies through more flexible routing and scheduling would require changes in the ICC's regulation of the trucking industry, a matter that will be discussed later in this section and also in Chapter 15.

FINANCIAL IMPACTS ON THE TRUCKING INDUSTRY.   The above operational changes could ultimately lead to increases in profitability and decreases in fuel requirements within the industry. Assuming that each of the present 2,622 (in 1975) radio-equipped firms have realized the expected 20 percent increase in efficiency implies a 3 percent increase for the industry as a whole. Since the gross revenue of regulated Classes I, II, and III carriers is on the order of $21 billion per year, and since the industry operates at about 4.8 percent gross profit (before taxes and dividends), this 3 percent gain in efficiency translate into a yearly increase in gross profits of about $29 million.[17] Each additional percent increase in efficiency obtained by the use of new technologies would result in higher gross profits of nearly $10 million per year.[18]

Energy savings, which is of growing concern, can also be calculated given the fact that each truck consumes about 2,800 BTU per ton-mile, or about one gallon of gas for every 52 ton-miles of travel. Table 6.2 shows the energy use of a typical nonradio-equipped vehicle (0 percent increased efficiency) and compares this energy use with that of vehicles 5, 10, 15, and 20 percent more efficient. This analysis uses the fact that the total volume of traffic *per vehicle* in the for-hire trucking industry is about 246,000 ton-miles. It also assumes that increased efficiency can be regarded as an expansion in vehicle utilization. The percentage gains in efficiency correspond to estimates of radio use by different types of trucking operations,[19] and Table 6.2 summarizes this information in terms of potential savings per vehicle per year. Considering that more than 700,000 trucks (ICC regulated and unregulated vehicles) are still not radio-equipped, the potential for energy savings is quite substantial.

ORGANIZATIONAL IMPACTS.   In the past, increasing the range of control allowed by communications has led to expansion of the areas served by pickup and delivery operations; local operations became "regional," extending their service to the limits of the available range. Introduction of the new technologies will probably cause a corresponding expansion of dynamically routed and scheduled P&D operations.

Broadening the market coverage of P&D operations would result in reduced numbers of short-haul, regular-route operations. Of course, these operations are

Table 6.2: *A Comparison of Energy Needs for Each For-Hire Vehicle With and Without Mobile Communications*

Assumed Increase in Efficiency and Load Factor	Possible Situation	Efficiency (10^3 Ton-Miles)	Gallons of Gas/Year	BTU/Year (Million BTU)	Fuel Savings Per Vehicle (Gal/Yr)
0%	Typical for-hire vehicle without mobile radio	246.1	4,395	689	0
5%	Regular-route prescheduled vehicle with radio	234.4	4,186	656	209
10%	Regular-route nonprescheduled vehicle with radio	223.7	3,995	626	400
15%	Irregular-route prescheduled vehicle with radio	214.0	3,822	599	573
20%	Irregular-route nonprescheduled vehicle with radio	205.1	3,663	574	732

necessary for linking collection and distribution points with shippers that are outside the range of local P&D operations. If the local coverage were expanded, shipments between points once outside service areas could be accomplished by a single vehicle, thus reducing transshipment delays.

A second possible organizational change might be caused by the ability of long-haul trucking firms to alter routes and schedules in order to serve requests just off their regular routes; this would also reduce the number of regular-route operations needed to provide complete market coverage.

By supporting recent trends in the movement of freight, communications could have a final type of organizational impact. Specifically, the capabilities of the new mobile cellular and multichannel trunked systems might make consolidation of the trucking industry more feasible. It has been suggested by a number of researchers that many services provided by the motor freight industry are duplicative; if the equipment and terminals of smaller operations were combined under a single control center, such duplication could be eliminated.

In summary, the trend in organizational changes that can be foreseen is toward a motor freight industry comprised of fewer, larger firms. However, when considering the potential changes that could be stimulated by the new technology, it must be kept in mind that almost every aspect of the industry's structure—the areas served by each firm, the routes traveled by long-haul truckers, the consolidation of firms, etc.—is controlled by federal and/or state regulations. Any organizational changes in the industry require approval from the ICC. Each of these suggested impacts should be viewed as potential changes that cannot be realized without action by the appropriate regulatory agency. Policy issues raised by the changes we have discussed are dealt with later in this chapter and also in abbreviated form in Chapter 15.

ANALYSIS OF MARKET PENETRATION OF THE NEW TECHNOLOGIES IN THE TRUCKING INDUSTRY

The market for new communication technology in the trucking industry will depend upon one of the following three "abilities" of a cellular or multichannel system: to decrease congestion within radio channels and hence transmission delays; to provide dispatching service at a lower cost than present mobile equipment; to perform functions not possible with present communications equipment.

Radio users interviewed for this study did not view channel congestion as a major problem. This is somewhat surprising because, in metropolitan areas, radio channels are sometimes shared by a number of users, and the resulting congestion can mean inefficiency within the trucking industry as in any other type of dispatch service. The previously cited work of H. A. Plotkin suggests that radio congestion is costly; we believe the use of a more spectrally efficient technology by truckers in areas of high radio congestion can be expected.

Lower cost would also facilitate the development of a broader market. The way in which charges will be made is important as well. The following is a summary of the costs used in this analysis:

CELLULAR, AUTOMATIC DISPATCH SYSTEM. Two estimates of the monthly cost of this technology are available. Motorola has estimated that service would cost about $60 per month per unit; AT&T gives a lower estimate, about $40 per month per unit. Both estimates are based on an assumption of 30,000 users, including dispatch operations and regular telephone subscribers. With greater market penetration, the system could be offered at a price only moderately higher than that of conventional telephone service. A range of $20 to $30 per month is conceivable for a very widespread system.

MOTOROLA'S 3-C TRUNKED SYSTEM. The estimated initial cost of fixed equipment for this system is about $600,000,[20] so that a substantial number of users are required for this system to be practical. Motorola feels the system could handle at least 2,000 mobile units per megahertz. The investment for fixed facilities would thus be about $300 per mobile unit. In addition, the cost of an individual 3-C mobile unit, including distribution costs, will be about $750. Total capital investment would therefore be approximately $1,050 per mobile unit.

PRESENTLY USED RADIO SYSTEMS. From the interviews we have conducted, it was found that the cost of present equipment is about: $3,000 for a single dispatcher base station; $1,000 for each mobile unit; and $100 per month for leased lines to an antenna (see Chapters 2 and 4 for more detail).

In order to compare these costs, one additional factor must be considered. The automatic dispatch service (ADS) proposed by AT&T would have a maintenance cost included in its monthly charge, while the 3-C and present systems incur an additional maintenance fee. One carrier we examined, for example, pays a maintenance cost of about $150 per month for approximately 20 trucks. Repair of base stations is more frequent and costly than that of the mobile units, and its cost will not depend on the number of mobile units.

Under the simplifying assumption of linearity, relationships between costs and the number of vehicles are plotted in Figure 6.3 for fleet sizes varying from one to twenty. Several tentative conclusions can be drawn concerning a possible market for these new technologies, based on economic advantages.

First, a cellular dispatch system priced at $50 per month would be advantageous for fleets of only one or two vehicles. Larger fleets, because they can amortize the capital expense over more units, and because the monthly charge of the ADS system increases sharply for bigger fleets, would be better off purchasing the conventional equipment or sharing the capital expenditure of the 3-C system.

Second, if priced at $20/month/unit, the ADS system would hold a considerable advantage over a conventional dispatch system, for any number of vehicles. The slope of the ADS cost line is $30/mobile/year less than the conventional

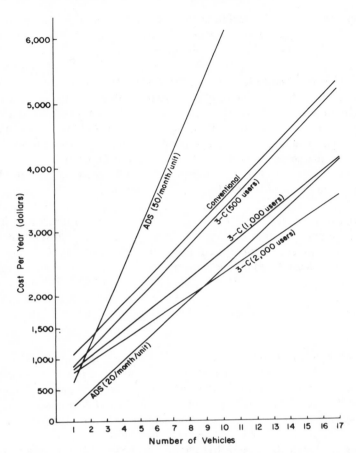

Figure 6.3: *A cost-per-year comparison of the cellular, MCTS, and conventional systems as a function of fleet size*

system, thus guaranteeing that its cost will always remain below conventional prices. However, this $20 charge per month and the previously quoted $50 per month per vehicle do not include any extra charges due to long-distance calls, which may be significant for long-haul truckers.

Third, at all three stages of sharing (500, 1,000, and 2,000 users) the Motorola 3-C system appears to be economically advantageous. One reason for this is that the 3-C system, after being paid for initially, does not require a monthly fee for leased antenna lines. Furthermore, the mobile units for the 3-C system would be $250 cheaper than present mobile units, presumably because the 3-C units require less power output, since the range needed to read one of the antennas is reduced by distributing the antennas throughout the service area.

And finally, for larger fleets, the 3-C system will be more economical than even the ADS system at high penetration rates.

Some Policy Implications and Regulatory Alternatives

Many agencies will influence the operation and use of communications in the trucking industry. Three particular agencies—the ICC, the FCC, and the U.S. DOT—will be faced with important questions of policy and regulation. The major policy considerations for each are outlined below.

REGULATION AND THE INTERSTATE COMMERCE COMMISSION. Two general regulatory issues now confront the federal government concerning the trucking industry—the question of continued regulation versus deregulation, and the issue of extensively consolidating the industry. These issues raise the following points that are important for the future of mobile communications:

1. If a regulation policy for the industry is to be continued, the ICC should consider easing its restrictions on allowed routes and service areas in cases where the potential for substantial efficiency improvements exists. The ICC should weigh the potential benefits of both present and future communication technologies in making its rulings.

2. Deregulating the industry would provide the greatest opportunity for improved efficiency through the use of communications. The elimination of route restrictions would allow over-the-road truckers to utilize the unlimited range of the new technologies. Relaxed service area restrictions of local pickup and delivery operations would allow these truckers also to take advantage of potential increases in communication range. Finally, deregulation of rates would enable the financial savings brought about by increased mobile communications use to be passed on to the shipper. Such factors should be taken into account when legislation dealing with deregulation of the trucking industry is considered.

3. If regulation is continued, the ICC should establish rates that reflect the savings allowed by mobile communications. Savings should be passed on to the shipper and to the general public.

4. The fact that the new technologies will facilitate more efficient control of vehicle fleets and coordination of goods movement tends to support a policy favoring consolidation of the industry. To make full use of the mobile-to-mobile capabilities and interconnection with the landline network, truckers should be allowed to coordinate shipper requests. Antitrust regulation could be amended to allow for consideration of cases in which economies of scale can be proven and truckers agree to reflect these efficiencies in their rates.

U.S. DEPARTMENT OF TRANSPORTATION POLICY

1. As mentioned earlier, the U.S. DOT has been actively pursuing the development of antihijacking systems. Given the fact that a cellular network might exist in major metropolitan areas and over interstate highways within 20 years, the DOT should consider experimenting with cellular technology as an antihijacking tool.

2. Research should be continued to determine how seriously driving ability is affected by the use of mobile communications, and whether legislation or new safety standards should be developed to ensure that the increasing number of radio-equipped truckers does not present a threat to the motoring public.

THE FEDERAL COMMUNICATIONS COMMISSION.

1. At present, the spectrum allocated to land mobile radio in the lower frequency bands cannot satisfy the demand. In its future decisions regarding the 900-MHz band, the FCC should continue to be concerned about the inefficiencies in the trucking sector and societal costs that result from communication deficiencies.

2. One of the greatest advantages of cellular and multichannel trunked technologies is the ability to reuse spectrum and thus provide more or less unlimited communication capacity. This scheme works best, however, for individual (nondispatch) users. Large radio-equipped truck fleets could face special problems. For example, a large number of mobile units attempting to contact a common dispatcher could saturate several channels and require long wait times; the FCC should consider adopting policies to reduce this problem. One scheme might be to require a specific mobile unit/base station ratio (e.g., a maximum of 30 to 40 mobile units/base station) to ensure that large queues do not develop. A balance between communication costs to the industry and the public cost of inefficient spectrum use needs to be established.

This completes our analysis of the impact of new mobile communication technology on the trucking industry. We now turn to consideration of passenger transportation, with special emphasis devoted to the development of demand-responsive public transportation systems.

NOTES

1. Paul Grey, "Prospects and Realities of the Telecommunications/Transportation Tradeoff" (Los Angeles, Calif.: Center for Futures Research, University of Southern California, September 1973); Jack M. Nilles et al., "Telecommunications Substitutes for Urban Transportation" (Los Angeles, Calif.: Center for Futures Research, University of Southern California, November 1974).

2. U.S. Department of Commerce, "Telecommunications Substitutability for Travel: An Energy Conservation Potential," prepared for the Office of Telecommunications by Charles E. Lathey, OT Report 75-58, Washington, D.C. January 1975.

3. William J. Hudson and James A. Constantin, *Motor Transportation: Principles and Practices* (New York: The Ronald Press Company, 1958).

4. Interstate Commerce Commission, "90th Annual Report" (Washington, D.C.: U.S. Government Printing Office, June 1976).

5. Personal communication with William Elder, Director of Communications, American Trucking Association, Washington, D.C., January 20, 1976.

6. William J. Hudson and James A. Constantin, "Micro-electronics Adds Clout to Management Control Plan," *Transport Topics,* March 31, 1975.

7. E. T. Bennett, "Metropolitan Area Microwave," brochure (Syracuse, N.Y.: Communications Xperts, Inc., 1975).

8. E. T. Bennett, "Security-Across the Dock Operation," brochure (Syracuse, N.Y.: Transportation Communications Associates, Inc., 1975); U.S. Department of Transportation, Office of the Secretary, "A Report to the President on the National Cargo Security Program" (Washington, D.C.: U.S. Department of Transportation, March 1977).

9. E. T. Bennett, "Digital Link Aims at Truckers," *Electronics,* September 19, 1974, p. 36.

10. The estimate of 15 to 25 percent efficiency was established through an American Trucking Association survey of 37 trucking companies. H. A. Plotkin, "Mobile Radio—A Pickup and Delivery Tool," *The Logistics and Transportation Review* 9, no. 4 (1973):357-66.

11. U.S. Department of Transportation, "A Report to the President."

12. U.S. Department of Transportation, Office of Transportation Security, "A Cooperative Approach to Cargo Security in the Trucking Industry" (Washington, D.C.: U.S. Department of Transportation, 1973).

13. U.S. Department of Transportation, "A Cooperative Approach"; "DOT demonstrates Electronic License Plate," *Electronics*, September 13, 1973, p. 50; U.S. Department of Transportation, Office of the Secretary, "Department of Transportation News," Washington, D.C., February 20, 1975.

14. U.S. Department of Transportation, "Department of Transportation News."

15. Hudson and Constantin, "Micro-electronics"; William J. Hudson and James A. Constantin, "On-line Systems Up-to-Date in Kansas City," *Transport Topics,* April 28, 1975.

16. American Trucking Association, *American Trucking Trends 1974* (Washington, D.C.: American Trucking Association, 1974), p. 3.

17. The calculation is as follows: revenue before communications use = \$21.00 $\times$ $10^9/1.03=$ \$20.39 $\times$ 10^9 per year. Therefore, the increased gross revenue due to communications is equal to \$610 million per year based on a 4.8 percent gross profit margin; this means a \$29,280,000 increase in gross profits per year.

18. This analysis assumes that the three percent efficiency can be considered analogous to a three percent increase in revenue. This is similar to saying that the industry would use its gain in efficiency to move three percent more freight.

19. Energy statistics for the trucking industry revealed the following. The typical motor carrier consumes approximately 2,800 BTU/ton-mile of goods moved. Also, combination trucks used about 8,860 $\times$ 10^6 gallons of fuel in 1973. Given the gallons of fuel consumed and that an estimated 505,000 $\times$ 10^6 ton-miles of goods were moved by the industry, one can conclude that the typical carrier uses about 0.018 gallons of fuel per ton-mile (or, in other words, the typical carrier gets about 56 ton-miles per gallon of fuel). Based on a study of 37 trucking companies, it was found that mobile communications use by irregular-route, nonprescheduled operators results in an average increase in efficiency of 20 percent. Hudson and Constantin, "Micro-electronics"; Hudson and Constantin, "On-line Systems"; Bennett, "Metropolitan Area Microwave."

20. Motorola, Inc., "Reply Comments," Formal submission to FCC Docket 18262, Motorola, Inc., Schaumburg, Ill., July 20, 1972.

Chapter 7
LAND MOBILE COMMUNICATIONS AND THE TRANSPORTATION SECTOR
Passenger Transportation

Arnim Meyburg
Russell Thatcher

DEMAND-RESPONSIVE TRANSPORTATION

Unlike the freight industry, "demand-responsive" passenger transportation has not been a longtime or large-scale user of communication services, with the single exception of taxicabs. We are emphasizing demand-responsive systems in this study because this type of passenger transportation is likely to become more common in the future, and the use of mobile radio is critically important to their operation.

In order to serve the public better and to expand the transit market beyond the "captive rider," a search for new concepts in passenger transportation was initiated in the mid-1960s. One concept that evolved was that of the "demand-responsive" system, transportation routed and/or scheduled to meet the specific requests of the customer. Ideally, door-to-door service would be provided between origin and destination, and the request for a ride would be answered with the least possible delay.

Although the concept of door-to-door, variable-schedule operation for widespread public use was first introduced in the 1960s, demand-responsive systems have a long established counterpart in taxicab service. Taxicabs are a substantial part of the total national transportation complex. For example, the 175,000 taxicabs operating in 1970 accounted for approximately 71 percent of all public transportation equipment, and in 1969 about 80 percent of travel by public transport vehicles (6,800,000,000 revenue vehicle-miles) was accomplished by the use of taxis.[1] One reason for their success is the high level of service and convenience provided, which is comparable to or exceeds that of the privately owned automobile in many circumstances.

Attempts to incorporate the taxicab's service quality into other means of public transportation were motivated by the fact that taxis were able to attract a substantial portion of the "automobile-dependent" population. But there were three problems in extending taxi systems to public transit use: the low vehicle productivities,[2] the consequently high fare needed to cover the operators' expenses, and local ordinances regulating fare structure and operating characteristics for taxi companies.

One attempt to remedy the high fare problem and still maintain a relatively high level of service is represented by shared-ride taxis. Organized shared-taxi systems are not yet as extensive in North America as they are overseas. Although a number of companies leave to the discretion of the driver (and the approval of customers) whether more than one party is to be served, standard taxi service is still designed to serve a single customer or party. Nevertheless, recent attempts at providing organized shared-taxi service have been successful in eight U.S. cities.[3]

Another type of system that represents an attempt to introduce the demand-responsive concept to the public transit industry is dial-a-ride, an alternative also referred to by names such as call-a-bus and tel-a-bus. As the name implies, customers receive service by phoning a central dispatching office and making known their desired destination and present location. The dispatcher will then assign the trip to that vehicle most able to meet the demand. Such dial-a-ride systems are designed to reduce costs substantially below those of taxi services by increasing vehicle productivity, and thereby perhaps ultimately becoming comparable in cost to conventional transit systems. Typically, small buses with a capacity of 10 to 20 persons are used: this greater passenger capacity involves more complex routing and scheduling functions than occur in taxi operations.

Several forms of dial-a-ride systems exist, which are defined by the routing arrangements used. They are:

MANY-TO-ONE SERVICE. Several origins are served and transportation to a single destination is provided. The one destination is usually a shopping center, work location, or a transfer station for other modes of transportation. (Figure 7.1.)

MANY-TO-FEW SERVICE. Where a number of high demand destinations exist, this service is employed to provide a doorstep service from several origins and access to a limited number of destinations. (Figure 7.2.)

MANY-TO-MANY SERVICE. This most extensive form of dial-a-ride operation will provide door-to-door transportation from any origin to any destination within the system's service area. (Figure 7.3.)

Another distinguishing feature of dial-a-ride services is the way customer requests are handled. Demands can be incorporated into the dial-a-ride system either dynamically, on a "call ahead" basis, or through subscription. The most complex of these is the dynamic operation, which enables customers to receive service soon after their phone request; wait time averages about 30 minutes. Because of the difficulties inherent in providing instantaneous service, many systems require that requests be made one day in advance. A second way of

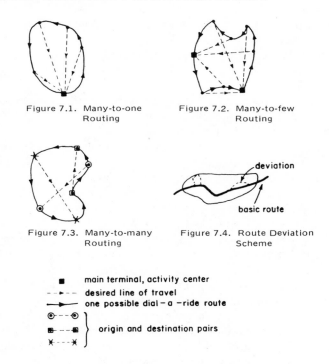

Figure 7.1.   Many-to-one
Routing

Figure 7.2.   Many-to-few
Routing

Figure 7.3.   Many-to-many
Routing

Figure 7.4.   Route Deviation
Scheme

■   main terminal, activity center
‒‒▸‒‒   desired line of travel
━▸━   one possible dial‒a‒ride route

◉‒‒◉ ⎫
■‒‒■ ⎬ origin and destination pairs
✕‒‒✕ ⎭

Figures 7.1 to 7.4: *Types of demand-responsive routing arrangements*

SOURCE: Adapted from U.S. Department of Transportation, Office of the Secretary and Urban Mass Transportation Administration, "Demand Responsive Transportation, State-of-the-Art Overview" (Washington, D.C.: U.S. Department of Transportation, 1974).

alleviating the complex dispatcher functions is to allow for subscription service where a patron registers an order to be picked up on a regular basis. Subscription is common for work or school trips.

Dial-a-ride represents a rapidly growing form of public transportation. The first system was established in Peoria, Illinois, in 1964, and since that time over 70 systems have been put into operation throughout North America, the majority after 1970. Although presently limited in terms of number of vehicles and passenger ridership, the future of dial-a-ride appears promising. It is particularly well suited for serving areas of low and medium population density and for providing special groups such as the elderly with transportation.

A final demand-responsive system to be examined is *jitney*. This system—which has the least amount of demand flexibility—operates on a "route deviation" basis. As illustrated in Figure 7.4, jitneys travel on a more or less fixed route that can be altered slightly according to customer demand. Also, the jitney is responsive in the sense that no stops are scheduled along the fixed route and customers are served by either flagging down the vehicle or, in more sophisticated systems, by phoning their request to a dispatch center. Unlike dial-a-ride, jitney service is limited to requests from persons within the immediate area of a fixed route.

Literature describing demand-responsive systems and providing information on costs, ridership, and types of service is extensive.[4] Information and research pertaining to the communication aspects of demand-responsive systems is much less abundant, however; only two studies treat communications as something more than a peripheral element.[5] Thus, to obtain information concerning communications uses, problems, impacts, and needs, we conducted a survey of demand-responsive systems in the United States and Canada during the course of the research described in this chapter. A total of 74 systems were surveyed by questionnaire. Information resulting from this survey will be used throughout the remainder of the chapter.

Communications Interfaces

Although each of the four system types (taxi, shared-ride taxi, dial-a-ride, and jitney) being considered has various communications requirements, two requirements are common to all: effective means of communicating with customers and effective means of communication between driver and dispatcher.

Standard (nonshared) taxi operations accomplish these two needs in the simplest fashion. Many customer requests are handled directly by the driver, as a substantial fraction of riders simply flag an empty cab. In these cases, the only mobile communication that may be useful would be a message from driver to dispatcher, stating that the taxi is in use and where it is going. When requests are phoned to the dispatcher, however, the vehicle fleet members must be paged by the dispatcher. Because a single vehicle does not service multiple requests and can be in only one of two states—occupied or unoccupied—the taxi industry has been able to employ a very simple driver/dispatcher communication scheme. Typically, the dispatcher will send out an "all-call," which specifies the demand to be met, and a driver who is free and near to the customer will respond. All-call capability has enabled the industry to avoid costly and complex dispatcher routing and scheduling functions.

Jitney operation requires only slightly more sophisticated means of communication. Again, many customers are picked up by the driver without intervention of a dispatcher. If calls do go through the dispatcher, though, an additional level of control is needed. With certain vehicles assigned to specific routes, it is necessary to know the general location of vehicles in order to determine whether a request can be served quickly by a nearby vehicle or whether the customer will be required to wait. Thus, drivers must be able to report their position to the dispatcher (and possibly also the number of passengers presently on board so that overloading will not occur).

Shared taxi and dial-a-ride systems require greater vehicle control and more sophisticated communication methods. In these systems almost all customer requests must be channeled through the dispatching center and relayed to the

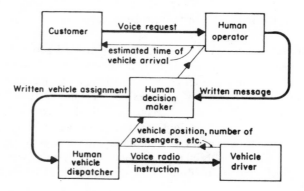

Figure 7.5: *Flow diagram of a manual control/voice communication operational scheme*

SOURCE: In part adapted from U.S. Department of Transportation, Transportation Systems Center, **Transportation Systems Technology: A Twenty Year Outlook** (Washington, D.C.: U.S. Department of Transportation, August 1971).

vehicle driver. Rather than allow vehicles to operate independently in the field, the dispatcher must now coordinate vehicles and customer demands.

Three levels of automation in dial-a-ride and shared-taxi operations have been identified in the literature; which one is used depends upon demand and the number of vehicles in the fleet. The simplest of these is the manual control/voice communication (M/V) level. In this mode, customer requests are accepted by a switchboard operator who writes down origin, destination, and other pertinent information. This information is passed to a human dispatcher who attempts to respond on the basis of information concerning each vehicle's location and occupancy rate. Once a vehicle has been chosen, the dispatcher relays two messages: one to the switchboard operator regarding the expected pick-up time, which is then passed on to the customer, and another to the radio dispatcher identifying the proper vehicle and the assignment to be given. This totally manual operation is shown in a flow diagram in Figure 7.5. It should be noted that information must also flow regularly from driver to dispatcher so that the latter can know the position and utilization of vehicles under his control.

Completely manual operation is limited to low demand or low density operations. Control of large vehicle fleets and hundreds of customer requests places an impossible burden on a human decision maker, causing inefficient and even erroneous assignments to be made. The second level of automation, labeled the computer control/voice communication (C/V) level, removes this decision-making limitation. As can be seen in Figure 7.6, the process remains essentially the same, except that the human decision maker has been replaced by a computer. Instead of writing down customer information, as in the M/V system, the operator punches this information onto a computer card and feeds it into a card reader. Programmed to set up optimal "tours" for each of the vehicles, the computer is able to handle a much higher demand.

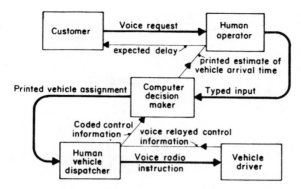

Figure 7.6: *Diagrammatical scheme of computer controlled/voice communication operation*

SOURCE: In part adapted from U.S. Department of Transportation, Transportation Systems Center, **Transportation Systems Technology: A Twenty Year Outlook** (Washington, D.C.: U.S. Department of Transportation, August 1971).

With this greater capacity for optimal decision making, the system is now limited only by the time necessary to feed information into the computer from both customer and vehicle elements, and by the time required to pass the results on to the customer and vehicle drivers. Although this problem can be alleviated by increasing the number of operators and vehicle dispatchers, a much more economical solution is represented by the final level of automation, defined as the computer controlled/digital communication (C/D) level. This completely mechanized flow of information employs digital communication technology at both the customer interface and the computer/vehicle interface. The user of the system is required to use a coded message for his location and destination (which means that push-button phones are needed), and each vehicle must be equipped with two-way digital capabilities, both to receive printed messages from and to feed information back into the computer.

As shown in Figure 7.7, this final level of automation will probably require the addition of voice capability between the vehicles and the customers. Emergency situations will call for the flexibility of human understanding, and unless an extremely complex customer/computer interface is designed, inquiries about the operation of the system will have to be addressed to a human operator. Customers who do not own push-button phones will also be required to pass information to the computer through a human operator, as in the C/V level.

Two additional "quasi-levels" that can be considered variations of the M/V and C/D levels also exist. These levels, the manual control/digital communication (M/D) and the computer control/voice-digital (C/V-D), are products of past or present technical limitations and are unlikely to survive. Finally, it should be noted that there are many variants of the computer-communication arrangement. Systems can involve dedicated or time-shared computers, remote or local siting, integrated or separate communication links.[6]

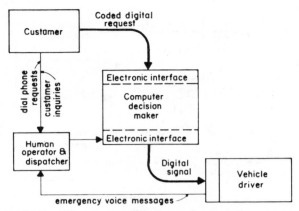

Figure 7.7: *Sophisticated scheme for demand-responsive systems operation: computer control/voice communication*

SOURCE: In part adapted from U.S. Department of Transportation, Transportation Systems Center, **Transportation Systems Technology: A Twenty Year Outlook** (Washington, D.C.: U.S. Department of Transportation, August 1971).

Communications Equipment Presently Available

A number of technologies are available for dial-a-ride communication use. These are indicated below in order of increasing sophistication:

1. *LANDLINE TELEPHONE/TWO-WAY RADIO.* This is by far the most common technology used at present; it consists of a phone or phones, usually with a number of trunked lines, and a simple two-way radio system operated in the present land mobile frequency ranges of either 150 or 450 MHz.

2. *COMBINED SWITCHBOARD/DISPATCHER SYSTEM.* Designed for small dispatching operations, where the functions of the operator and dispatcher can be combined, this system (developed by Motorola) is used in a number of dial-a-ride operations. The control console, which resembles a large multiple-extension telephone, can handle up to four remote control functions and is capable of connecting regular wireline phones to mobile units through a voice-coupling device.

3. *RADIO SYSTEM WITH SELECTING CAPABILITIES.* Many operations prefer to direct the transmission of messages to specific vehicles, so that assignments can be picked up much more easily and drivers are distracted only when necessary. In this case, a slightly more sophisticated two-way radio system is needed that requires more channel space than systems relying on all-call.

4. *TELEPHONE/ONE-WAY TELEPRINTER SYSTEMS.* Mobile teleprinter systems in combination with a manual, telephone-oriented customer interface

represents the next level of present technological sophistication. Digital systems with visual displays in mobile units are also now available.

5. *TWO-WAY DIGITAL CAPABILITY.* This technology provides a keyboard in mobile units allowing the driver to transmit digitally coded messages; it is used in only two demand responsive systems (some New York taxis and Mississauga, Ontario, taxis.) Its advantage is that vehicles can be linked directly with the routing and scheduling computer. See Figure 7.7.

6. *AUDIO-RESPONSIVE UNITS.* Although not presently used in demand-responsive systems, technology is available to accomplish a mechanical, customer communications interface. Audio-response units react to the various tones of a push-button telephone and can be used to relay the customer's coded message directly to a scheduling computer.

Finally, we wish to note an unusual use of paging for control purposes in the Rochester, New York, dial-a-ride operation. Each time a driver arrives at a new pickup or delivery spot, he activates a tone-paging system. Since each vehicle is given a unique signal tone, the dispatcher is able to keep track of the progression of vehicles through their assigned "tours." This technology, though it is used in only one case, illustrates the efforts being made to devise suitable systems for the important vehicle/dispatcher control interface.

Digital vs. Voice Communications

It has been suggested that the new mobile cellular system be built with digital as well as voice capabilities. This section compares the relative merits of voice and digital technologies as applied to demand-responsive transportation systems. John E. Ward, discussing these two communication technologies, states that:

> Investigation of traditional voice radio dispatch techniques has indicated that in a dial-a-ride system, a human dispatcher using voice radio could handle approximately 100-200 demands (passenger service requests) per hour, or about 5-10 vehicles, and that this message traffic would completely utilize one voice radio channel (or channel pair). Although voice radio could be a feasible and economic means of vehicle communications in small-scale dial-a-ride systems (10-20 vehicles), the requirements for a separate dispatcher and radio channel for each 10 vehicles indicates that voice radio is not a feasible solution for larger systems. The large number of dispatchers would be an economical burden and perhaps more important, the requisite number of radio channels are just not likely to be available in the congested land-mobile frequency bands.

> Digital transmission systems, which can make much more efficient use of the voice-channel bandwidth, are now coming to the fore in police, transportation, fire, and other public communications services, both because of the severe spectrum congestion and also because they provide

faster, more reliable communications, including hard-copy printout. In a dial-a-ride system, for example, a digital system operating on a single channel pair at 1000 bits per second could transmit 3600 40-character messages per hour, sufficient to direct about 180 vehicles, and simultaneously poll (scan for information from) the vehicle fleet at a rate of 40 vehicles per second for driver inputs or responses.[7]

Although this statement indicates that digital communications would be highly desirable in systems with more than 20 vehicles, the benefits to be gained from a faster, more reliable means of communication are not limited to large-scale systems. In 1972, the Ford Motor Company, in conjunction with Motorola and the Rochester-Genesee Regional Transportation Authority, sponsored a test of digital communications effectiveness in the Batavia, New York, dial-a-ride system, covering a trial period of 17 days.

Defining efficiency as the percentage by which total communications time is reduced, the report concluded that the digital teleprinter was between four and ten percent more efficient. Moreover, the vehicle productivity increased from a previous average of eight or nine passengers/vehicle-hour to a high of 11.2 passengers/ vehicle-hour. Improvement in operational efficiencies was not the only benefit, however. The teleprinter's production of a hard-copy print of each customer request provided an efficient means of system monitoring and reduced manual paperwork. It was also noted that utilizing the teleprinter was likely to improve safety because the drivers could receive messages at any time, without their attention being diverted.

At the time the teleprinter was installed, Batavia was operating with only five demand-responsive vehicles. The efficiencies reported by this test should therefore be considered very conservative. If extrapolated to larger dial-a-ride systems, with greater message traffic, it is not unlikely that efficiencies of 15 percent could be achieved in systems with ten to 20 vehicles.

If cellular technology were to provide digital capability, it is probable that the combination of improved operating efficiencies, increased safety, lowered cost, and absence of channel crowding would favor the acceptance of this new technology in the demand-responsive passenger market.

The Impacts of Present Communication Technology

First, it must be emphasized that pure dial-a-ride or shared-taxi operations depend heavily on mobile communications to achieve reasonable efficiencies. It is possible to operate these systems without such technology by using subscription service or requiring vehicles to return to the dispatching center for new assignments, but the service can become very inefficient. The need for effective mobile communication technology was particularly evident in two responses to our survey by managers of systems that did not use two-way radio. In general, communications is seen to have an impact on demand-responsive systems in the following areas:

PRODUCTIVITY AND OPERATING EXPENSE. Vehicle control, provided by an efficient driver/dispatcher link, can lead to more efficient routing and scheduling and can, therefore, increase vehicle utilization and decrease operating costs. Also, well-designed communication systems, such as a combined operator-dispatcher console, can eliminate unnecessary labor costs and substantially reduce operating expenses.

CUSTOMER LEVEL OF SERVICE. Use of mobile communications can greatly reduce the delay between placement of a phone request and the vehicle's actual arrival time. Because much of the demand-responsive systems' success is based on their competitiveness with the automobile, a high quality of service can influence the degree of customer acceptance.

SAFETY (DECREASED). Constant use of two-way communications could be severely distracting to the driver. In the case of voice assignment, the driver is required to manually record each new address and name. This hazard can be reduced through the use of digital equipment, but the driver must still read the printed message. Two-way digital schemes would require drivers to code messages manually for transmission and could eliminate the benefit of digital communications in this impact area.

SAFETY (INCREASED). Some dial-a-ride systems are used exclusively for transporting elderly and handicapped persons. Systems not completely devoted to the disadvantaged have also found their elderly and handicapped ridership to be large. With accidents and illness high among this population group, mobile communications become a useful tool in cases of emergency. They are also useful in dealing with other emergency situations such as traffic accidents and mechanical breakdowns.

Questions concerning all of the above impacts were included in the survey of demand-responsive systems. Operators were asked whether their communication scheme had no effect, some effect, or a significant effect in each impact area. Responses of the 42 systems that replied to these questions are tabulated in Table 7.1. To permit a comparison of aggregate responses in each impact area, an average value was calculated by assigning weighting factors of zero in the case of no effect, one for some effect, and two for a significant effect. The importance of mobile communications is reflected in a majority of the responses; in Table 7.1 all average values exceed 1.0, and three of these (productivity, wait time, and aid in emergency situations) approach a maximum value of 2.0.

. Although a decrease in operating expenses can result from increased vehicle utilization and productivity, the former scored well below the average value associated with vehicle productivity (1.13 vs. 1.83). This outcome was due to the high cost of communications equipment. These responses seem to indicate that although improved operating efficiencies are realized, the cost of mobile communications equipment offsets this gain to a significant degree. Also the category "safe travel" did not seem to represent as tangible or direct an impact area as those categories that could be more readily associated with everyday operations.

Table 7.1: *Perceived Effects of Communications Equipment on Selected Impact Areas (Replies from 42 Systems)*

	Perceived Effect				
Impact Areas	No Effect	Some Effect	Signif. Effect	N.A.	Average Value
Vehicle productivity	0	7	35	0	1.83
Customer wait	0	7	34	1	1.83
Operating expenses	8	18	13	3	1.13
Emergencies	1	7	34	0	1.79
Safe Travel	6	23	11	2	1.13

Communication Needs of Present Systems

Present communication methods have two important limitations: (1) capacity restrictions of manually dispatched, voice controlled methods, which are used in all but seven demand-responsive systems in operation; and (2) spectrum limitations that affect the availability of channels required to serve various levels of demand.

One must consider the consequences of these limitations for both the customer/switchboard operator and dispatcher/driver interfaces. Technology used in the customer communication interface limits the level of demand that can be served. With conventional telephone equipment and several trunked lines, an upper limit of about 200 demands per hour has been established.[8] This figure is based on the assumptions that three to four switchboard operators constitute the maximum number of call-takers that can coordinate information, and that a single operator is capable of handling 50 to 75 demands per hour for conventional taxi operations and about 30 to 60 demands per hour in shared-ride systems. The latter are likely to have a lower demand capacity because more information is required to pass between the customer and the operator.

The dispatching functions of a demand-responsive system, if done manually, present an even more severe limitation.

> In shared taxi systems where one man does both the scheduling and dispatching, studies have shown that this limit is about 100 demands per hour, allocated amongst a maximum of about 20 vehicles. . . . For conventional taxi service, a dispatcher can cope with perhaps 120 demands per hour, and possibly as many as 150 vehicles. But with this many vehicles he is not dispatching them all. Many are flagged on the street by customers. A dispatcher can handle perhaps 40 vehicles if they are totally centrally dispatched.[9]

Thus it appears that for dial-a-ride systems, a likely dispatching limit for one dispatcher exists at approximately 100 to 200 demands per hour, or about five to ten vehicles.[10]

Few presently operated demand-responsive systems have average hourly demands that exceed the limitations suggested above. However, at peak hours traffic, as in any other form of public transportation, tends to be several times greater than average demand. Some demand-responsive systems have found it necessary to limit peak-hour service to subscription riders, or to switch from a many-to-many mode of operation to a many-to-one or many-to-few type of service, thereby easing the routing and scheduling problems. Peak-hour demands can present dispatch capacity and response speed problems in systems with five or more vehicles.

Nearly half of the systems presently operating use five vehicles or less, and only 12 of the 77 systems have a fleet of greater than 20 vehicles. Probably the biggest reason for this is the fact that many systems are used for transportation of special population groups (e.g., handicapped, elderly) and the capacity is limited to what can easily be handled by only a few vehicles. It is also likely, however, that expansion of many operations is restricted by the dispatching and vehicle control problems.

A second limitation placed on the growth of present demand-responsive systems results from the crowded land mobile radio spectrum. A report prepared by J. K. Associates observes that: "In shared-ride systems, depending on the vehicle productivity, the form of service, the size of vehicles, etc., the number of vehicles per channel might reach a maximum of about 30, or perhaps be as low as 10."[11]

For a conventional taxi system, as many as 150 vehicles may be operating on one channel pair (but not all of these vehicles are "voice dispatched," since many customers simply flag down a vehicle). Few presently operated dial-a-ride systems have been allocated more than one channel pair by the FCC. Because the 150-MHz and 450-MHz bands are congested, limitations on channel allocation could become another major restriction on present dial-a-ride systems. For efficient operation and conservation of spectrum, digital communication systems have been suggested for dial-a-rides with more than 20 vehicles.

Our survey contained a number of questions concerning present operational problems and present and future communication needs. Only 15 of the 42 systems responding indicated that no operational problems had been encountered with present systems. This and other prominent responses are summarized

Table 7.2: *Frequency of Operational Problems with Present Communications Equipment (42 Systems Responding)*

Problems	No. of Systems Experiencing Problem
No operational problems	15
Breakdowns, unreliable equipment	7
Interference	5
Overloaded channels	4
Dispatching problems	3
Quality of transmission (dead spots)	6

in Table 7.2, along with the frequency with which each of these problems was mentioned.

In addition to these more frequent responses, operational problems unique to single systems were: "unreliability" of CB radio for use in dispatching; handicap of having only one-way transmission; too frequent use of mobile-to-mobile capability for personal conversation; and inefficiencies caused by the lack of mobile communications.

Speculation as to dispatching and channel problems in larger demand-responsive systems is substantiated when the occurrence of dispatching and capacity-related operational problems are cross-tabulated with vehicle fleet size. It was observed that all seven of these problems occur in systems with seven vehicles or more, and five of them occur in systems with 14 vehicles or more.

Two other questions dealt with the need for better communication capabilities. Table 7.3 shows the respondent's level of satisfaction with abilities of present communications equipment. Answers are disaggregated in Table 7.4 to show the communication needs as a function of the number of vehicles operated. The range necessary appears to be independent of the fleet size (as would be expected) and would probably vary with service area. The need for both capacity and response speed, however, increased with larger systems. Only those systems that indicated the need for the abilities listed appear in the table. Unfortunately, the disaggregated table entries involve very small numbers. A final question in our survey asked operators whether vehicle productivity, customer wait time, operating expenses, aid in emergency situations, and level of safety in travel could be further improved with the introduction of "better communications equipment." The disaggregated responses are presented in Table 7.5; average values were calculated as in Table 7.1. Table 7.5 indicates that "room for improvement" with better communications is perceived in almost half of the responses. Again, impact areas that are associated with everyday operation scored well above the more indirect impacts.

An interesting result was that "room for improvement" in operating expense ranked second only to vehicle productivity. Many systems responding gave the general impression that equipment expense was a significant burden. An inquiry into the capital and operating costs of communications equipment produced the following results: capital cost of equipment typically ranged from $1,200 to $1,700/vehicle; and operating cost was between about $15 and $50/month/vehicle (these figures would include the costs of the base equipment divided

Table 7.3: *The Number of Systems Indicating a Need for Better Communications Abilities*

Abilities	Not Needed	Presently Needed	Will Be Needed
Range	26	5	12
Response speed	21	5	17
Capacity	25	4	14

Table 7.4: *Communications Abilities Needed, Disaggregated for Different Vehicle Fleet Sizes*

Abilities	Number of Vehicles					
	Needed Presently			Will be Needed		
	<5 (16 systems surveyed)	5-10 (13 systems surveyed)	>10 (14 systems surveyed)	<5	5-10	>10
Increased range	2	1	2	4	3	5
Increased response speed	0	2	3	4	7	6
Increased capacity	0	1	3	3	5	6

among the vehicles). Although these costs are comparable to those borne by the trucking industry (see Chapter 6), they tend to be more of a burden to demand-responsive systems for which capital costs are generally lower. Since low capital investment is one of the positive features of this new form of transportation, high costs of communications can become a major deterrent to wider acceptance and growth of such systems.

System operators also responded to a question concerning the percentage of the total costs attributable to communications using two different bases. About half of the respondents provided communications costs as a percentage of initial system costs (i.e., other hardware), and about half seemed to be calculating communications costs as a percentage of ongoing expenses (i.e., wages, fuel, administrative costs, etc.). This led to the following ranges of cost estimates: between 12 percent and 20 percent when calculated as a fraction of initial capital cost; less than one percent when considered as a percentage of on-going expenses. The very high percentage indicated in the first category supports the conclusion that communications costs represent a substantial burden for this type of passenger transportation.

A review of several studies and the results of our survey indicate that the needs of present shared-ride demand-responsive systems include:

Table 7.5: *Additional Perceived Benefits in 5 Impact Areas*

Benefits	Room for Improvement				
	No Room For Improvement	Some Room For Improvement	Significant Room For Improvement	N.A.	Average Value
Vehicle productivity	16	17	6	4	0.744
Customer wait	17	18	4	4	0.667
Operating expenses	16	17	5	5	0.711
Emergencies	26	9	3	5	0.395
Safe travel	26	9	2	6	0.351

1. *A LESS EXPENSIVE FORM OF DISPATCHING.*   This applies especially to systems with less than five vehicles. While voice communications are adequate for their dispatching functions, the expense incurred is very high.

2. *A RELIABLE, INEXPENSIVE FORM OF AUTOMATED ROUTING AND SCHEDULING.*   Although this capability is only necessary in larger systems (greater than 20 vehicles), dispatcher requirements in the peak hours have been found to be a problem for systems with as few as seven vehicles.

3. *GREATER VEHICLE COMMUNICATING CAPACITY.*   This can be accomplished in two ways. If voice dispatching techniques are retained, more channels are needed. A better solution would be to use digital communications, because digital equipment uses the spectrum more efficiently and can be linked to an automated routing and scheduling unit.

The needs of conventional taxi systems appear to be met by present-day technology; with the exception of large systems (greater than 150 cabs), a single dispatcher and one radio channel pair appear to be adequate. Large fleets, however, have the same needs as those listed for shared-ride systems.

Communication Needs of Future Systems

Table 7.6 summarizes the responses of 42 systems to a question concerning plans for expansion. Over 71 percent of the systems responding had plans for future expansion, and a significant fraction were considering extensive growth. Yet as more systems move up into the 20-plus vehicle fleet category, they are likely to experience serious problems of communication capacity and speed that are associated with large systems.

For shared-ride systems to become effective means of public transit, much larger vehicle fleets are required. Table 7.7 suggests the level of demand and

Table 7.6: *Likely Growth of Demand-Responsive Systems*

Future System Expansion Plans	Total in Each Category	Percentage
No expansion	12	28.6
Increase in service area or fleet size up to 25%	19	45.2
Increase in service area or fleet size of 25-50%	5	11.9
Increase in service area or fleet size of over 50%	6	14.3
Total responses	42	100.0

number of vehicles needed to reduce private automobile travel by 14 percent, 33 percent, and 51 percent, respectively, through the use of shared-ride systems. The systems required would need hundreds of vehicles and would have to service several thousand demands per hour. If shared-ride systems are to be expanded to this level, present routing, scheduling, and vehicle and customer communication systems are not practical.

In 1969, planners at MIT outlined the requirements for a C/D level large-scale dial-a-ride system for the U.S. Department of Transportation.[12] Routing and scheduling was accomplished using a computer, and both the proposed customer and vehicle/dispatcher communications interfaces were digital. A rather elaborate vehicle-communication system was developed that included a communications coupler, separate from the routing and scheduling computer, to handle the tasks of message formatting and transmission, polling (scanning) of vehicles for responses, and error detection and control. Two radio channel pairs would be needed—one for digital transmissions to and from the vehicle fleets and one reserved for emergency voice transmission.

The MIT study also reviewed the possibility of incorporating automatic vehicle monitoring (AVM) technology into future dial-a-ride systems. Presently proposed monitoring systems are characterized as fixed-route proximity systems. Transmitters are stationed at intervals along transit routes and each is assigned a location number that is transmitted continuously. As vehicles pass these

Table 7.7: *Estimated Demand Densities and Vehicle Fleet Sizes for Future Dial-A-Ride Systems*

Demand Densities		Future Estimates		
	Present	Conservative	Medium	High
Urban Trips/Capita/Annum				
By private automobile	410	350	275	200
By fixed transit only	85	100	150	200
By para-transit* (including fixed transit connections)	5	50	75	100
	500	500	500	500
% para-transit	1%	10%	15%	20%
Private automobile reduction		14%	33%	51%
Para-transit Estimates				
Daily trips per 10,000 population		1,452	2,258	3,065
Peak demands/hour/10,000 population		218	339	460
For a community of 100,000 population				
Peak demands/hour		2,200	3,400	4,600
Number of vehicles in service at peak demand**		180-330	290-500	390-680
For a community of 600,000 population				
Peak demands/hour		13,100	20,300	27,600
Number of vehicles in service at peak demand**		1,000-2,000	1,700-3,000	2,300-4,100

SOURCE: Canada, Ministry of Transport, "Para-Transit Dispatch Systems," Report prepared for the Transportation Development Agency by J. K. Associates, Toronto, Ontario, 1975, p. 6-6.

NOTE: Based on reducing automobile travel by 14%, 33%, and 51% (assuming 550 trips/capita/annum).

*Para-transit refers to demand-responsive systems.
**Variations are due to routing arrangements chosen (e.g., many-to-one would require the least and many-to-many the greatest number of vehicles).

markers, the location numbers are picked up and transmitted to a central control unit. Vehicle monitoring would provide for direct interval surveillance of all vehicles in operation and could improve vehicle assignments made by a routing and scheduling computer.

Several technical problems remain to be solved, however, before AVM systems can be combined with C/D dial-a-ride technology. First, digital equipment is much more sensitive to variation in radio signals than are voice communication systems, and this may result in message errors. More reliable digital systems need to be developed before AVM will find widespread use in demand-responsive passenger transit systems. Second, because dial-a-ride systems do not utilize fixed routes, present AVM designs are not directly applicable to them. On-board locating equipment is required, and although technologies for independent "navigation" systems have been proposed, none have been tested in actual use. And third, cost requirements for integration of digital communication and monitoring systems could be prohibitive. Based on present technology, these systems would have to function separately and therefore would be purchased as separate components. Also, as was suggested in the MIT study, a voice system should be maintained for emergency transmissions. It is likely that such elaborate communication and control facilities will be too expensive for all but the largest and most intensively used dial-a-ride systems of the future.

A final communication need of future systems is a method for coordinating dial-a-ride systems with fixed-route, fixed-schedule systems. Since demand-responsive systems seem to be best suited to medium and low density operations, they could provide "feeder" service to existing transportation facilities; several systems (Regina, Alabama, Rochester, New York, and New Orleans, Louisiana, for example) are predominantly this type. With much larger fleets, communications between fixed-route, fixed-schedule drivers and dial-a-ride drivers (i.e., mobile-to-mobile capacity) may be required to accomplish smooth transfer of passengers.

The preceding sections have been concerned with demand-responsive passenger transportation systems. Transportation systems with predetermined routes and schedules do not usually employ mobile communications technology. Recently, however, several fixed-route, fixed-schedule bus operations have found that while mobile communication is not a necessity, it can be used to improve service. For example, traffic congestion or weather conditions can cause buses to fall behind their schedules; mobile communications can facilitate the taking of corrective action by system managers. Increased safety through more rapid response to emergency situations is a second example. Driver satisfaction can be increased because the communication link reduces their isolation from the parent organization. Automatic monitoring of the mechanical function of transit vehicles, such as engine temperature, oil level, and air pressure, constitutes a further benefit the mobile radio may provide these types of transit systems.[13]

POTENTIAL IMPACTS OF NEW COMMUNICATIONS TECHNOLOGY ON PUBLIC PASSENGER TRANSPORTATION

The 1960s and 1970s have been a period of innovation in public transportation in the U.S., much of it stimulated by actions of the federal government. The Urban Mass Transportation Act of 1964, and the many subsequent amendments to this bill, have provided considerable assistance to transit operations while calling for extensive research in the transportation field. Continued pressures for innovation within public transportation would appear to enhance the possibility of new communication technologies being adopted.

As is the case within the trucking and private passenger transportation areas, use of cellular or multichannel trunked (MCT) technology by public transit systems will be affected by a number of variables. Although widespread consumer use of cellular technology is not imperative for its adoption by the mass transit industry, its utilization could improve interaction with customers. Parallel developments in computer technology would also readily complement new mobile communication technologies: full exploration of capabilities such as greater communication capacity and vehicle monitoring could be enhanced by computerized routing and scheduling algorithms and/or computer control technologies, for instance. Greater industry integration would facilitate more flexible communications between drivers and dispatchers of different transit systems; interconnection with the wireline network and mobile-to-mobile capabilities provided by the new technologies would also be highly advantageous. Costs and capabilities of these new communication services are obviously important. Finally, when analyzing the potential consequences of implementing mobile communications, it is important to distinguish between the many different types of public transportation systems, since impacts vary significantly according to the type of operation.

Operational Efficiencies

Precise control of vehicle fleets is perhaps the greatest single determinant of operational efficiency outside of 'management and operator skills and vehicle hardware technology. Control reduces "bunching," enables schedules to be maintained, and provides quick response to nonroutine situations. Achieving good vehicle control in bus operations has been very difficult. Control systems that have been used include conventional two-way radio and the newly developed automatic vehicle monitoring systems. Two-way radio systems, most notably the Motorola Metrocom system, have been extremely beneficial to intraurban bus systems. For example, installation of a two-way radio system (Metrocom) in San Diego, California, resulted in improved schedules and fleet control, and increased ridership by 50 percent in the first year of use.[14] While benefits of two-way communications have been recognized, use has been limited by cost. The Syracuse (New York) Centro bus system reported paying approximately $450,000 for a Metrocom system to control 150 buses. Deployment of AVM systems has also been limited because of high cost. Subscribing to a common carrier system, such as cellular or MCT, capable of incorporating the functions of an AVM system into a two-way radio network and offered at a reasonable monthly rate would seem to be a viable alternative for intraurban bus fleet control.

Intercity bus systems could benefit from new technologies if mobile communications covered a greater range. This technical barrier has limited the usefulness of mobile communications for intercity operations. Intracity systems, such as dial-a-ride, jitney, and shared-ride taxi systems, could benefit from vehicle monitoring and increased dispatching capacity and speed. Since most taxis now operate in large cities where frequencies are very overcrowded, increased capacity would alleviate a major problem for present conventional taxi operations. Vehicle monitoring, however, would have little immediate impact on the industry because complex routing and scheduling of vehicles are not used.

All systems would gain from lowered costs of communications equipment. Specifically, the monthly charge for cellular service would eliminate the high capital cost of equipment. Shared-ride taxi systems and dial-a-ride systems, which are of necessity radio-equipped, would experience the greatest benefit. If services on the cellular system were only 50 percent more per month than a fixed telephone, the economic impact would be felt by all demand-responsive operations. If, however, the charge were substantially higher, closer to $50 per month, large dial-a-ride and jitney services would probably opt for the outright purchase of a communications system, especially if federal and state money were available for purchasing equipment. Furthermore, high costs would cause shared-ride taxi systems to continue using their conventional equipment, and conventional taxi systems would either remain totally nonradio-equipped or would continue to rely on the all-call form of dispatching. The only systems that would benefit from a cellular technology given a high monthly charge would probably be very small operations, for which even a $50/month charge would be more reasonable than purchasing a base station, a leased line to an antenna, and mobile units.

The interconnection of cellular devices with the landline network may reduce the "no-show" problem experienced in demand-responsive operations. Some customers call the dispatching center and then appear late or fail to appear when the vehicle arrives. At present, the driver must handle such problems by communicating with the customer via the switchboard operator. The cellular system will permit direct communication between driver and customer.

Even if a cellular technology were not employed in the dispatching of demand-responsive vehicles, one very important benefit would still result from public use. Customers would be able to phone in their requests without having to be at home or to use a pay phone. Of course, a fully portable system would eliminate virtually all access restrictions to demand-responsive transportation systems.

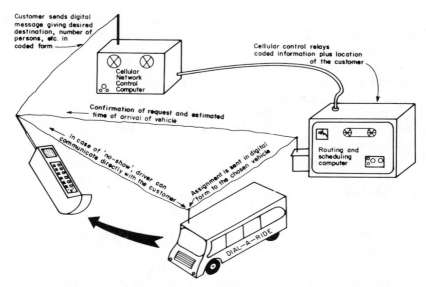

Figure 7.8: *A C/D level dial-a-ride operation utilizing a cellular mobile phone system. (The ability to receive customer location from the "cellular network control computer" and the ability to have drivers in direct communication with customers are benefits that would be unique to a cellular technology.)*

Cellular technology will also facilitate the incorporation of the most advanced automatic and computerized dispatch and control methods. No system operating today has reached the computer controlled/digital communication level of operation because of problems in developing reliable computer algorithms for automated control. The potential for a C/D operation would be increased by deployment of the cellular system. Through the use of digital coding, a mobile phone could be used to relay information between a vehicle and a computer control center without the intervention of a dispatcher. The customer could enter a coded message giving desired destination and present location, which could then be fed into the routing and scheduling computer. Elimination of dispatcher and switchboard operators would drastically reduce the operating cost of demand-responsive systems. This type of totally automated operation is illustrated in Figure 7.8.

Market Coverage

Increased control and operational efficiencies in interurban, intraurban, and demand-responsive bus systems could result in mass transit services being made available to a much larger number of people. Fixed-route bus systems might be able to offer more diverse routing arrangements if greater vehicle control could be achieved. By covering wider areas, an increased ridership could be attracted.

Broader market coverage would be generated by increasing the number of demand-responsive operations. Such systems are especially suited to serving the handicapped and elderly, and more systems are needed for this purpose. They are also well suited for areas with moderately low population densities and thus enable transportation to be offered to rural areas, small cities, and suburban areas. In metropolitan areas, demand-responsive systems can be used as feeder services, connecting suburban and fringe areas to city-based, fixed route, fixed-schedule systems. The mobile-to-mobile capability could facilitate the transfer of passengers.

Safety and the "Transit Image"

Many Americans view public transportation (especially subway systems) as unsafe and unpleasant. Reports of high crime rates in transit systems have led many city inhabitants to shun public transportation. As a result, older subway systems have had to hire special security personnel and newer transit systems are using closed-circuit television to discourage crime and ease the fears of riders. Central computerized monitoring, in combination with radio-equipped security forces, is also being used.

Widespread use of hand-held cellular mobile telephones would extend the communications network mentioned above. If a "panic button" were to be provided as a special feature of the mobile phone, its activation could alert the security-personnel dispatchers to emergency situations. Locating the incident could be facilitated by special receivers associated with the transit network.

On-Board Communications: Use by Patrons

In the preceding discussion, use of new communication technologies by patrons of transportation facilities has been limited to reporting and surveillance functions. But it would also be possible to use communications for conducting personal and job-related business. Indeed, if automobile users were able to communicate while traveling, mass transit systems might be disadvantaged if similar options were not available.

The potential for conducting conversations on public transit vehicles would appear to be limited to intercity rail and bus systems. The noise level and crowded condition of many subway systems would inhibit a phone conversation in a subway car, and privacy of phone conversations would be virtually zero unless special booths were provided. Intercity transit vehicles, though, could easily provide facilities for mobile phone conversations. In fact, many Amtrak and Trans-European Express trains have a limited number of booths containing telephones connected with the regular landline network. The provision of some form of telephone service on future intercity trains (or buses) seems highly desirable; the possibility of connecting a portable unit to an outlet seems feasible.

Mass Transit Versus the Automobile

While it is impossible to provide quantitative predictions of future shifts in automobile and mass transit ridership that might accompany deployment of new communication technology, some general impacts can be suggested.

1. The automobile user can travel to a destination by any number of routes and make any number of intermediate stops; in contrast, a transit user is limited to the route taken by the transit vehicle. Use of the automobile, therefore, makes it possible for the mobile phone user to change routes immediately in response to messages received during the trip. Because the transit user has much less flexibility, mobile phone technology will enhance the individual mobility allowed by the private automobile.
2. Intracity transit systems are unlikely to be able to provide individual communication capabilities. They will be disadvantaged by the new technologies being widely deployed in automobiles.
3. The amount of patronage that urban transit systems might lose as a result of the above considerations would probably be limited. A large part of present transit ridership consists of "captive riders" who, for one reason or another, do not have access to a car. Another group prefers to avoid the particular frustrations of automobile use in a major city. The remaining transit riders are not likely to be "mobility-oriented," which is evidenced by their use of transit, and would not be prime candidates for mobile telephones.
4. While urban transit operations may not lose a substantial portion of their ridership, they may encounter difficulty attracting additional patrons. Mobile communication technology would tend to increase the dependence of present automobile users on their cars.
5. Mobile communications might provide intercity transportation with a competitive advantage. Freed from driving responsibilities, travelers could use mobile phones to make their transit time more productive.

Some Policy Implications

THE FEDERAL COMMUNICATIONS COMMISSION. The potential inefficiencies caused by large radio-equipped truck fleets using new technologies were discussed in the previous chapter. The same concern would apply to large bus fleets. Spectrum could be "wasted" on queues of drivers waiting to contact a dispatcher. Again, the specification of a required mobile unit/base station ratio would seem appropriate.

THE U.S. DEPARTMENT OF TRANSPORTATION.   The possible advantages that mobile communications would offer private travel should be examined carefully for their potential impact on mass transit ridership. Demonstration projects that utilize the full potential of mass transit efficiencies afforded by mobile communications should be considered. Arrangements might be made between industry and large demand-responsive and fixed-route, fixed-schedule bus systems to test cellular and MCT technologies in these transit systems.

Potential Market for New Communication Technologies in Demand-Responsive Transportation Systems

Table 7.8 summarizes the responses to questions in our survey concerning the need for new (innovative) communications equipment. It indicates that demand-responsive operators are certainly not completely satisfied with their present communications facilities. Of the 42 respondents, five indicated immediate plans to purchase new communications equipment, and 16 indicated that new communications equipment is needed.

The questionnaire provided a brief description of the cellular system and the multichannel trunked system. Because of the brevity of the description, the possible capabilities of these new technologies were emphasized rather than their technical specifications. Expected costs were not stressed so that responses concerning usefulness could be obtained irrespective of a particular system's cost.

Table 7.9 presents the responses concerning the new systems. Vehicle monitoring was perceived as being most useful, reflecting the high level of control problems. Of the 42 systems responding, 37 percent recognized value in at least one of the new technologies for at least one area of potential benefit. Looking at each system separately, 28 percent of the operators recognized the cellular system as having benefits in at least one category, and 14 percent indicated that MCTS could aid in improving capacity, communication speed, vehicle monitoring, or quality of transmission.

In order to gain a deeper understanding of factors that may cause a positive attitude toward the new technologies, three cross-tabulations were performed. The first of these attempts to correlate a recognition of the new technologies'

Table 7.8: *Attitude Toward the Purchase of New Communications Equipment*

Need for Equipment	Number of Systems
1. Not needed now or in the future	6
2. Not needed now but will be needed	20
3. Needed now, but too expensive	6
4. Needed now, but right technology is not available	5
5. Needed now and plan to buy	5
6. Not applicable, system ceased operation	1

Table 7.9: *The Perceived Usefulness of Cellular and Trunked Technologies*

Abilities	Present System Is Suitable	MCTS Could Help	Cellular System Could Help	Both Could Help	N.A.
Capacity	27	3	5	1	7
Speed	24	5	6	1	7
Monitoring	23	4	9	1	6
Quality	25	4	6	1	7

potential usefulness with size of fleet. Table 7.10 presents these data. To gain a measure of the "positive response rate" for each category of vehicle fleet size, one must compute the ratio of the number responding positively to the total number of systems in each size category (it will be remembered that smaller demand-responsive systems are predominant). The column on "number of possible responses" was obtained by multiplying the number of systems in each category by four, since a positive response is only one of four possible responses. A rather good correlation between size and perception of potential benefit was

Table 7.10: *Responses Concerning Perceived Usefulness of the Cellular and MCT Systems*

Abilities (according to technology)	Number of Vehicles in the Fleet					
	0-5	6-10	11-15	16-20	21-25	26-30
Cellular technology could aid:						
capacity		2	2			1
speed		3	2			1
monitoring	1	6	2			
quality	1	3	2			
MCTS technology could aid:						
capacity			1	1		2
speed	1		1	1	No systems in this category responded	2
monitoring			1	1		2
quality						
Both MCTS and cellular technology could aid:						
capacity	1					
speed	1					
monitoring	1					
quality	1					
Total responses	7	14	12	3		10
No. of possible responses	64	52	20	8		16
Percentage of total responses	10.9	29.6	60.0	37.5		62.5

Table 7.11: *Cross-Tabulation of Abilities Needed and Response to the Acceptability of Cellular and Trunked Technologies*

Abilities that are or will be needed	Cellular Could Aid	MCTS Could Aid	Both Could Aid	T O T A L	No. of Systems in Each Category	%
Increased range of communication is now or will be needed	5	3	1	9	17	53
Increased capacity of communication is now or will be needed	6	3	1	10	18	56
Increased speed of communication is now or will be needed	8	4	1	13	22	59

obtained. The dividing line between the positive and negative responses to these new technologies seems to be ten vehicles; this would tend to support the conclusions drawn earlier regarding the relation of fleet size to communications problems.

Another cross-tabulation correlates expressions of specific communication needs with perceptions of the new technologies' usefulness. The results given in Table 7.11 indicate a rather high level of recognition of the new technologies' usefulness among organizations with expressed needs for communications and seem a good indicator of the potential for cellular and MCTS utilization by present demand-responsive systems. It seems safe to assert that between 50 and 60 percent of the systems whose needs are not met with present technology would turn to either cellular or trunked technologies if they were made available.

Variance in the Market According to System Type and Abilities Provided

While taxi systems were not covered in our survey, some comment is desirable on the use they are apt to make of new mobile communication technology. As mentioned before, the communication needs of conventional taxi and route-deviation jitney operations are not nearly as complex as shared-ride taxi and dial-a-ride services. Taxi systems do not require increased communication capacity until the vehicle fleet size is well over 100 vehicles. Since the all-call capability is an effective communication method in taxi dispatching, the acceptance of new technologies would also depend heavily on its inclusion; for this reason, the MCTS would be preferable for taxi services, unless the prohibition against cellular systems providing all-call service is removed. Finally, the inclusion of digital capabilities in the cellular system would greatly affect its use among larger systems. The trend toward digital transmission and its benefits over voice communications is quite clear for this portion of the transportation industry.

In summary, cellular systems seem to be best suited to: (1) small-scale shared-ride systems because of its low capital cost; (2) large shared-ride systems for which present technology is inadequate to handle capacity and control problems; and (3) large-scale taxi systems. The degree of adoption by (2) and (3) would depend on whether digital capabilities are provided in automatic dispatch systems (see Chapter 6).

Trunked systems would seem to have the same general markets, although their acceptance among larger systems would probably be less since the cellular system has superior potential for vehicle control and monitoring. The 3-C system (described in Chapter 4) may have a substantial market among small-taxi systems because all-call would be available, and the potential for savings through sharing a communication network would be quite high.

Having completed our analysis of mobile communications' usage in two sectors of the transportation industry, we shall conclude this chapter by discussing several topics that relate to its use by private citizens and commercial users. The topics include: motorist aid and information systems; safety hazards of mobile communication devices; traffic control and some impacts of mobile devices on travel patterns.

HIGHWAY MOTORIST AID AND INFORMATION SYSTEMS

For several years, the U.S. Department of Transportation has sponsored costly demonstration projects in an attempt to develop effective communication systems for use on highways. For the most part, such efforts have utilized stationary rather than mobile technology. The recent surge in CB radio purchases, however, has brought increasing attention to the potential of in-vehicle radio devices. Indeed, with CB radio on the verge of heavy penetration in the automobile market, it is possible that private automobiles could soon represent the greatest user of mobile communications in the transportation sector, so that use for highway aid and information might then take on added importance. Consequently, we shall briefly describe and analyze a number of both stationary and mobile highway communication systems in order to identify communication services needed for safe highway travel and the potential market for mobile communication technology.

Communication systems that complement automobile travel can be classified in two general categories: motorist aid systems, and traffic information and control systems. Motorist aid systems deal with needs for assistance, such as changing flat tires, correcting mechanical or electrical failures, providing fuel and towing services; they are also used to summon ambulance and fire equipment to highway accidents. Traffic information and control communication systems provide the motorist with knowledge of road and traffic conditions, so that hazardous conditions or congested areas can be avoided and traffic can be guided for optimal flow rates.

TELEPHONE AND RADIO CALL-BOX SYSTEMS AND ROADSIDE DETECTOR SCHEMES. One type of motorist aid that is becoming increasingly common employs roadside devices, such as call boxes or telephones. A coded call-box system generally employs radio transmission to a central receiving console, usually police operated. Each box has from two to four coded buttons, all boxes having buttons for police and vehicle service, while some include fire and ambulance. Four examples are presented below.

1. An emergency call system, tested in New Jersey, uses two-wire line buried along the roadway and signal devices placed every 200 feet.[15] Pressing a simple push-button switch activates a central police monitor. The two-wire system was not found to be significantly cheaper than other more sophisticated systems; it had relatively high maintenance and operation costs, including personnel costs.
2. An eleven-mile stretch of a Houston, Texas, freeway was equipped with four-button call boxes at quarter-mile intervals. This system was unique in that charges were made for services. There was apparent reluctance to use the call boxes, either because of the fees or out of simple ignorance, and the majority of stopped drivers obtained help by other means. However, on an elevated portion of highway the usage rate was almost three times greater than that of the entire eleven-mile portion.[16]
3. Roadside telephones generally use leased lines and are spaced at intervals of a half to one mile. One such phone system, connected to the state police, was successfully tested in a two-and-one-half-year experiment in Michigan.[17] An extrapolation of costs experienced with the test system gave an estimated cost of $25 per call, without accounting for economies of scale that would be present in a more widespread system.
4. FLASH, which stands for Flash Lights and Send Help, is a system that has been developed and tested in Florida.[18] Motorists flash their lights at marked detectors to inform police of disabled vehicles. Equipment consists of signs informing motorists of the system, roadside detectors capable of detecting headlights during day or night (spaced up to eight miles apart), a roadside computer to sort out signals and decide which are valid (one per detector, connected by landlines), and a central console at police headquarters (connected by phone lines). Several vehicles have to flash their lights within a given time span before action is taken on a particular incident. Test results showed that driver waiting time was reduced from a mean of 134 minutes before FLASH to 18 minutes. Driver comprehension and usage were considered satisfactory.

The costs of the Michigan, Houston, and New Jersey systems were roughly comparable on a per mile basis: installation costs ranged between $10,000 and $15,000 per mile. Cost expressed on a per station basis showed much greater variation—$197 in New Jersey, $1,830 in Houston, and $4,835 in Michigan. Maintenance costs ranged between $60 and $150 per mile/month. Costs of operating the three systems, including personnel, could not be compared because the figures presented did not refer to the same categories. No costs were given for the FLASH system, though it was described as "economical." For any of these systems to be effective, however, they would have to be extended beyond their present locations, and there are no estimates available for such extensions, other than a general consensus on the high cost.

The systems vary in their impact on police manpower. Those systems that enable the motorist to specify the need (e.g., the telephone and coded call-box systems) place fewer demands on law-enforcement personnel; 50 to 80 percent of the motorists involved in these demonstrations needed services of a nonpolice nature. The New Jersey and Florida systems, on the other hand, required police to respond to all calls and identify the problem. The Florida experimenters found that patrolling the 50-mile stretch of highway tested necessitated increasing police manpower by one man-shift during each 24-hour period.

Only the effectiveness of the FLASH system was quantified. The Michigan and New Jersey systems did not arrive at any figures for average waiting time, but in some cases motorists who used the aid system waited longer than those who did not. In New Jersey, response time of the service vehicles was often slower than expected. Experience in Michigan demonstrated, as expected, that where self-help was not feasible, driver waiting time was reduced when the phone system was used.

Most of the experiments involved observers monitoring stopped vehicles to see what efforts had been made to gain aid. Options available to drivers included the motorist aid system, walking or hitching, public phones, self-help, or help from a passing motorist. In the Houston coded call-box experiment, only 13.5 percent of drivers stopped used the call box; of the nonusers, 31.8 percent were unaware of the system and 34.9 percent objected to the cost. The Michigan roadside telephone experiment recorded the following use levels: in summer, 28 percent of motorists stopped on the freeway used the aid phones and 52 percent of motorists needing aid on other roads used the phones; in winter, the comparable percentages were 41 percent and 50 percent, respectively.

It is apparent that there are significant problems with aid systems that rely on some form of roadside apparatus. While the cost of providing widespread systems would be high, perhaps the more serious problem is the low utilization of the experimental systems. It was established that a major cause of nonusage of the Michigan and Houston systems was unawareness of the system's existence; this could be alleviated through increased publicity. Nevertheless, a major portion of stopped motorists did not use the aid systems by choice, either because of cost or other considerations. Given these patterns of use, it would seem questionable whether the large expenditures required for these systems would be justified.

CB RADIO NETWORKS AND PATROLS.   Citizens Band radio is presently the most common motorist aid system using an in-vehicle device. A major CB system was set up experimentally in Detroit, Michigan, sponsored initially by General Motors.[19] One hundred employees of the City of Detroit and GM who generally traveled the same highway were given CB sets for their cars, free of charge. At first one base station was set up, but soon remote receivers connected by phone lines were added to improve range and reception. The object of the test was to determine how effective CB-equipped motorists would be in reporting incidents along the highway. Initial responses were quite favorable, and all participants were satisfied. When the service was expanded, the number of calls increased greatly, with a majority (up to 90 percent) coming from motorists who were not part of the originally organized group but who possessed their own CB sets. Although coverage included the entire urban area, most calls concerned freeway incidents. Over a two-year period, the approximate distribution of the reasons for calls being made was: 40-45 percent reporting stalled vehicles; 20-30 percent requesting information; 20 percent notifying of accidents; and 15-20 percent calling for other problems. The base operator in the Detroit test served as an intermediary between motorists and authorities, and relayed only those calls that required some kind of response. Fifty to 60 percent of all calls received were reported to authorities; the remainder either did not require a response or were duplications. The most recent statistics for the Detroit system showed 900 calls were received per month in 1971.

As a result of the demonstrated effectiveness of CB radio networks and patrols, a national volunteer CB organization (REACT) was formed.[20] Subsequently, an experiment involving the Ohio branch of REACT and the Ohio State Police was established that used Channel 9, the emergency channel specified by the FCC. During the first year of operation, approximately 60 percent of all calls concerned incidents that were potentially hazardous, 18 percent were calls requesting information, and 22 percent were "other" incidents. Accidents were reported for freeways and interstates (48 percent), city streets (35 percent), and secondary roads (11 percent). The initial reaction by participants and police agencies was favorable, and the program was expanded.

The cost of a CB system aimed at aiding motorists is reasonable. The major expenditure is for central monitoring equipment. Individual CB units are relatively inexpensive, though they do require a motorist's investment. Nevertheless, most of the drivers in the Detroit experiment indicated that they planned to purchase their own units at the conclusion of the test.

Because most CB systems rely on volunteers, explicit manpower costs to any public service agency are very low. More importantly, demands on police manpower are not greatly increased, since the central monitoring station screens the calls. In addition, CB use is not limited exclusively to highways, as is the case with the roadside motorist-aid systems; it can operate wherever CB-equipped motorists are within range of the central monitors.

There are technical problems associated with CB usage, however. CB communications are limited in their range, susceptible to interference, and subject to problems of illegal use. Strictly enforcing the utilization of Channel 9 for emergency purposes only would help reduce some of these problems.

In both tests, initial fears that volunteers would not be effective or regular in their activities proved to be unjustified. Concern over the dependence on volunteers can be alleviated by training and/or paying those who man the central monitor, as the Detroit system did.

How effective two-way radio or CB systems are in reducing average waiting delay has not been established. One study based on the Detroit CB system reported that detection delay was reduced by an average of 17 minutes.[21] Another study cites a mean reduction in urban notification delay of only one minute (12 percent) when 20 percent of all vehicles are two-way radio equipped.[22] Detecting an incident depends upon the coincidental passing of a radio-equipped vehicle; because of the randomness involved, detection delay might still be significant. Notification delay, however, might be reduced. The effectiveness of a CB system is clearly dependent upon the level of market penetration—the more vehicles that are equipped with CB, the faster incidents will be reported.

Traffic Information and Control Systems

AERIAL TRAFFIC SURVEILLANCE AND AM/FM RADIO REPORTS. Current commercial traffic reports are generally provided as a service by radio stations and vary in their accuracy and usefulness. An experiment in Buffalo, New York, suggested that an airborne observer broadcasting over one commercial radio station was cost effective if one compared the estimate of travel time saved with cost of the service.[23] In Houston three stations broadcast reports based on information drawn directly from the police, without any airborne observers of their own.[24] Although listeners used the information, they wanted greater accuracy and timeliness in the reports. A survey was conducted that showed the mode of communication preferred by drivers: changeable signs and commercial radio broadcasts were each favored by 45 percent, while 5 percent each favored television and the telephone.

Current advisory and emergency broadcast systems suffer from several deficiencies. First, because such broadcasts are often provided as a service or as a promotion by commercial radio stations, their frequency and the accuracy of their reports vary considerably. Second, it is impossible for a single observer to cover an entire region reliably, and stations usually cannot afford multiple observers or coverage during nonrush hours. Thus, most of the time commercial broadcasts are unavailable or are not useful to very many drivers. A final problem is that people traveling in unfamiliar areas have no way of knowing which stations provide traffic advisory broadcasts. Means of reducing the latter problem are incorporated into proposals for new systems in Germany and Japan.[25]

HIGH-SPEED CONTROL SYSTEMS.   There are two basic types of high-speed control systems. General surveillance entails determining overall traffic flow, density, and speed for the purpose of optimizing "throughput" (vehicles/lane/unit of time). Freeway ramp access control is specifically designed for limited-access roads, where vehicle entry is controlled by entrance ramp signals to improve overall throughput.[26]

General surveillance systems currently employ some automatic detection devices—optical speed and position detectors, sonic detectors, pressure plates, and loops underneath or alongside the road. Other detection and surveillance techniques include airborne observers and remotely controlled television cameras.[27] In the most extensive system proposed, the Comprehensive Automobile Control System being developed in Japan,[28] a central computer will use automatic detection devices along the road to determine traffic density, and will automatically query vehicles at each intersection to determine their destination. The computer will generate route guidance information for each vehicle based on this information, current data on time and conditions, and past statistical data.

Freeway ramp access control systems are of two different types. In the first type, vehicles are allowed to enter the highway at preset time intervals, based on previous traffic patterns; this system is not responsive to unexpected changes in traffic levels. The Harbor Freeway in Los Angeles, California, among others, uses such a system.[29] The second type—gap-acceptance merging control—is more sophisticated. It uses a computer controller and automatic detectors to measure gaps in the traffic flow on the outside lane of the highway. When a gap larger than a preset value appears, the controller releases a vehicle(s) stopped on the ramp, timing the release so that the car will enter the acceptable gap. The Gulf Freeway in Houston uses this system.[30]

CHANGEABLE MESSAGE SIGNS.   Changeable message signs, which can reflect different traffic conditions, offer more flexibility than a static sign but also require some form of control system. The signs range from simple systems that can display only one message to flexible lamp matrix signs that can display an unlimited number of messages.[31] Most sign systems use fixed-wire links, connected to some central terminal. The Varicom sign, made by 3M, is adaptable to either hard wire or radio control.[32]

Safety: Benefits and Hazards

At present, the need for increased motorist aid and emergency services is still unsatisfied. Roadside call boxes have been installed just on selected major highways and thus offer only a partial solution. Since such systems do not provide immediate notification, a stranded motorist may have to walk to the nearest installation and may cause additional hazard on busy, high-speed roads. Citizens Band radio networks provide the driver with direct contact to emergency aid services but appear to be labor intensive. They would require "listening posts" approximately 40 miles apart in order to provide reliable, complete road coverage, because of the limited range of CB equipment. Without 24-hour listening posts, total reliance on other CB owners is necessary.

The development of CB motorist aid services will reduce the need for other communication devices to improve traffic safety. If the CB unit becomes an automobile option as inexpensive as an AM/FM radio unit, and if a network of aid stations is provided to cover major highways, any impact on driving safety that might be derived from adding a mobile telephone system would be small. If, on the other hand, CB motorist aid networks remain merely demonstration projects, introducing mobile telephones in combination with a cellular network on major highways could have a considerable impact on automobile emergency aid provision. For example, a toll-free number could be provided in a given area to accept emergency calls; the central listening station would then notify the closest appropriate emergency agency.

The benefit of providing quick responses to emergency highway situations is large. It has been reported that for a ten-mile stretch of highway "the number of emergency stops to be expected will vary up to 160 per day (depending on average daily traffic)."[33] Reducing inconvenience is another substantial benefit that cannot be readily quantified. Quicker responses could also save much suffering and loss of life. These matters will be discussed in more detail in the next chapter; an attempt to quantify the potential benefits will be found in Chapter 10.

While utilization of mobile communication technology is likely to increase safety in many areas, it has also been suggested that widespread use of mobile communications might present new driving hazards. Such concern is reflected in a State of Connecticut statute passed in 1957, which stated that "the mobile telephone cannot be used [by the driver] while the vehicle is in motion."[34] Although haphazard introduction of this technology into vehicles might have dangerous consequences, several potentially risky situations could be avoided by correct product design, careful equipment installation, and proper use.

The location and mounting of transceivers poses the most apparent hazard associated with vehicular communications. Secure mounting of a unit is particularly important, since loosely bolted radio components can become lethal projectiles in a collision.[35] It is equally important to have units mounted in a position where driver visibility will not be reduced. The Batavia (New York) dial-a-bus system has several recommendations regarding the location of their teleprinter units.[36]

The possibility of audio distraction is a second potential concern. Using the pervasive auto radio as a roughly analogous situation, one might speculate that audio stimulus does not significantly affect driving safety. However, since two-way radios require more user involvement than the broadcast radio, additional research would be needed before any conclusions could be drawn. States already have statutes prohibiting the use of earphones and headsets while driving.[37]

A final concern is the operational convenience of a communication device while driving. Questions arise about driver performance when dialing or holding a microphone. Such situations might easily be avoided by substituting omni-directional, dash-mounted mikes for hand-held ones, or by designing units so that abbreviated phone numbers could be accepted, thereby easing the dialing task. These product modifications are only a few examples of how altered technological designs might alleviate potentially dangerous situations. Manufacturers and regulators must give serious attention to the safety problems posed by new mobile communications equipment; rigorous standards and an enforcement program will probably be needed, and they can be introduced at the initial stages of the technology's development.

The Consumer Product Safety Commission has major responsibility for addressing safety questions associated with consumer use of mobile communications equipment. In situations involving in-vehicle use, however, such responsibility may be shared by the National Highway Traffic Safety Administration (U.S. Department of Transportation), if mobile communication devices are construed as motor vehicle accessories.

Some Policy Implications

The U.S. DOT and the FCC will be faced with a number of policy issues. Some will become more important as new cellular and MCT technologies approach the implementation stage, but there are also concerns that need to be addressed at present.

THE U.S. DEPARTMENT OF TRANSPORTATION.   The National Highway Traffic Safety Administration of the U.S. DOT should plan, develop, and enforce safety standards for mobile communications use by motorists. This issue is of immediate concern due to the rapidly increasing number of CB radios. Because of the many highway assistance and informational benefits, the policy should not discourage motorists from purchasing mobile communications equipment; instead proper use and installation should be emphasized. Policies regarding the development of motorist aid and information services should take into account the rapidly developing cellular and MCT technologies. If CB networks are established on major highways, their structure should allow for easy integration of future technology. A decrease in the "cost of travel" could increase the level of automobile use. Thus, the U.S. DOT should also consider the implications for fuel demand and urban sprawl; pricing mechanisms to keep automobile travel stable might be needed, including increased gasoline taxes, higher commuter tolls, etc.

THE FEDERAL COMMUNICATIONS COMMISSION.   The FCC should work in conjunction with the U.S. DOT to ensure that present communication systems are compatible with foreseeable future technology. Licensing of CB motorist aid volunteer groups and networks should stipulate that consideration be given to the transition from CB systems to CB/cellular systems.

Impacts on the Value of Time

The concept of the "value of time" is used extensively in transportation economic studies to express the value of increasing or decreasing travel delays in financial terms. Often, in studies of highway construction proposals, the calculated dollar value of time saved becomes the dominant benefit considered to justify costs. Estimated values of time are multiplied by the expected change in travel time to yield a dollar cost or benefit.

These estimates are typically based on "opportunity cost": the time spent in transit is valued according to an individual's judgment of the worth of that time's best alternative use. The issue of assigning appropriate values to different types of travel time has concerned transportation researchers for almost fifty years. Values ranging from 30¢ to $5.78/hour for the work trip have been suggested in the literature;[38] these values are 14 percent to 132 percent of the wage rate. Computing values of time for a nonwork trip is even more controversial.

The value of travel time will be significantly affected by the motorist's potential ability to conduct business via mobile communication devices while in travel. Hence, the widespread use of mobile communications will require a change in the way travel delay costs are derived from the value of time. The communication device will not change the total time in transit; multiplying the value of time by the change in trip time will result in zero benefit and, consequently, this type of cost procedure will no longer be useful. Because mobile communications permit travel time to be used for business and social purposes, however, they permit a partial "recovery" of the opportunity cost of travel. Thus travel time can no longer be treated as completely separate from business or social time.

Changes in costs associated with increasing or decreasing travel times would have two very important implications. First, if the cost associated with increased travel time is effectively reduced by mobile communications, travel patterns (and specifically trip lengths) will be affected. Second, the division of those traveling on various transportation modes will change. For example, if mobile phone capabilities are provided in automobiles but not in public transit vehicles, perceptions of relative "costs" could result in decreasing utilization of public transit vehicles and an increase in the use of automobiles.

Impacts on Land Use, Travel Patterns, and Energy Use

History illustrates the important influence of transportation on urban structure. Better transportation reduces travel time for a given distance and allows

Table 7.12: *Example of Postulated Maximum Travel Times*

Residential Trip to	Postulated Maximum Travel Time (Minutes)
Local market	7
General market	21
City center market	63
Work site	84

SOURCE: U.S. Department of Transportation, Office of Systems Analysis and Information, "The Role of Travel Time Limits in Urban Growth," prepared for the Strategic Planning Division by Walter D. Velona (Washington, D.C.: U.S. Department of Transportation, 1973).

people to reside farther away from work and service locations. If mobile communications significantly improve traffic control and reduce costs associated with travel time, they will contribute to further outward expansion of cities.

Walter Velona proposed that there are observable limits to the time people will spend traveling for different purposes; his estimates are based on present perceptions of the value of time and travel data for the Chicago area.[39] Table 7.12 summarizes his calculations. Such limits can be translated into impacts on urban structure. For example, Hans Blumenfeld suggests that the maximum growth of cities appears to be limited by a one-hour travel time from the city center.[40]

Total volume of passenger travel could also be affected by mobile communications use. Volume of travel can be broken down into two components—number of trips and trip lengths. In the former case, reducing travel times and cost could lead to an increase in the number of trips made. But trip lengths

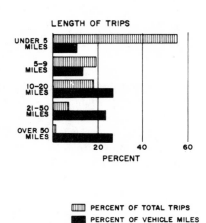

Figure 7.9: *Trip lengths and vehicle-miles*

SOURCE: Motor Vehicle Manufacturers Association, **1973/74 Automobile Facts and Figures** (Detroit, Mich.: Motor Vehicle Manufacturers Association, 1974), p. 37.

could increase as well with mobile communications use. Figure 7.9 shows the typical length of trips made by automobile both as a percentage of total trips and as a percentage of total vehicle-miles traveled. Potentially, widespread introduction of mobile communication technology could increase the percentage of long-distance trips and decrease the number of shorter trips.

Perhaps the most significant second-order impact of mobile communications use is the energy consumption that might result from increased automobile travel. Because automobile travel accounts for about 54.2 percent of the total energy consumed by transportation and about 13.6 percent of all energy consumed in the U.S.,[41] even small shifts in the total number of vehicle-miles traveled can be significant. The following figures illustrate this point:

—In 1973, passenger cars traveled a total of approximately 1.0×10^{12} miles.[42]
—Each one percent increase would mean an extra 1.0×10^{10} vehicle-miles per year.
—Using the 1972 figure for automobile fuel efficiency (13.5 miles per gallon[43]), a one percent increase in total vehicle-miles traveled would result in approximately 7.5×10^7 extra gallons of gasoline being consumed per year.

Thus, small changes in travel patterns can have serious consequences for energy consumption.

In this chapter we have frequently mentioned the role of mobile communications in emergencies. The next chapter takes up this subject again and discusses communications use by organizations that provide public safety services.

NOTES

1. U.S. Department of Transportation, Office of Assistant Secretary for Policy and Internal Affairs, *1972 National Transportation Report* (Washington, D.C.: U.S. Government Printing Office, 1972).

2. Vehicle productivity is usually defined as the number of person-trips carried by a single vehicle per hour. Vehicle mileage is less significant in productivity considerations, since the vehicle is assumed to be operating within a specific service area with a relatively uniform average speed and, consequently, a relatively uniform travel distance.

3. The cities were: Davenport, Iowa (p)   Madison, Wisconsin
                    Ft. Leonard Wood, Missouri (p)   Merced, California
                    Hicksville, New York (p)   Petersborough, Ontario
                    Lowell, Massachusetts (p)   San Diego, California (p)
The notation (p) indicates that the system is privately owned and operated.

4. Transportation Research Board, "Demand Responsive Transportation Systems and Services," *TRB Special Report 154* (Washington, D.C.: National Research Council, 1974); U.S. Department of Transportation, Office of the Secretary and Urban Mass Transportation Administration, "Demand Responsive Transportation, State-of-the-Art Overview" (Washington, D.C.: U.S. Department of Transportation, 1974).

5. John E. Ward, "Vehicular Communications for a Dial-a-Ride System" (Cambridge, Mass.: MIT Urban Systems Laboratory, March 1971); Canada, Ministry of Transport, "Para-Transit Dispatch Systems," report prepared by J. K. Associates, Toronto, Ontario, 1975.

6. For details, see Highway Research Board, "Demand Actuated Transportation Systems," *Special Report No. 124* (Washington, D.C.: National Academy of Sciences, 1971), p. 45.

7. Ward, "Vehicular Communications," p. 1-1. Note that researchers at Bell Telephone Laboratories feel that at 900 MHz, the typical rate of 1,000 bits per second would be significantly smaller.

8. Ward, "Vehicular Communications."

9. Ministry of Transport, "Para-Transit Dispatch," sect. 4.2.

10. Ward, "Vehicular Communications."

11. Ministry of Transport, "Para-Transit Dispatch."

12. "CARS (Computer Aided Routing System): A Prototype Dial-A-Bus System" (Cambridge, Mass.: MIT Urban Systems Laboratory, 1969).

13. Motorola, Inc., "Radio Spectrum Smoothes Schedules, Soothes Drivers for CNY Centro, Inc. in Syracuse" (Schaumburg, Ill.: Motorola, Inc., Communications Division, 1975).

14. Frequent advertisement in various transportation magazines.

15. Eugene F. Reilly et al., "Two-wire Emergency Call System," *Highway Research Record 450,* 1973, pp. 1-2.

16. Merrell E. Goolsby and William R. McCasland, "Use of Emergency Call-Box System on an Urban Freeway," *Highway Research Record 358,* 1971, pp. 1-7.

17. Walter J. Roth, "Study of Rural Freeway Emergency Communication for Stranded Motorists," *Highway Research Record 303,* 1970, pp. 74-88; Walter J. Roth, "Rural Freeway Emergency Communications for Stranded Motorists: Final Phase Report," *Highway Research Record 358,* 1971, pp. 8-16.

18. Ivor S. Wisepart, "Designing the First Flash Installation," *Highway Research Record 303,* 1970, pp. 89-94; Bernard Adler et al., "Evaluation of the First Flash Installation," *Highway Research Record 402,* 1972, pp. 28-36.

19. Clark E. Quinn, "A Highway Communication System for the Motorist: The Case for Two-Way Radio," *Highway Research Record 402,* 1972, pp. 9-15; Herbert J. Bauer et al., "Response to a CB Radio Driver Aid Network," *Highway Research Record 279,* 1969, pp. 24-39.

20. Robert M. Chiaramonte and Henry B. Kreer, "Measuring the Effectiveness of a Voluntary Emergency-Monitoring System in the Citizens Radio Service," *Highway Research Record 402,* 1972, pp. 16-22.

21. Ibid., p. 17.

22. U.S. Department of Commerce, "Land Mobile Communications and Public Policy," prepared by Systems Application Inc. (Springfield, Va.: National Technical Information Service, 1972).

23. Highway Research Board, "Surveillance Methods and Ways and Means of Communicating with Drivers" (Washington, D.C.: National Academy of Sciences, 1967).

24. Conrad L. Dudek et al., "Evaluation of Commercial Radio for Real-Time Driver Communications of Urban Freeways," *Highway Research Record 358,* 1971, pp. 17-25.

25. "Traffic Information Broadcasting," *Wireless World,* May 1973; Johann H. Buhr et al., "Design of Freeway Entrance Ramp Merging Control Systems," *Highway Research Record 279,* 1969, pp. 137-49.

26. Buhr et al., "Design of Freeway Entrance."

27. Kenneth G. Courage, "Some Electronic Measurements of Macroscopic Traffic Characteristics on a Multilane Freeway," *Highway Research Record 279,* 1969, pp. 107-18.

28. Automotive Traffic Control Design Center, "Automotive Traffic Control System," Tokyo, Japan, 1972.

29. Leonard Newman et al., "An Evaluation of Ramp Control on the Harbor Freeway in Los Angeles," *Highway Research Record 303,* 1970, pp. 44-55.

30. Courage, "Some Electronic Measurements."

31. Irwin Hart, "Changeable-Message Signs," *Highway Research Board Special Report 129,* 1971, pp. 26-28.

32. Lawrence A. Powers, "A Varicom Traffic Control and Communication System," *Highway Research Board Special Report 129,* 1971, pp. 29-32.

33. Highway Research Board, "Motorist Aid Systems," *National Cooperative Highway Research Program Report 7* (Washington, D.C.: National Academy of Sciences, 1971), p. 3.

34. Connecticut General Statute, sect. 14-259.

35. W. D. Nelson, "The Influence of Vehicle Add-ons, Alterations, Maintenance, and Use on Vehicle Safety" (Paper presented at the International Automotive Congress, Detroit, Michigan, January 1973).

36. Karl W. Guenther and R. G. Augustine, "Evaluation Report–Radio Teleprinter Test, Batavia, New York" (Dearborn, Mich.: Ford Motor Co., Transportation Research and Planning Office), pp. 41-43.

37. California Vehicle Code 27400.

38. Peter R. Stopher and Arnim H. Meyburg, *Transportation Systems Evaluation* (Lexington, Mass.: Lexington Books, 1976), Chapter 4.

39. U.S. Department of Transportation, Office of Systems Analysis and Information, "The Role of Travel Time Limits in Urban Growth," prepared for the Strategic Planning Division by Walter D. Velona (Washington, D.C.: U.S. Department of Transportation, 1973), p. 4.

40. Hans Blumenfeld, "Selected Essays," in *The Modern Metropolis*, ed. Paul D. Spreiregen (Cambridge, Mass.: The MIT Press, 1967).

41. U.S. Congress, Office of Technology Assessment, "Energy, the Economy and Mass Transit" (Washington, D.C.: U.S. Government Printing Office, 1975), p. 14.

42. Motor Vehicle Manufacturers Association, *1973/74 Automobile Facts and Figures* (Detroit, Mich.: Motor Vehicle Manufacturers Association, 1974), p. 37.

43. U.S. Congress, Office of Technology Assessment, "Energy, the Economy and Mass Transit," p. 14.

Chapter 8
LAND MOBILE COMMUNICATIONS AND THE DELIVERY OF PUBLIC SAFETY SERVICES

Sara Edmondson

Introduction

This chapter reviews and examines potential uses of cellular technology in the public sector. Two potential primary impacts of the cellular system, the reduction of total response time to a client and the capacity to stimulate integration of communication systems into public agencies, will be considered. The possibility of second and third order consequences, involving the organizational structure of the service agencies and the quality of service provided, will be examined somewhat speculatively. Public services in which mobile communication is already an integral part of their performance, such as police, fire, and emergency medical services, are the most likely users of the new technology. Since most empirical evidence on the use of mobile communication technology comes from analyses of these three services, potential impacts are examined in depth only in these areas.

Mobile communication technologies are used by a public agency to provide information linkages within the organization and to those served by it. Three major types of linkages should be distinguished: communication between citizens and service providers; communication among service providers; and communication among citizens. The following assessment of the cellular system's potential impacts on the public sector has been based on an examination of how technologies affect these communication linkages, and an analysis of the attitudes and behaviors exhibited by citizens and service providers when using existing communication technologies.

A limitation of the analysis, which is especially relevant to the treatment of response time, but which characterizes most statistical analyses of public agency operations, must be pointed out. Current procedures and strategies used in the delivery of public services are largely based on the subjective and cumulative of agency administrators and supervisors. For example, the belief that reduced response time will improve service outcomes is a common assumption based on intuition and experience. Our analysis examines whether this assumption can be substantiated by social science studies, which use quantitative and less subjective methods of assessment. As we shall see, however, these studies have shortcomings of their own, and opinions differ on whether they should be given more weight than the subjective experience of agency personnel. Clearly both require serious consideration when new strategies are being devised and resources allocated. Whatever the limitations, the studies surely demonstrate that the relation between response time and service outcomes is complex, providing a desirable counterweight to oversimplified notions that have been used in the past to promote new procedures and new technologies in the public service area.

POTENTIAL USERS OF MOBILE COMMUNICATIONS

Within the wide range of services offered to clients by public agencies, five classes of services have some identifiable use for mobile communication technologies: education, social services, health care, police, and fire. Systematic studies of the potential application of telecommunications have been undertaken in the education and social services sectors, but no attempt has been made to identify uses of mobile communication technologies per se. Ambulance services in the health care sector, police departments, and fire departments, on the other hand, have been using mobile communications for a number of years. These "public safety services" are the most likely market for new developments in these technologies, since dispatched fleets of mobile units are central to their organizational form.

Education and Social Services

In the education sector, there has been strong resistance to using technology as a substitute for face-to-face exchange; it has been argued that this will have a dehumanizing effect on students. Despite this resistance, experiments with new technologies, such as computer assisted instruction, have proceeded. Some analysts expect to see a trend in this direction, even the development of large-scale "tele-systems" similar to those currently being developed for health care delivery. Such expectations are based on increasing investment in technological innovations by governments, a greater emphasis on cost effectiveness, and an acceptance of behaviorist learning theory by educators.[1]

While these trends may indicate a potential market for some large-scale telecommunication systems in schools, including cable television and satellite receivers, they do not imply any special need for mobile communication technologies. There have been some attempts to adapt mobile units for instructional use with dispersed populations in rural areas,[2] but these services do not affect the majority of students or personnel.

There are two types of potential applications of telecommunications for delivering social services, such as family counseling provision, family income supplements, and rehabilitation services. First, use of mobile communication devices could provide a caseworker access to client information stored in a computerized information system at the state and federal levels. In addition, information and referral systems for human services could serve mobile as well as stationary clients.[3] A second possibility is implicit in a proposal for an integrated system of social service delivery and scheduled transportation services.[4] Such a system would require support functions for processing information, scheduling, and accounting. Three kinds of transportation would be needed: fixed schedule, demand-responsive, and negotiated schedule. The transportation aspects of these uses have been analyzed in the preceding chapter.

Police Services

Mobile communication technologies are widely used by police departments, fire departments, and other organizations providing emergency health care services. Within police departments, communications command and control centers house the base station from which mobile units are dispatched.[5] A typical center processes citizens' requests for service received through public, private, and emergency telephones; citizen's alarms; automatic detector alarm systems; and police radio. For low levels of demand, a single person can act as switchboard operator, complaint operator, and radio dispatcher. When demand increases, the functions are divided and allocated to separate personnel. Upon further demand, additional personnel are added, and components can be automated. Most, however, are manually operated, using conventional two-way mobile radio.[6]

Communications between the command and control center and the mobile units accomplish the following functions: (1) dispatch instructions to the field forces; (2) provide tactical communication between supervising personnel and field forces; (3) communicate status among members of a department; (4) disseminate information among members of a department.[7]

Police departments in some cities and metropolitan areas use highly sophisticated systems. Several are experimenting with computer assisted dispatch and mobile digital terminals, and departments in New York City and Chicago are testing hand-held portable radios for all foot patrolmen.[8] Because channel congestion is a difficult problem in heavily populated districts, technological advances that could reduce channel crowding and improve the functioning of the system would be likely to find a market here, if implementation costs were acceptable.[9]

Health Services

A program sponsored by the federal government for development of emergency medical services (EMS) is stimulating a trend toward central dispatching of ambulances and the use of patient monitoring devices that can transmit data on clients from a mobile unit. This may increase the use of mobile communications in the health care sector. Previously, a command and control center for ambulance services existed only for systems provided by police and fire departments.

Prior to EMS systems, four types of emergency transportation existed: private ambulance companies, private voluntary organizations, hospital-based ambulance services, and public ambulance services. A survey taken for the U.S. Department of Transportation indicates that the majority of ambulance services are privately owned.[10] Of the 1,763 municipalities sampled, 36 percent had ambulances operated by funeral homes, 31 percent by private companies, 26 percent by the police, fire, or other governmental departments, 14 percent by hospitals.

Most independent ambulance services seem to be poorly equipped with mobile communication capacity. A survey of ambulance operators conducted in 1970 by the State of California indicated limited communications capability: only 17 percent of the ambulance providers, operating in just a few counties, had connections of any kind with county communications centers, and only 11 percent had two-way mobile radio connections with these centers. About 31 percent had some type of radio communications with the police or sheriff. The ambulance operators felt that public agencies should supply two-way radios, since most operators could not afford adequate equipment. Another study of rural ambulance services in North Carolina, conducted in 1965, found that 50 percent of the ambulances were not equipped with two-way radios, and suggested that similar conditions existed in other states.[11]

The Emergency Medical Services System Act of 1972 (PL 93-154) encouraged counties and municipalities to organize ambulance and other emergency medical services in a more systematic and comprehensive way.[12] One provision of the act required that there be communication between ambulances, hospitals, and related public agencies, and funds supplied for purchasing equipment. It also specified the inclusion of a centralized communications command and control system similar to that used by police. In addition, federal guidelines stress that the command and control center should be the locus for coordinating information and resources between local police, fire, highway patrol, civil defense, and other services, particularly during large-scale emergencies. For counties and municipalities that institute federally funded emergency medical systems, therefore, present patterns of underutilizing electronic communication will change. Thus, the health care sector is a potential market for mobile communication technologies, especially if the federal government continues its stimulation.

One of the primary reasons for providing mobile communication capacity between ambulances and hospitals is that data on the status of an emergency victim can be transmitted from the scene of the incident to a doctor or a specialist at a distant medical facility, using recently developed telemetry devices. Emergency medical and ambulance services could thus be important users of mobile equipment that provides for both voice and telemetric communication.

Fire Services

Fire and police departments were the first public safety services to utilize emergency notification systems. Originally, fire departments had strategically located alarm boxes; current systems can indicate the exact or approximate location of the fire, depending on their technical sophistication. Also, most fire departments receive notification of fires by telephone. The National Fire Protection Association argues that a dual notification system is necessary because the telephone system can be destroyed by fire and is overloaded during emergencies. The association also claims that it is difficult to obtain precise locational information over the phone. Recent experience suggests that these arguments are no longer persuasive, however. The street alarm boxes in New York City produce 90 percent of the false alarms, while only 9 percent are received by telephone. High rates of false alarms are causing many cities to contemplate discontinuing the use of their municipal fire alarm systems, placing reliance on conventional telephones instead.[13]

The need for sophisticated base-mobile communication in fire trucks is not as essential as for squad cars and ambulances. Strategies to improve the operations of fire departments have focused on the optimum siting of fire stations, and on alternating the numbers of trucks stationed at the facilities, rather than on improving communication between the crew in route to a fire and the base station. Nevertheless, some trucks are equipped with two-way radios so that crews can be contacted when they are out of the station.

Once at the site of a burning building, however, firemen need communication with each other. Walkie-talkies are currently used, but their transmission capacity is limited. A lightweight, easily accessible device with transmission capacity in a variety of locations might help firemen to coordinate their actions and assist each other when they are endangered.

Use of Mobile Communications by Citizens

Most clients currently rely on fixed communication technologies, predominantly the conventional telephone system, to notify agencies of a need for service. But the telephone's effectiveness is seriously limited because it is not always readily available at the scene of an incident. While this may not be essential to a client for education or human services, it is very important in emergency situations. The existence of a hand-held portable or vehicular mobile communication device could provide a potential client with rapid access to a service agency. The use of Citizens Band radio to report traffic accidents to police is an example of such usage.[14] Thus, cellular system devices are likely to be used by citizens for emergency notification, as well as for other personal and business purposes.

A new trend in the formation of voluntary citizens organizations suggests that citizens might use mobile communication technologies to facilitate their own public service activities, either in cooperation with or independently of public agencies. A recent study of locally based citizens' groups estimated that about 900 patrols are operating in major metropolitan areas, of which four general types were identified: building, neighborhood, social service, and community protection patrols.[15] Building patrols protect specific dwellings and groups of private residential compounds, surveying areas that usually receive minimal attention from local police. Neighborhood patrols may cover many blocks and may not have well-defined boundaries. Since they cover streets and public areas, they frequently coordinate their activities with local police. Social service patrols are building or neighborhood patrols that perform such functions as providing ambulance and rescue service as well. Community protection patrols, in addition to performing functions similar to those previously mentioned, also monitor police activities. The monitoring is carried out because of fear of harassment by police. Obviously, the surveillance routines carried out by these groups can be facilitated by using base-mobile and mobile-to-mobile communications.

IMPACT OF MORE INTENSIVE USE OF MOBILE COMMUNICATIONS

Introduction of the cellular system could intensify these utilization patterns of mobile communication technologies just discussed, thereby causing a number of effects on the operations and organization of public safety agencies and their relationships with their clientele. In order to analyze this complex set of interactions, we shall separate the effects, somewhat arbitrarily, into two groups. The first will concern modes of operation: effects on response time and demand for service will be analyzed with the assumption that the technology does not cause changes in the way the service is organized. The second will concern organization: the potential of the technology to cause important restructuring of service agencies and their service areas will be explored. The significance of the former is clear from the previous discussion. Organizational questions are also necessary to an assessment of cellular technology, though, because the technology is capable of providing all the communication linkages needed for a centralized and integrated system. Separating the effects into these two groups is desirable, because a different kind of analysis is required for each.

Communication Linkages within Service Agencies:
The Effect of Reduced Agency Response Time

One of the primary purposes of a standard dispatch system is to minimize the time required to reroute mobile units in the field. Centralizing several independent dispatch systems in a given service area—by means of computerized automation and the use of digital transmission for routine messages—can further reduce response time. The use of cellular technology is likely to alleviate channel congestion, and thus conceivably achieve a response time shorter than any currently available.

One major potential effect of shortening response time is that the outcomes of emergency services could be improved. As mentioned earlier, this is a common intuitive assumption of public safety service providers and analysts. In the case of fire, with its progressive character and threat of spreading, there is no question that this assumption is reasonable. With regard to police and medical emergencies, however, the assumption is not justifiable in every case, and the relationship is more complex. Studies concerning the effects of response time on arrest rates for certain classes of crimes indicate that a functional relationship exists. But these studies are not conclusive, since they have not been able to account for the possibility of intervening variables. Evidence from studies of the relationship between the response time of ambulances and reduced mortality rates is fragmented and even less conclusive.

ARREST RATES.   One of the first studies to consider the influence of response time on arrest rates was conducted by Herbert Isaacs for the President's Commission on Law Enforcement and Administration of Justice.[16] Figure 8.1 displays the various increments into which total response time may be divided. Isaacs' study actually analyzes increments two through eight, that is, the time required to process a request for service at the command and control center and the time required for the mobile unit to travel to the scene. The delay between the time the incident occurred and when the agency was notified was not analyzed in this study.

The study was based on 4,704 cases occurring in January 1966 that were chosen from the files on two field divisions at the Los Angeles Police Department. Calls that were judged by communications center personnel not to concern crimes, such as drunk and vagrancy arrests, traffic incidents, and aid to injured persons, were excluded from the sample. Isaacs found that delay both at the communications center and in field travel time were related to a drop in arrest rates. Also, he found that giving priority to emergency calls reduced response time for emergency calls by one half, and that this reduction corresponded to an increase in arrest rates. He concluded that both communications center and field response time should be minimized.

Isaacs noted that these findings do "not directly imply that faster response time will result in more arrests," since a controlled experiment would be required to establish a direct causal relationship. He does suggest, however, that

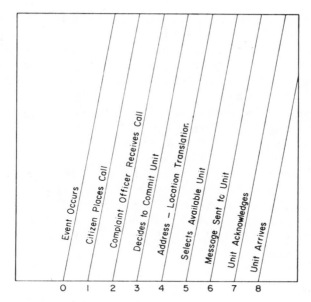

Figure 8.1: *Response time increments*

SOURCE: President's Commission on Law Enforcement and the Administration of Justice, **Task Force Report: Science and Technology,** prepared by the Institute for Defense Analyses (Washington, D.C.: U.S. Government Printing Office, 1967), p. 91.

they confirm the intuition that reduced response time is significant for effective police performance, and justify the investment to automate the command and control center communication support functions. But one of the most serious limitations of analyses based on frequencies only is that arrest rates could be increasing due to a number of factors other than response time. For instance, a special campaign against drug offenders or prostitutes may have occurred during the same period, manpower allocations could have changed, or some other unexamined variable may have been producing this effect.

An analysis of the effect of response time on arrest rates by Calvin Clawson and Samson Chang of the Seattle Police Department leads to similar conclusions.[17] This study is based on a sample of 5,875 calls for service received between April 1974 and March 1975, each of which was identified as a crime-in-progress requiring rapid response. The calls were classified into five different crime groups and an analysis of the relationship between communications center processing time, field travel time, and total response time was made for each. Although no statistically significant relationship was found between communications center processing time and arrest rates, a significant relationship was found (at both the 0.90 and 0.95 degree of confidence) between both travel time and arrest rates, and between total response time and arrest rates. It should be noted that the travel time and total response time analyses were conducted on a sample of 507, since complete time intervals were available in only that number of cases.

Extending their analysis, Clawson and Chang fitted their total response time and arrest rate data to a curve that illustrates that as response time decreases, the percentage of arrests increases. Since the resulting function is very similar in form to that found in the Isaacs study, they adjusted Isaacs' data to permit direct comparison with their own and fitted a curve to both data sets. The two curves are plotted in Figure 8.2.

The functional relationships exhibited by both data sets, conducted on different populations, using different sampling procedures, with an eight-year time interval, are surprisingly similar. A regression analysis on the two data sets produced a coefficient of correlation of 0.95.

As Clawson and Chang carefully point out, this analysis does indicate that "shorter response times and travel times are related to a higher percentage of on-scene arrests."[18] But, as is true of the Isaacs study, while it does not *conclusively* demonstrate that reduced response times could bring about higher percentages of arrests—because of the possible influence of uncontrolled

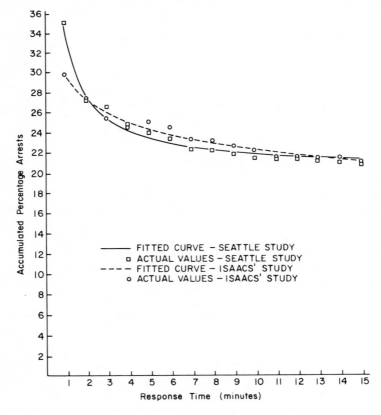

Figure 8.2: *Accumulated percentage arrests (Seattle study and Isaacs study)*

SOURCE: Seattle Police Department, "Impact of Response Delays on Arrest Rates" (Paper prepared by Calvin Clawson and Samson Chang for the Inspectional Services Division, Seattle, Wash., September 1975), p. 21, graph 3.

variables—it does support the intuitive assumption that improved communication technologies could contribute to increasing the effectiveness of police agencies.

The slope of the curves relating arrest rates to response time is small for delays that exceed a few minutes. This result is reasonable, since one would not expect a large effect until the response time could be made comparable to the time it takes for the criminal to escape from the immediate vicinity of the incident. Indeed, the data and fitted curves suggest the possibility of significant increases in slope for total response times that are less than three or four minutes.

EMERGENCY AMBULANCE SERVICES. Determining the importance of response time for emergency medical ambulance services is even more difficult. Thomas Willemain concludes from his study that the relationship between reduced response time and improved patient outcomes is not always consistent or positive, partly because it is difficult to select an adequate measure for patient outcomes. Efforts to develop such indicators have encountered a number of problems, which Willemain discusses in his monograph, "The Status of Performance Measures for Emergency Medical Services."[19]

While it is clear in some types of emergencies, such as heart attack cases, that minimizing response time is a desirable objective, it is not so in others. There are two circumstances, for instance, in which a more lengthy response time would prove better in the long run. First, if a more detailed and necessarily longer interrogation of callers were used in order to assign priorities for service, the waiting time for the most urgent cases might be reduced. A second circumstance would involve patients for whom the best equipped medical facility may be preferable to the nearest. For cases of this type, increased travel time may be associated with reduced death rates,[20] and response time can be a misleading measure of effectiveness.

Nevertheless, if response time is shortened, it seems reasonable to infer that improved outcomes are likely to result in some cases. These may take the form of lower death rates (though they may be reflected in higher morbidity rates) or, alternatively, of reduced morbidity rates for cases that would not have resulted in death. For example, a study cited in a Systems Applications, Inc. (SAI) analysis of mobile communications concluded that fatalities and disabilities from automobile accidents could be reduced by 18 percent in emergency vehicles equipped with notification devices.[21] Clearly, it is desirable to eliminate any delays that are not consciously used to provide improved service to the client.

ALTERNATIVE STRATEGIES. Yet improvements in communication technologies are not the only techniques available for reducing the response time of mobile units. Plans have been developed for determining both the optimum location of ambulances and police vehicles within a specific jurisdiction, and the optimum dispatching policies required to minimize response time.[22] While these models have not been widely accepted by practitioners, they suggest an alternative approach that should be considered.

Each of these approaches, however, assumes that reducing response time is the optimum service goal. Experiments with a team policing program, involving patrolling strategies that resulted in longer response times but improved relations between citizens and police, suggest an alternative approach.[23] A mobile communication system, not requiring rapid response capability, would be useful for squad coordination and management in this approach to service delivery. Other alternative crime prevention strategies for police services may have comparable communication needs.

Still, the evidence we have presented suggests that shortening response time may result in improved outcomes in a fraction of police and medical emergencies, with the improvement obviously varying in different situations. It is important to note that alternative strategies exist for reducing response time, and other means are available for improving service delivery that might be less costly and more effective than the major investment in mobile communication technology. Indeed, if only the service agencies, as opposed to clients, are equipped with the new sophisticated mobile communication devices, it is unlikely that decreases in response time will result in substantially improved service in most situations. On the other hand, if citizens were equipped with mobile devices, the possibility exists for a reduction in the notification or reporting delay. This topic will be discussed in the next section.

Communication Linkages Between the Service Agency and the Citizen: The Effect of Reduced Reporting Delay

Widespread deployment of a cellular system would have impacts not only on the internal communications of public agencies, but also on the linkages between agency and client.

Herbert Isaacs' analysis of response time for police departments defined total response time as the time elapsed between an event's occurrence and the squad car's arrival at the scene of the incident. His study examined the nature of delays only from the time the police department received a call to the time a squad car arrived. The study did not examine the causes of delay between the time the emergency occurs and the time a call for service is placed. Lack of an available telephone or other warning device must contribute to some degree, particularly for incidents in remote places and for automobile accidents. But like the communications center and field travel time components mentioned previously, it may be difficult to assess the impact of reduced reporting delays on service outcomes. The widespread availability of portable or vehicular mobile communication devices might substantially reduce the time required to report an emergency, yet it may also increase the demand for service. Other impacts, such as lessening the anxiety and confusion that a victim may experience prior to receiving assistance, which are harder to quantify, may also be stimulated by the increased deployment of mobile communication units.

The only study dealing with the causes of delay in reporting emergency incidents to police is currently being conducted in Kansas City, Missouri, by the Kansas City Police Department. Preliminary results indicate that the client reporting increment is the longest interval in the total time necessary to respond to a service request.[24]

A study conducted by Mogielnicki, Stevenson, and Willemain examines the reasons that clients tend to delay reporting medical emergencies.[25] It provides some insight into factors other than the availability of communication technology that contribute to the length of the client reporting interval.

The study involved 181 emergency incidents in which patients were treated in the Cambridge Hospital emergency room in Cambridge, Massachusetts, during April and May 1973. The primary means of emergency medical transportation in Cambridge at that time was the fire department's rescue squad. Since in most cases the patient himself did not call the rescue squad, and the caller was unknown or could not be reached, only 63 persons constituted the final sample of rescue squad users. Patients who entered the emergency room during the same time period but did not call the rescue squad were also interviewed; these 23 patients were identified as "nonusers." For all cases, an assessment of need for immediate medical care and seriousness of incident was made by two staff physicians and a head nurse. Because the sample is small and drawn from only one city, the results of this study can only be considered illustrative.

In most of the sample cases, client reporting time exceeded communications center and field travel time. For users of the rescue squad, the findings suggest that factors other than the availability of communication are the primary determinants of reporting delay, since only two percent of the users attributed their delay to the unavailability of a telephone. The majority of the sample, 66 percent, responded "called as fast as I could." Because of the unfortunate ambiguity of the wording, it is not clear whether close proximity to a telephone could have reduced the client reporting time significantly. Conclusive evidence would require additional research using more precise questioning.

A much larger percentage of nonusers indicated that communication technologies would have been helpful in shortening the reporting delay. Their responses suggest that confusion and uncertainty resulting from an unclear or inaccurate perception of their medical condition may have been the predominant factors in their avoidance of the rescue squad: 31 percent did not want to get help unnecessarily; 18 percent did not recognize the problem as an emergency; and 8 percent did not know what to do. In addition, 22 percent of the sample thought that transportation other than the rescue squad was faster, and 17 percent were unaware of its existence. It seems likely that offering some type of medical diagnostic and counseling service at the dispatch center would help reduce response time for cases of this kind. Forty percent of the nonusers indicated that they would have used a publicized telephone number to obtain medical counseling if one had been available.[26]

Access to such a service through a hand-held portable or vehicular telephone might further decrease client reporting time and increase demand for service within the potential client population. Experience with the 911 emergency telephone system, which will be considered below, suggests that convenient telephone access to emergency services greatly increases service demands. A secondary impact of mobile access to emergency medical counseling might be a reduction in clients' anxiety and uncertainty.

Integrated Communications with Service Agencies: The Effects of Centralization

Most studies of optimum location and dispatching procedures indicate the advantages that can result from centralized dispatch operations.[27] Centralization can reduce response time, while providing economies in resource allocation, bettering utilization of spectrum, and improving coordination of field units during large-scale emergencies. Some municipalities and counties have begun to experiment with centralized dispatch, integrated communication systems, and the new computer assisted and digital technologies. Evidence from studies of these systems will give some insight into the cellular technology's potential to stimulate comparable centralization and integration in public service agencies.

A primary organizational benefit of centralizing dispatch operations for emergency services is that some activities can be coordinated without the use of radio communication. This can directly conserve radio spectrum, which is congested in many localities at the present time, and facilitate the coordination of spectrum use by police, fire, and ambulance services. The benefits of the coordination are highest during wide-area emergencies, which place the greatest demand on emergency resources and radio spectrum.[28]

Centralized dispatching usually requires fewer dispatchers to perform the same workload. (For small police departments, however, the loss of record-keeping, receptionist, and secretarial services, which the dispatcher perhaps also provided, may offset this benefit.) Another advantage is that the cost of sophisticated equipment can be shared by several departments. Moreover, problems caused by radio interference can be overcome, since dispatchers can speak to each other without using radio communication.[29]

An evaluation of a centralized dispatch system used by police departments in three suburban Chicago communities showed improvement on a number of performance indicators when compared with the operations of three independent dispatch centers.[30] Communications center processing time for all three communities decreased after centralization. In the largest communities, radio system delay time was reduced by almost two-thirds. Telephone ringing time before a call was answered decreased from 8.2 seconds to 6.8 seconds, and the probability that a caller would receive a busy signal dropped from 0.60 to virtually zero. Total annual cost did increase by 7.5 percent—annual cost, including personnel, was $311,488 for the three individual systems and $344,658 for the centralized system—but cost per complaint dropped from $5.73 to $4.35, since the centralized system could handle more requests for service.[31]

In terms of technical parameters, computer assisted dispatch and mobile digital terminals can be used by dispatch systems of any size. The currently high cost is restricting their use, however, to municipalities with large fleets or cooperative, centralized systems. For example, the Huntington Beach, California (population 150,000), Police Department estimated spending to be $22,920 less per year, based on an analysis of man-hours saved in dispatch operations and office report preparation.[32] A comparable analysis of a centralized dispatch system used by three police departments in Illinois estimated annual savings of $168,840.[33] Savings over a five-year period, including adjustment for a projected population of one million, were estimated at $825,750. Total implementation cost was estimated to be $1,841,500. Costs for mobile digital devices are comparable.[34]

The improvements in performance that a manual centralized dispatch system provides, the high implementation cost of computer assisted dispatch and mobile digital devices, and the continued competition for radio spectrum are factors that favor centralizing dispatch operations and sharing existent communication technologies on a metropolitan or regional basis. These factors are reinforced by federal initiatives as expressed by the National Advisory Commission on Criminal Justice Standards and Goals, the Emergency Medical Services Systems Act of 1973, and the Highway Safety Act of 1966, which advocate comprehensively planned and coordinated communication systems for public safety agencies. Introducing the cellular system would be likely to further the trend toward system integration, since its operational capabilities are particularly well suited to the requirements of integrated systems.

Existing studies on the centralization and decentralization of public services can provide some insight into the impact integration has on service quality and effectiveness. First, March and Simon make an observation which suggests that integrating mobile communication functions may be more successful than attempts at other kinds of coordination. They point out that "the division of work that is most effective for the performance of relatively programmed tasks need not be the same as that which is most effective for the performance of relatively unprogrammed tasks."[35] This implies that the work of dispatchers, which tends to be repetitive or relatively programmed, need not be organized in the same manner as the work of the field personnel, who respond to each incident as the circumstance demands. Thus, dispatch functions could be centralized without reducing the opportunity for face-to-face encounters between clients and field personnel, and without sacrificing spontaneity of action in unanticipated situations. In the past, restricting the personal contact and flexibility of agency personnel affected the quality of service provided, particularly for police.[36] Providing all personnel with two-way communication would greatly facilitate this organizational strategy, because centralized coordination of decentralized units would be possible when needed.

The second point March and Simon make is that "it is possible under some conditions to reduce the volume of communication required from day to day by substituting coordination by plan for coordination by feedback." This would involve putting greater emphasis on preestablished schedules, for work that is relatively routine, instead of relying on communicated information.[37] In dispatch operations, the highly repetitive support services of information processing and status monitoring can be automated. Furthermore, a centralized dispatch center would obviate the need for coordinating one or more dispatchers by feedback over radio channels. The work of March and Simon suggests that this type of reorganization is more likely to be tolerated than attempting to centralize and integrate the activities of police patrolmen and paramedics.

Fifteen years ago, the organization of most public services had become highly centralized. The more recent strategy to decentralize these services resulted from criticism that the agencies lacked responsiveness to their clients' needs. Client dissatisfaction was most demonstrable with public safety services, where client-agency interactions were frequently face-to-face and under stress. A recent study of service decentralization has shown that while this strategy did not measurably improve the service outcomes, it did tend to improve client attitudes toward police patrolmen and firemen.[38] Thus the cellular system may enable public agencies to benefit from the administrative advantages of centralized direction and control, and at the same time maintain whatever advantages are derived from personal interaction between clients and field personnel.

Integrating the communication system in public safety services may have some negative consequences, however. A highly centralized and interconnected communication system could increase public domination by the executive branch of the government, so vividly characterized in science fiction and social criticism.[39] Because of its monitoring capability, the cellular system, if widely deployed, might provide federal agencies with more extensive electronic surveillance over private citizens.

A third major impact that increased use of the most sophisticated technologies could entail would be greater influence of the federal government over local and regional service delivery. Currently, the high cost of most of these technologies precludes adoption by local governments without federal financial support. This fiscal dependency on the federal government could contribute to the decline of local initiative.

Integrated Communications between the Service Agency and the Citizen: Organizational Responses to Increased Demand

It seems likely that the widespread availability of portable or vehicular communication devices may substantially increase the demand for service. Unless resources are also increased, service agencies will have to develop some method of rationing response to this demand. Recent experience with the 911 emergency telephone system is relevant to this problem. The extraordinary increases in both legitimate and inappropriate requests for emergency service in New York City that followed the installation of the 911 system may be equaled or exceeded if the cellular system extends the integration to mobile units.[40] Consequently, the client reporting interval may be longer if the demand results in queuing on agency switchboard lines. This in turn may necessitate the use of more screening and priority systems by public agencies to rank requests for service.

Stevenson and Willemain conducted an analysis of screening processes for emergency medical services.[41] They note that there are two major problems confronting a screening or priority system. One, noted earlier, is that some clients with severe medical problems bypass the ambulance system entirely. The other is that some citizens request ambulance service for nonemergency transportation, thus reducing the service's ability to respond to serious incidents when all ambulances are in use.[42]

Stevenson and Willemain analyze two types of screening processes intended to resolve these problems. One type reserves certain vehicles and services for cases identified as "true" emergencies. The second type serves all calls with the same vehicles, but ranks them for service by placing cases identified as "true" emergencies at the head of the queue.

The analysis concluded that either type of screening is apt to improve service when demand is heavy, providing that only a small percentage of cases are "true" emergencies and that the probability of misclassifying a patient's condition is very small.[43] It is assumed the screening personnel can distinguish "true" emergency cases from nonemergency cases with a high level of certainty. The analysis does not specifically address the subjectivity problem in perceiving the level of urgency or the issue of whether the service provider or the client should resolve any difference of opinion. A citizen's exaggeration of the seriousness of the problem may be due to a real difference in his perception, or it may be an attempt to obtain service from an agency that has been unresponsive to his needs—particularly if he is located in a high risk, low income neighborhood.

This issue has some important implications for mobile communication technology. On the one hand, when resources are limited, the institution of screening procedures may be the only equitable response to increased service demand possibly stimulated by the availability of communication technology. On the other hand, a screening procedure could be utilized by a service agency to deny reasonable requests for service, and to avoid investing additional resources to expand service capacity. Stevenson and Willemain anticipated this possibility; they advise that "at heavy system loading, only if the probability of misclassifying an emergent patient is [well] below 20 percent is screening preferable to system expansion as a way of meeting increases in demand."[44] Should screening procedures be instituted, however, citizens who become dissatisfied with the process might find a means to bypass it entirely and contact the emergency mobile unit directly. The nature of cellular technology could facilitate this tactic, which could disrupt the centralized management system of service agencies. Citizens might also respond by forming voluntary associations to provide emergency service protection independently of the public system. This approach will be examined below.

A high diffusion of portable mobile communication devices in the consumer market could therefore have two major impacts on citizen-service agency interactions. The immediate availability of a communication channel to public safety agencies could both assist in substantially reducing response time in the client reporting interval, and contribute to an increase in service demand, which may overload public safety agencies operating at a given service capacity. Agencies would have to develop policies for handling this potential increase in service demand, either by reallocating resources or deflecting a certain percentage of the demand.

Effects of Improved Communications Among Citizens: The Role of Citizen Patrols

The third fundamental communication linkage that a high deployment of the cellular system is likely to affect is communication among citizens. The proliferation of citizen patrols in metropolitan areas suggests the use of mobile communication devices to enhance patrol operations could have indirect effects on the services that are provided by public safety agencies.

The formation of citizen patrols has been motivated by a number of factors. The existence of some patrols may be a reaction to irresponsible police practices, or a reflection of a real or imagined need for protection from the police. Community patrols, especially social service patrols, may be part of a movement toward community control and self-help that began in the 1960s, but they may also manifest opposition to "big government" and the invasion of public authority into every sphere of life.[45]

EFFECTS ON SERVICE DELIVERY.   Cooperation between public safety agencies and patrols is most likely to be achieved by those patrols who represent an interest in revitalizing resources at the local level. If cooperative activities are productive, public safety agencies might consider forming voluntary associations as an alternative to making heavy capital investments in sophisticated technologies to deal with increased service demand.

Patrols that are formed in opposition to encroaching governmental intervention may be less interested in joint activities with local agencies. More apt to encounter indifference from public agencies, these patrols may be permitted to operate independently, within legal limits. Some of these activities may be less than innocuous, however, since their motivation is a potential source for vigilantism; the availability of low cost mobile communication technology could facilitate activities of this kind. This motivation for forming patrols also suggests that citizens might institute a set of parallel emergency services should a trend toward centralization of service delivery result in a lack of responsiveness to their needs.

ORGANIZATIONAL RESPONSES TO CITIZENS' INITIATIVES.   Public agencies may respond to these citizens' associations with hostility and defensiveness, a cooperative and exploratory attitude, or simple indifference. To the extent that the existence of some patrols represents hostility and fear of the police, police departments may be more likely to respond with defensive, combative strategies, countering the patrols' monitoring practices with more extensive surveillance of their activities. The police might be able to locate any mobile unit within the dimensions of one cell through the cellular system's vehicle location capacity. If deployment of the technology is very high, cell size might ultimately be reduced to several blocks in some areas. This monitoring capability, coupled with the legal authority to engage in tactics that may harass certain patrols or obstruct their activities, would provide police with extraordinary opportunities to effectively deactivate citizens' groups that are perceived as threatening. However, it could also be used to violate the civil rights of groups that are operating legally and serving a beneficial social purpose. These possibilities should be considered when decisions about the permissible use of the cellular system's monitoring capacity are made.

We do not wish to suggest that widespread availability of these new communication devices will foment hostility between police and neighborhood groups, foster the formation of community-oriented voluntary associations, or reinforce the resentment of government intervention into the private domain. Any combination of these attitudes may have motivated the formation of citizens' patrols; the technology merely facilitates their activities.

ADOPTION OF THE CELLULAR SYSTEM IN THE PUBLIC SECTOR

Political, Organizational, and Bureaucratic Influences

The full impact of the possibilities discussed here will not be felt until the cellular system is extensively utilized by public agencies and has fairly wide deployment in the consumer population. In any event, many of the impacts would not be substantially different from those caused by heavier use of technologies that have already been introduced. Because of its efficient use of spectrum, cellular technology may be the only means of increasing utilization by citizens or agencies, although in the latter case the adoption of technologies based on trunking may be sufficient. If its costs become competitive with standard two-way radio systems, and the additional capabilities of computer assisted dispatch and digital transmission can be included at a reasonable cost, the cellular system may displace currently available systems. An examination of trends in the deployment of centralized and multiservice dispatch systems, computer assisted dispatch, mobile digital devices, and the 911 emergency telephone system reveals no distinct pattern emerging for the adoption of any of these innovations on a national scale. Adoption is progressing, but it is occurring on a regional or local basis.[46]

The number of individual dispatch centers is still greater than the number of centralized ones. While there is a definite trend toward centralized dispatch and service integration in some states, these states have not introduced the most sophisticated technology—such as computer controlled dispatch or mobile digital systems—suggesting that factors other than the technologies' cost effectiveness may affect the degree of telecommunications planning for and adoption of sophisticated technologies that occurs. Some evidence to support this hypothesis can be found in a survey that shows respondents perceive "political factors" are a major constraint in telecommunications planning.[47] This evidence also tends to reinforce the view that the decision to adopt a technology within a public agency tends to be influenced as much by bureaucratic goals as by service goals. One study reports:

> Thus, even though a particular innovation can improve services, if it does not serve bureaucratic goals, it is not likely to be incorporated. Similarly, some innovations may be incorporated because they serve these goals, even though little service improvement has been produced.[48]

A study of organizational interdependence suggests that police, fire, and emergency medical services may pool their resources to form centralized dispatch systems so that specialized services, such as mobile coronary care units, can be provided at a reduced cost to the individual agency.[49] The study also suggests that motivation to participate in a joint program might come not only from the opportunity to improve service, but also from the potential to increase one's prestige by being associated with a large-scale operation involving sophisticated technology. Therefore, leaders may be willing to sacrifice some autonomy for increases in staff and other resources. While there are strong organizational motives to avoid such exchanges, which might result in loss of autonomy and necessitate more internal coordination, the need for more resources makes such exchanges more attractive.[50] Obviously, many of the factors influencing cooperative programs will be strongly dependent on local conditions.

Evidence from attempts to implement the 911 emergency telephone system nationally, on the other hand, brings up some organizational factors that may impede the universal deployment of an integrated system. The overlapping of public safety agencies' service area jurisdictions and the noncoincidence of telephone company service boundaries with service area jurisdictions are concomitant problems. For example, calls for 911 originating within the boundaries of a telephone company central office can be directed to only one answering point. But when that central office boundary overlaps two or more local government jurisdictions, a call originating in the overlap zone could be received by an agency that does not serve the caller. The routing of calls to the appropriate jurisdiction can be simplified technologically by using capabilities inherent in the Automatic Number Identification facilities currently employed by phone companies for billing purposes. Unfortunately, the cost of providing these capabilities may be prohibitive in areas with less sophisticated telephone systems.

The desire of individual service agencies to maintain autonomy will constrain many efforts to integrate the activities and responsibilities of public agencies. Apart from institutional factors, the personnel's perceptions of self-interest can compound the problem. For example, some agency heads may discount or not understand the longer term bureaucratic benefits discussed earlier; losing responsibility or autonomy may become the dominant factors. Such considerations may motivate some to resist interagency cooperation, even though the objections may be voiced as threats to "local interest."[51]

A second constraining factor is the accountability of agency administrators to their local constituents or legislatures. While the potential benefits of improved service access might enhance the police, fire, or medical systems, poorly administered services would clearly damage the agency's standing. Observing the performance of other local agencies could influence the willingness of an agency's director to undertake cooperative activities.

The complexity of organizational factors that affect any decision to adopt or implement a new technology makes it very difficult to predict confidently how the cellular system will be used by public agencies. It should be noted, however, that some of the cellular system's most profound effects on the public sector would result from mobile communications being employed to contact service agencies or to facilitate the operations of citizen service associations. Thus, public agencies may be pressured to adopt the cellular technology by clients who are already using it.

SUMMARY AND CONCLUSIONS

Since the operational capabilities of cellular systems are similar to mobile communication technologies currently available to public safety services, their large-scale adoption is not likely to have internal impacts upon an agency that would differ substantially from a more intensive use of available devices. While the use of cellular systems might improve response time rates over those attainable with a conventional transmission system, its effect on service outcomes will vary with the service provided. There is no conclusive statistical evidence that decreases in response time would consistently improve service outcomes, as they are currently measured.

Because its initial implementation cost could be expected to be high, especially with add-on capabilities, the cellular system is likely to stimulate the current trend toward centralizing and integrating communications for public safety services. There is a danger that increased centralization of the service delivery structure could lead to lessened responsiveness to clients' needs and a loss of autonomy by local service agencies. Given the nature of dispatch operations, however, it may be possible to centralize base station operations only and to maintain a decentralized deployment of mobile service personnel. The cellular system would greatly facilitate a strategy of this kind, as its superior spectrum efficiency would enable all personnel to be equipped with mobile transceivers. Since increased contact between clients and service personnel seems to improve the attitudes of both, local agencies might be willing to sacrifice some autonomy for improved community relations, if this strategy is feasible.

Nevertheless, the cellular system could have some negative effects on internal agency operations. One is that an increased use of sophisticated technologies may create cleavages between the specialized personnel, who must be hired to operate it, and their administrative superiors, whose authority may reside in bureaucratic rank rather than knowledge of operations. Resulting conflicts could impede effective operation. A second potentially erosive effect is that the high implementation cost of the technology could preclude its being adopted by communities with modest resources without federal cost sharing. The consequence could be increased fiscal dependency on the federal government at the expense of locally initiated and supported activities.

More profound effects on the service delivery system could be anticipated if cellular devices become widely available in the consumer market at a reasonable cost. Both citizen access to public safety services and citizen initiated community protection associations would be enhanced by portable or vehicular mobile communication devices interconnected to the landline telephone system. Reductions in total response time and increases in demand for service could be expected. Part of the increased demand would no doubt be caused by clients who are currently bypassing the emergency transportation system. The uncertainty and confusion expressed by these clients about the nature of their emergency and the appropriate action to take suggests that professional counseling over a publicized emergency telephone number, such as the 911 system, would improve their access to service and reduce their anxiety and confusion.

Experience with the 911 emergency telephone system suggests, however, that many of the requests that may be stimulated by new mobile communication devices may be inappropriate to the emergency service functions of the public safety agencies. Consequently, agency manpower could be occupied with non-emergency requests, while other more serious cases were left unattended. This situation is apt to necessitate a screening or priority system to rank requests for service.

The cellular system, if regulation and technology permit, might provide clients dissatisfied with a screening process with an alternative route to service. Citizens with a mobile transceiver and the phone number of the appropriate mobile vehicle might find a means to contact it directly, bypassing the agency communications center and disrupting the fleet management system. This possibility, if uncontrolled, could have profound effects on relationships between local service agencies and clients. Problems of this kind could lead to serious disagreements over the agency's control of and accountability for service quality.

Interaction between service agencies and clients would also be affected by the improved surveillance and monitoring capability that the cellular system could provide the citizen initiated community protection patrols. Groups hostile to local services may be able to escalate their activities, while those seeking to coordinate their activities with public agencies may supplement existing services and provide an alternative to heavy investment in sophisticated equipment. Others wanting to operate independently may either provide additional services outside the domain or capacity of public agencies, or create a parallel system of services as an alternative to publicly provided assistance. The use of mobile communications both by clients, to improve their access to services, and by citizens' associations, to facilitate their operations, will result in competition with public agencies for radio spectrum until cellular technology relieves spectrum congestion.

The limitations of the evidence from which these conclusions have been derived are evident. In the analyses of response time, sample sizes were often unacceptably small and drawn from only one metropolitan area; there is no data available on total response time for even a single agency in one locality. Classification schemes for arrests and medical emergencies are subject to question, the problem of choosing acceptable indicators for service outcomes is unresolved, and the total set of variables affecting the outcomes of police and emergency medical services has not yet been specified, let alone controlled for. It is therefore clear that additional research should be undertaken before conclusive decisions can be reached about the most productive areas for intervention. The weight of the evidence suggests, however, that resources should be devoted to questions concerning improved access to service, potential for increased demand, and activities of local citizens' organizations, rather than to analyzing how to improve the internal communication capacity of service agencies.

If it can be demonstrated that a reduction in the client reporting interval could contribute to improved service outcomes, then it might be advisable to provide high risk and low income clients with mobile communication devices and to institute a telephone counseling service for emergency victims. A second consideration is that if the increased availability of communication technology can stimulate inordinate increases in service demand, such as occurred in New York City with the 911 system, then perhaps the entire notion of "system overload" ought to be reexamined. The massive demands for assistance may indicate that the potential client population for public services of all kinds is being seriously underserved. Instead of responding to overload with priority systems aimed at deflecting cases defined as "nonemergent" by an agency's professional authority, planners might investigate the social and economic sources of emergencies that develop continually and devise preventive, as opposed to reactive, strategies in order to reduce the demand pressure. Reactive strategies may lead citizens to create alternative emergency service structures that either circumvent or disrupt the existing system.

Another alternative to be considered is that the proliferation of citizen patrols seeking cooperative relationships with public safety agencies may reduce the need for large-scale, sophisticated technologies, such as computer assisted dispatch and mobile digital communications. Management strategies might be developed to balance the patrol and ambulance workload between the established service system and the voluntary and, sometimes intermittent, citizen patrols. The type of technological development to be encouraged in this instance would be two-way, mobile-to-mobile, spectrum efficient, low cost communication, as opposed to large-scale information-processing support systems. The allocation of spectrum between public safety agencies and Citizens Band communication might also need to be reconsidered.

A final implication is that careful consideration should be given to the complex problem of resource allocation before heavy investments in large-scale cost effective technology are made at the expense of human contact and responsiveness in delivering services that inherently involve stressful situations. Nowhere in the literature or in this analysis has a comparison been made between the relative contributions to service outcomes of increasing the manpower and utilizing physical technology. Illumination of this problem would require an entirely new research agenda.

In the four preceding chapters, we have organized our discussion of the use of mobile communication technology around the role it plays in the activities of organizations. But no discussion of mobile communications would be complete without giving attention to the extraordinary growth in personal uses, as manifested by the Citizens Band phenomenon. This is the subject of the next chapter.

NOTES

1. B. E. Robinson and R. P. Morgan, "A Delphi Forecast of Technology in Education with Implications for Educational Satellite Development" (Paper prepared for the 4th Annual National Educational Technology Conference, San Francisco, Calif., March 1974).

2. See, for example, "Teaching by Telephone," *Nations Schools and Colleges* 1 (November 1974):15; "Any Telephone," *Nations Schools and Colleges* 93 (March 1974):50.

3. James J. O'Neill, C. G. Paquette, S. Polk, and F. L. Skinner, "Testing the Applicability of Existing Telecommunication Technology in the Administration and Delivery of Social Services" (Springfield, Va.: National Technical Information Service, April 2, 1973).

4. William K. Linvill, "A Concept for Improving Human Service Delivery," Engineering-Economic Systems, Stanford University, Palo Alto, Calif., December 6, 1974.

5. President's Commission on Law Enforcement and the Administration of Justice, *Task Force Report: Science and Technology,* prepared by the Institute for Defense Analyses (Washington, D.C.: U.S. Government Printing Office, 1967).

6. For a detailed account of the functions performed by command and control centers, see Associated Public Safety Communications Officers Inc. *Police Telecommunications Systems,* prepared by IIT Research Institute for the Law Enforcement Assistance Administration (Washington, D.C.: U.S. Department of Justice, 1971).

7. Ibid., p. 103.

8. See U.S. Department of Justice, "Application of Computer Aided Dispatch in Law Enforcement," prepared by Robert L. Sohn et al. for the National Criminal Justice Information and Statistics Service (Washington, D.C.: U.S. Department of Justice, December 1975); and U.S. Department of Justice, "Application of Mobile Digital Communications in Law Enforcement," prepared by Robert L. Sohn et al. for the National Criminal Justice Information and Statistics Service (Washington, D.C.: U.S. Department of Justice, May 1975).

9. IIT Research Institute, *Illinois Police Communications Study* (Springfield, Ill.: Associated Public Safety Communications Officers, Inc., December 1969).

10. U.S. Department of Commerce, "The Economics of Highway Emergency Ambulance Services," prepared by Dunlop and Associates (Springfield, Va.: National Technical Information Service, January 1969).

11. Keith A. Stevenson, "Emergency Ambulance Services in U.S. Cities—An Overview" (Boston, Mass.: Operations Research Center, Massachusetts Institute of Technology, August 1969), pp. 9-10.

12. U.S. Congress, Senate, 93rd Cong., 2nd sess., *S2410,* November 16, 1973.

13. Clark Whelton, "A Radical Solution to Our Deadliest Problem," *Village Voice,* March 29, 1976, p. 9.

14. See, for example, Corwin Moore, "Report on the Michigan Emergency Patrol" (Paper prepared for the Highway Safety Research Institute, University of Michigan, Ann Arbor, Mich., January 20, 1976).

15. Robert K. Yin et al., "Patrolling the Neighborhood Beat: Residents and Residential Security" (Santa Monica, Calif.: The Rand Corporation, March 1976), pp. 61–74.

16. President's Commission, *Task Force Report,* p. 88.

17. Seattle Police Department, "Impact of Response Delays on Arrest Rates" (Paper prepared for the Inspectional Services Division, Seattle, Wash., September 1975).

18. Ibid., p. 22.

19. Thomas R. Willemain, "The Status of Performance Measures for Emergency Services" (Cambridge, Mass.: Operations Research Center, Massachusetts Institute of Technology, July 1974).

20. Ibid., p. 8.

21. U.S. Department of Commerce, "Land Mobile Communications and Public Policy," prepared by Systems Applications, Inc. (Springfield, Va.: National Technical Information Service, 1972), vol. I, p. 221.

22. For a review of these studies, see Hyrum Plaas et al., "The Evaluation of Policy-Related Research in Emergency Medical Services" (Springfield, Va.: National Technical Information Service, August 1974).

23. Kent W. Colton, "Computers and the Police Revisited: A Second Look at the Experience of Police Departments in Implementing New Information Technology" (Cambridge, Mass.: Operations Research Center, Massachusetts Institute of Technology, October 1974), pp. 41-50.

24. Response Time Study, Kansas City Police Department, Kansas City, Mo., March 1977.

25. Peter R. Mogielnicki, Keith A. Stevenson, and Thomas R. Willemain, "Patient and Bystander Response to Medical Emergencies" (Cambridge, Mass.: Operations Research Center, Massachusetts Institute of Technology, July 1974).

26. Ibid., p. 17.

27. See, for example, E. S. Savas, "Simulation and Cost Effectiveness Analysis of New York's Emergency Ambulance Service," *Management Science* 15, no. 12 (August 1969):B-608-B-627; and William K. Hall, "The Application of Multifunction Stochastic Service Systems in Allocating Ambulances to an Urban Area," *Operations Research* 20, no. 3 (May-June 1972):558-69.

28. IIT Research Institute, *Illinois Police,* vol. II.

29. U.S. Department of Justice, "Muskegon Co. Central Police Dispatch," prepared for the Law Enforcement Assistance Administration by Abt Associates, Inc., January 10, 1975.

30. B. Ebstein, F. C. Back, and B. I. Marks, "Central Dispatching System: Design Test and Implementation" (Chicago, Ill.: IIT Research Institute, September 1973), chap. 3, pp. 2-46.

31. Ibid., sect. 3.3.3.; for a study of ambulance services, see E. S. Savas, "Simulation and Cost Effectiveness."

32. U.S. Department of Justice, "Application of Computer Aided," p. 62. Huntington Beach has a combined police and fire department system.

33. Ibid., p. 63.

34. U.S. Department of Justice, "Application of Mobile Digital," pp. 44-57.

35. James G. March and Herbert Simon, *Organizations* (New York: John Wiley & Sons, Inc., 1958), p. 158.

36. Ibid., p. 160.

37. Ibid., pp. 160-62.

38. Robert K. Yin and Douglas Yates, "Street Level Governments: Assessing Decentralization and Urban Services" (Santa Monica, Calif.: The Rand Corporation, October 1974), Chapters 4 and 9.

39. For example, see Bertram Gross "Friendly Facism, A Model for America," *Social Policy,* November-December 1970, pp. 44-52.

40. See National Service to Regional Councils, "Emergency Telephone Communications Workshop: Summary of Proceedings" (Springfield, Va.: National Technical Information Service, March 1971); Kenneth Bordner and J. Spencer Houston, "A Study of the Single Emergency Telephone Number" (Philadelphia, Pa.: The Franklin Institute Research Laboratories, March 1970); *The New York Times,* July 10, 1973, August 13, 1973, and July 31, 1975.

41. Keith A. Stevenson and Thomas R. Willemain, "Analyzing the Process of Screening Calls for Emergency Service" (Cambridge, Mass.: Operations Research Center, Massachusetts Institute of Technology, September 1974).

42. Ibid., pp. 43-76; Albert J. Reiss, *The Police and the Public* (New Haven: Yale University Press, 1971), pp. 75-76.

43. Stevenson and Willemain, "Analyzing," p. 43.

44. Ibid., p. 44.

45. Yin et al., "Patrolling," p. 38.

46. U.S. Department of Justice," A Review and Assessment of Telecommunications Planning in the 50 State Planning Agencies," prepared for the Law Enforcement Assistance Administration and the Associated Public Safety Communications Offices by Booze, Allen, and Hamilton, Inc. (New Smyrna Beach, Fla.: APCO, Inc., November 1, 1975), vol. 2, pp. 166-69.

47. Ibid., pp. 180-81.

48. Robert K. Yin et al., "A Review of Case Studies of Technological Innovations in State and Local Services" (Santa Monica, Calif.: The Rand Corporation, February 1976), p. 161.

49. Michael Aiken and Jerold Hage, "Organizational Interdependence and Intra-Organizational Structure," in *Human Service Organizations,* eds. Yehezkel Hasenfeld and Richard A. English (Ann Arbor: University of Michigan Press, 1965), p. 6077.

50. Ibid., p. 589.

51. For an analysis of the complex psychological motives for maintaining a tight organizational management style, see Victor Thompson, *Modern Organizations* (New York: Alfred Knopf, 1971), Chapter 8.

Chapter 9
PERSONAL USES OF MOBILE COMMUNICATIONS
Citizens Band Radio and the Local Community

Cary Hershey
Eric Shott
Howard Hammerman

INTRODUCTION

American society is currently experiencing the synergistic effects of two of its most important consumer technologies, the automobile and the telephone. With Citizens Band (CB) radio, people now have a readily available, low cost, two-way communication technology for personal use. And the explosive growth in its use has attracted national attention. Journalists, commercializers, and policymakers are eagerly attempting to understand the phenomenon. They are asking why, suddenly, more than 500,000 persons are applying each month to the Federal Communications Commission (FCC) for CB licenses, and how we arrived at ten million licensees, an estimated 22 to 25 million sets, and 25 or more million users.[1] Is CB a fad, like the hula-hoop, or does the tremendous growth in CB reflect a more fundamental need or desire for electronic communication between people on the move? What does the boom in CB suggest about American society, and what are its implications for the way we live and work? What can be inferred about the future of personal mobile communications from a study of CB usage patterns?

While all of these questions regarding the mass popularization of a communication technology rooted in our mobile culture are very important, they are difficult to address with a high degree of confidence. By the time social scientists begin studying a "mass event" like CB with appropriate methods and adequate resources, the event is often on the wane; the study becomes retrospective, and its utility for effective public policymaking is reduced. This is especially true in the area of consumer technologies. Furthermore, social scientists have focused predominantly on work and deemphasized leisure, although leisure studies themselves emphasize economic and productivity issues. "New species of commodities (do-it-yourself programs) reflect the modern fragmentation and mutual displacement of work and leisure, and the emergence of new synthetic structures as yet unanalyzed."[2] Citizens Band radio falls squarely into this framework of leisure activity. Consequently, it is not surprising that one student of CB reports that, "in conducting anything but a cursory examination of the CB problem, one is struck by the scarcity of available data on the operating habits of CB licensees."[3]

Our aims are more modest than attaining a comprehensive study of the social forces and impacts associated with Citizens Band radio. The goals of this study were to undertake the acquisition of data on a limited scope regarding personal uses of mobile communications that would be adequate to explore certain public policy questions and to clarify the problem for future research. Not only were resources unavailable for a more comprehensive study, but also without a first assessment of issues to be investigated, as in this effort, it is questionable whether one should embark on a comprehensive research course.

We shall begin this chapter with a brief overview of the nearly twenty-year history of the Citizens Band radio service, including the shifting stance of the FCC toward the service. We will then place the CB phenomenon into the context of community sociology in order to understand one type of CB usage and to facilitate analysis of the factors that underlie its extraordinary and unexpected popularity. Finally, we will infer some social implications and consequences for policy from our analysis of the CB phenomenon, and speculate on their relevance to the future of mobile communications in general and the 900-MHz cellular system in particular.

REGULATION OF CITIZENS BAND SERVICE

In 1959, which marked the end of the first twelve years of CB, there were only 49,000 licensees. This was one year after the FCC reassigned a portion of the spectrum from amateur radio operators, or "hams," and inaugurated a new class of CB service (Class D), in an effort to promote the use of radio by private citizens for their personal use. Reallocation of the spectrum made it possible for manufacturers to build quality CB sets at a price that the average consumer could afford, around $150 to $250. In 1962, four years after Class D service was introduced, the number of licensees reached 300,000, a sixfold increase. By 1965, the number of licenses had increased another two and one half times to 745,000, and the FCC was beginning to express considerable alarm about the state of the service. In its 1965 *Annual Report,* the FCC stated that " 'widespread rule violations . . . threaten the continued usefulness of the service' " and that " 'illegal practices are seriously impairing the legitimate use of the service.' "[4] Surprisingly, there was no significant growth in the number of CB licensees during the 1967–1973 period, with the number stabilizing near 800,000; there is no satisfactory explanation for this stabilization.

The rules in force from 1958 through late 1975 permitted any person 18 years or over, or any business owned and operated by citizens, to obtain a Class D license. In contrast to the amateur service, no tests were required. Rules were adopted limiting use to the transmission of substantive and purposeful business or personal messages; hobby use, which is frequent within the amateur radio band, was prohibited. The FCC intended CB to be a readily accessible, two-way personal communication technology for the average American. And the public responded most enthusiastically to the opportunity, especially during the past few years after the media focused attention on the service's use by truckers and other cultists. The oil embargo-truckers strike demonstrated how effectively CB could be used.[5]

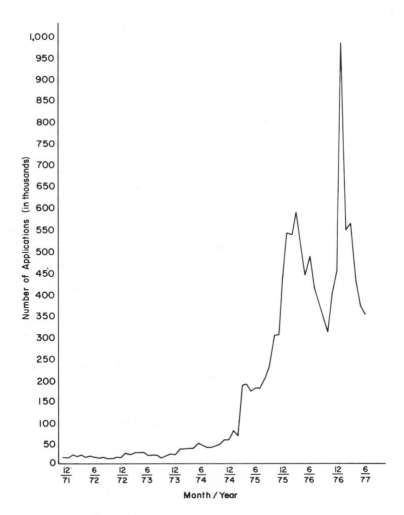

Figure 9.1: *Class D Citizens Radio Service Monthly Applications (November 1971-June 1977)*

As Figure 9.1 shows, from 1971 through late 1973 the number of applications for CB licenses approximated 20,000 per month. But, after the truckers strike in late 1973, the number of applications increased to about 40,000 per month in 1974, and then climbed dramatically throughout 1975 to reach a peak of 590,000 in December. Then, until October 1976, applications dropped off, presumably in part because of confusion over new types of CB units and the FCC's regulatory response to these options. In November, however, applications began increasing again, and by January 1977 they had reached an all time high of 980,253. During February and March 1977, there were almost 600,000 applications per month.

Not surprisingly, keen dissatisfaction among both CB users and the FCC accompanied the almost exponential growth in the number of users. Experienced CB users and those who employed CB in their businesses were especially annoyed as their channels became overloaded and misused. Numerous discussions such as the following appeared in the media:

CB'ers are, as a rule, less disciplined than other radio operators; the Citizens Band is frequently cluttered with overlapping and interfering transmissions. The messages are often inconsequential—breaks for radio checks and requests for the time of day, and, whatever their message, many CB'ers seem to feel no compunction about breaking in on someone else's conversation. As a result, the Citizens Band can be a mass of garbage, practically useless, especially around urban areas. All the blather and illegality are infuriating to serious CB'ers, particularly to those oldtimers who can remember when they had to do a lot of changing of channels to find anyone else transmitting.[6]

The FCC found itself under extreme political pressure to do something about "the CB mess." Deterioration in the service was attributed not only to the explosive growth but also to poor enforcement of rules by the FCC, resulting in excessive monopolization and congestion of channels. An authority on CB has described the FCC's enforcement dilemma in the following terms:

FCC rules require the licensee to identify his station with an FCC-assigned call sign at the beginning and end of each series of transmissions. The reason for requiring this identification is to enable FCC enforcement personnel to identify any station that may be in violation of any of the multiplicity of other FCC rules applicable to CB licensees. It was not long after the service was created, however, that CB operators discovered that by not giving their call sign they could significantly decrease their chances of being apprehended while operating in violation of other rules, and thus the vast majority of CB communications today is conducted by stations which address each other by pseudonyms, such as "Red Rooster," "Superman," and "Cannonball."[7]

According to FCC regulations, most users are required to limit conversations to five consecutive minutes. Other rules, in force until late 1975, prohibited discussion of CB equipment, since the commission viewed such subjects as not being "necessary personal communications"; the violation of this rule also served to increase channel congestion. Similarly, prohibiting the use of CB as a hobby was an attempt to preserve the congested channels for more "necessary personal communications."[8]

The FCC responded to complaints and increasing political pressure by partially deregulating the CB service. In late 1975, the commission promulgated a series of more permissive rules that, among other things, removed the restriction on hobby use and "chitchat" conversations, although they did keep in force rules prohibiting obscenity and limiting the length of conversations to five consecutive minutes. The present FCC regulatory stance appears to be moving toward liberalization and deregulation in an effort to design a personal radio service that will satisfy users. The Citizens Band may have been deregulated because the commission came to the conclusion that there was no effective means of and perhaps no compelling reason for regulating it.

Before turning to our data on CB communication, we shall discuss some structural changes in American society that may be related to the upsurge of CB usage in local communities. Understanding these changes has been useful in guiding our search for hypotheses concerning the reasons for the explosive and continued growth in CB.

COMMUNICATION AND THE STRUCTURE OF COMMUNITY

In the past century, America has changed from a rural, agrarian country with a few dense settlements to a large, industrial society with many urban centers. The importance of the family[9] and the community[10] have declined. Communities of interest have increasingly replaced communities of natality as social and geographic mobility has accompanied economic growth. According to the 1970 U.S. Census, 47 percent of the American population moves once every five years;[11] 20 percent of the mobile population are moving to another state.[12] The alienation and tensions produced by this transience and rootlessness have been studied by several researchers.[13]

The spatial character of American cities has also changed. From the dense, closely packed "walking city" typified by Boston,[14] most cities have undergone rapid "spread" in the form of suburbanization, and large-scale agglomerations have increased the distance between home and workplace. As a result, the average American spends more and more of his time in transit.

This trend toward mobility has affected our values. Whereas prestige had once been achieved through family reputation in a particular community, the advent of the automobile culture created an additional, more mobile crystallization of prestige. Much more than a transportation necessity, the car has become ingrained in our courtship patterns, can reflect our social class, and marks our progress toward the generally accepted goals of the American dream.

The automobile has also served as the ends and means for a new kind of mass culture, which has its own consequences. The mobility facilitated by the auto increased the user's isolation from his fellows in some circumstances and decreased it in others. Furthermore, this mobility has permitted the formation of "communities of interest" that extend over large geographic areas and, in doing so, has probably inhibited the development of more locally based ones.

We began our research on CB with the assumption that it allows the achievement of what Melvin Webber called a "community without propinquity" by providing a means of enhancing and/or creating communities of interest in an era when people are increasingly mobile.[15] CB also allows the user to overcome social barriers that might be impenetrable in other types of social interaction. For example, a "newcomer" to the CB community can listen to the flow of conversation and choose when and with whom he will communicate. The use of the "handle," or code name, protects the individual's anonymity. In this way, the CB user is free to project whatever image he chooses without jeopardizing his "real life" position in the community. Additionally, the use of a special jargon or "lingo" on the CB further tends to decrease the likelihood that the "real" identity of an individual will be discovered. If these assumptions are correct, we would expect a large portion of CB communications to be concerned with the ritual of establishing the user's "CB identity." This can center on the code name, but may also include location, occupation, and hobbies.

A community of interest is based largely on commonality.[16] Certainly the one thing all CB users have in common is the equipment they use. By assuming that CB is a means for establishing community, and not strictly for performing instrumental tasks, we would expect that a frequent topic of discussion would be the comparative benefits and performance of the various brands of CB equipment, their modes of deployment, and their care and maintenance. On the other hand, were CB used merely as a means of instrumental communication, we would expect the bulk of the messages to be concerned with information and advisories useful to the users' "real life" identities.

In other words, we need to determine whether CB is used to transfer information useful in achieving "real life" goals, or if it is used as an end in itself. Finding that CB communications are mostly instrumental and informative would lead us to conclude that the social implications of this technology are narrowly focused on facilitating people's activities while in a mobile state. However, demonstrating that CB is used mostly for expressive and socializing purposes would provide us with strong evidence that acceptance of this technology is associated with a need for closeness and a sense of attachment for a portion of American society.

If we find that CB is used predominantly for the latter type of communication, we are forced to ask the next question: Why is there this need to communicate, to be in touch? After all, radio communication has been with us for some time. Our image of the traditional "ham" radio operator is one of a person concerned with the technical aspects of radio communication itself. He communicates not simply for the content of the communication, but as part of a hobby, to establish the range of his equipment, for instance. And ham operators are very concerned with the protocol and procedures of radio communication. Thus amateur radio operators can be considered an "electronic neighborhood," but one that is fairly exclusive in terms of the expense of equipment, formal rules, and the insistence that every member of the "neighborhood" have a high degree of technical knowledge. We hypothesize that the interest in CB communication comes from a different source. CB can also be looked upon as an "electronic neighborhood," but this one is highly populist, requiring little technical competence. If this is true, then we would expect that CB users would be only minimally concerned with the formal rules of CB communication. While CB users might use discussion of equipment and range as a subject of conversation, their interest is mainly in communicating per se.

It is important to examine the characteristics of Citizens Band radio in order to fully appreciate its social meaning. First, CB is nonprivate: while communications can be directed to another specific user, all users understand that their conversations are being monitored or "sandbagged" by unknown third parties. CB is participative: unlike television, this technology invites the participation of the listener, and each user can perform in the daily drama of its chatter. CB is anonymous: it is difficult to locate or detect a particular user unless he wants to be detected. CB is extremely portable, and it has a limited range: because of restrictions on power output defined by federal law, communications are limited to a range of less than ten miles; in densely populated areas the effective range is even shorter.

Citizens Band radio has been extensively marketed as a device that can be used to aid travelers in emergency situations. Certainly it is used for this. But CB may also serve as a means of dealing with daily stress and boredom. As it becomes more difficult for the average worker to associate his own creativity with the finished products of his labor, it can be expected that he will seek other avenues for expressing his individuality. One can see the recent interest in recreational vehicles or the popularity of customized vans in this light, and CB communication may be congruent with this pattern. Each user is allowed to become his own broadcaster—through widespread participation in CB chatter, the user can make his handle as well known in the local community as Walter Cronkite's is in the national community. Therefore, we can expect highly personalized communications as participants announce their handle, hobbies, activities, and opinions to the growing CB public.

These hypotheses lead to another expectation: we would expect CB communications to be of at least moderate length. Unlike military or industrial use of radio communications, we suggest that local community CB is mainly *affective* in character, not *instrumental*. As such, the communications should be full of personalization, shadings of meanings, etc., all of which would make them lengthier.

Finally, we expect to find CB used as other subcultural symbols are used—as a way of defining the boundaries of a community of interest. Thus we anticipate transmissions to announce CB-oriented meetings and to relay information about other CB users in the community; to maintain and extend a social network; to track down other members and leave messages of the type, "Tell X that I was looking for him." In this context, CB can be seen as a telecommunications counterpart to the neighborhood tavern.[17] The CB allows participation and contact with others in a casual, nonprogrammed manner.

CONTENT ANALYSIS OF CB COMMUNICATIONS

Given these considerations, a content analysis of actual CB messages was undertaken in order to understand how CB radio is being used by the general public. Content analysis, a widely accepted technique for studying message content in communications research, has been described as "a research technique for the objective, systematic and quantitative description of the manifest content of communications."[18] The main aspects of the method are: (1) clearly and explicitly defined categories of analysis used to classify the content; (2) methodical classification of all relevant material in the analyst's sample; and (3) use of some quantitative technique in order to provide a measure of the importance and emphasis in the material of the various ideas found.[19] This research method permits a classification of the content of actual transmissions on CB channels.

CB is currently used in three rather distinct ways. The highest priority use is for emergency communications and is considered so important that the FCC has allocated a special channel (9) just for emergency transmissions. A second category can be termed highway or trucker use. Immortalized in popular song, these conversations generally occur between long-distance travelers on interstate highways and major traffic routes, as well as within local areas. Transmissions usually concern traffic conditions, "smokies," and other subjects that temporarily link the travelers together. The third category, which has received little attention, can be termed local use. Referring to citizens' use of CB within their local community as they commute to work, shop, or socialize in or outside of their homes, this latter type of more general CB use will be the focus of this study.

The goal of the research was to determine the type of messages transmitted over a *local* CB channel during a one-week period. It was decided that the conversations to be analyzed would be limited to those that *originated* on the calling channel (Channel 10, the local "home base") of the area studied. The research itself was carried out in two stages: stage one concentrated on developing a sampling frame of communications, while stage two involved collecting most of the data. However, data collected during stage one does provide information on the frequency, type, and logistics of CB transmissions.

During stage one, a coder listened to a standard CB mobile unit for 15 minutes out of every hour between the hours of 8:00 A.M. and midnight for seven consecutive days. The coder counted the total number of communications, including the number of completed, directed communications, that is, a communication in which an identified caller reaches an identified receiver by calling him on the "call" or "home" channel. From the resulting distribution of completed, directed communications, a quota sample was drawn—with probability proportional to size—to yield 55 "picks," where a "pick" is a specified hour on a specified day.[20] Clearly the more popular hours (such as 5:00 P.M. to 6:00 P.M.) had a higher probability of being selected than the less popular hours.

In stage two, the coder was directed to monitor the first four conversations occurring in each hour, selected according to the procedure outlined above. The conversations were recorded on standard cassettes.

Monitoring for stage two occurred during late June 1976, a period marked by substantial interference. As a result, reception was often very poor and the monitoring was extended over a three-week period. Out of a total target sample of 220 conversations, 205 were actually recorded, yielding a completion rate of 93 percent.

Coding was accomplished following standard procedures; a comprehensive scheme was developed that allowed the coder to classify up to 20 "mentions" per conversation. A "mention" is defined as a subject topic, so that if a conversation began with the CB users discussing the quality of their reception and then moved on to a discussion of the weather, two mentions were coded. In addition to recording the content of the communication, the duration was timed and noted.

The Frequency and Logistics of Citizens Band Communications

The objective of the first stage of research was to count the total number of communications made on the call channel being studied to establish a sampling frame. A communication was defined as: words spoken over the CB initiated by any party and terminated by (1) lack of contact with the desired calling party, (2) the end of a conversation between two or more parties, (3) the end of a broadcast message, or (4) the response to a broadcast message. The following types of communications were noted:

1. *BROADCAST, SELF-IDENTIFICATION.* These were messages that simply announced presence on the airwaves. A typical message was: "This is the Red Rider, standing by on the side of the channel."

2. *BROADCAST, OTHER.* This category encompasses all the other broadcast messages, that is, messages directed to any available listener and not to a specific CB user. Requests for the correct time, announcements of speed trap locations, etc., would fall under this heading.

3. *DIRECTED, COMPLETED.* These messages are specifically directed from one user to another and have reached their mark. The actual conversations

initiated or arranged on the call channel were often carried out on one of the other channels. An example of this type of communication might be: "This is Red Rider, calling Tiny Tim." "Tiny Tim here, Red Rider, take her up to 15."

4. *DIRECTED, INCOMPLETE.*   These are messages directed from one CB user to another that were not completed because there was no response.

Table 9.1 shows that of the 11,284 communications[21] in our test week, more than half were unsuccessful attempts by one CB user to reach another specific user. One-third of the calls were unsuccessful attempts by one user to reach another party. Broadcast messages, of either the "self-identification" or "other" variety, comprise only a small fraction of the total number of communications (seven percent and eight percent, respectively). Thus the preponderance of *local* CB communications do not represent a series of attempts, for whatever reason, to make contact with unknown parties but are purposeful efforts to communicate with a specific CB user, known at least by handle to the communicator. This finding is consistent with the hypothesis that local CB usage patterns conform to a community model, since one would hardly expect a CB user who is part of a "neighborhood" to broadcast to strangers when he wants to talk to a friend. The objective of the second stage of this research was to determine the content of these directed, completed communications.[22]

Table 9.2 presents the overall distribution of communications by the day of the week, a factor that appears to have little influence on the distribution of total communications. One exception should be noted. A high proportion (24.6 percent) of the directed, completed communications took place on Saturday, which is generally a nonwork day when many people spend a good deal of time in their cars.

The distribution of communications by the time of day is presented in Table 9.3. Clearly, the early evening hours (between 4 P.M. and 8 P.M.) are the most popular time to use the CB. The most popular one-hour time period is from 5:00 P.M. to 6:00 P.M. (not shown in Table 9.3).

From this distribution, there is an indication that CB is used mainly in connection with leisure time activities, since these are the hours when teenagers are out of school and most adults are driving home from work.

Table 9.4 presents the mode of transmission (origin-destination) of CB communications derived from our stage two sample and is sorted into three

Table 9.1: *Number of Communications by Type, 25% Sample (Stage One)*

Type of Communication	Frequency	Percent
Broadcast, self-identification	190	6.7
Broadcast, all other	219	7.8
Directed, completed	938	33.3
Directed, incomplete	1,474	52.2
Total, all types	2,821	100.0

Table 9.2: *Distribution of Communications by Day of the Week, 25% Sample (Stage One)*

Day of the Week	Total Communications		Directed, Completed Communications	
	Frequency	Percent	Frequency	Percent
Sunday	371	13.1	111	11.9
Monday	402	14.3	140	14.9
Tuesday	334	11.8	112	11.9
Wednesday	431	15.3	135	14.4
Thursday	372	13.2	95	10.1
Friday	442	15.7	114	12.2
Saturday	469	16.6	231	24.6
Total	2,821	100.0	938	100.0

Table 9.3: *Distribution of Communications by Time of Day, 25% Sample (Stage One)*

Hour of the Day	Completed Directed Communications	Total Communications
	Percent	Percent
8 A.M. to 11:59 A.M.	20.0	22.3
12 noon to 3:59 P.M.	22.4	22.0
4 P.M. to 7:59 P.M.	35.2	33.2
8 P.M. to 11:59 P.M.	22.4	22.5
Total	100.0	100.0
Number	938	2,821

Table 9.4: *Distribution of Communications on Local Channel by Mode of Transmission (Stage Two)*

Origin-Destination	Frequency	Percent
Base-to-base	52	25.4
Base-to-mobile/mobile-to-base	56	27.2
Mobile-to-mobile	12	5.9
No information	85	41.5
Total	205	100.0

categories: base-to-base, base-to-mobile (or mobile-to-base), and mobile-to-mobile. Although 41.5 percent of the local communications could not be categorized according to mode of contact, we found that 25.4 percent of the total communications could be classified as base-to-base, and only 5.4 percent of the communications were in the mobile-to-mobile mode. While this estimate may be inaccurate, because mode of contact was difficult to determine, the finding is extremely suggestive. The high proportion of communications in the base-to-base mode indicates that in many instances the CB radio may be used similarly to the conventional landline telephone. It is also interesting that only 5.4 percent of the communications *on the local channel* were mobile-to-mobile (i.e., between two CB users traveling in their cars).[23]

Data collected on the duration of conversations and channel switching provide additional information on the logistics of CB communication. The average directed, completed CB conversation in our sample lasted two and one-half minutes; the shortest was 15 seconds in length, while the longest lasted almost 29 minutes. Examining Table 9.5, we find that more than half the conversations were over in less than two minutes, and 91.2 percent of the conversations were well within the FCC's five-minute maximum duration limit. By comparison, the average conversation is longer than many messages occurring among other classes of mobile communicators (with the exception of amateur operators) and is about the same length as the average local telephone call.

The research design called for monitoring all directed, completed calls on the "home channel" (Channel 10) and following the conversations to other channels if the parties switched. The data given in Table 9.6 show that 68.8 percent of the

Table 9.5: *Distribution of Communications by Duration (Stage Two)*

Duration	Frequency	Percent
Less than one minute	55	26.8
One to two minutes	58	28.3
Two to three minutes	38	18.6
Three to four minutes	22	10.7
Four to five minutes	14	6.8
More than five minutes	18	8.8
Total	205	100.0

Table 9.6: *CB Channel Switching (Stage Two)*

Channel Switching	Frequency	Percent
Parties did not switch channels	139	68.8
Parties switched channels	63	31.2
Total	202	100.0

CB users who originally made contact on the "home channel" stayed on that channel throughout the conversation, while 31.2 percent switched from the "home channel" to one of the 22 other available channels.

This data indicates that a large minority of CB communicators may have wanted to ensure greater privacy by utilizing a channel that was less popular or at the very least, to lessen the possibility of being interrupted by third party "breakers." It is also possible that the "switchers" were motivated by good CB etiquette to vacate the calling channel so that others would be able to use it. Whatever the degree of channel monopolization occurring on the home or other channels, however, the conversations themselves are clearly two-party communications. In 89.1 percent of the monitored communications, two CB users engaged in a conversation were not joined by a third party. In only 11 percent of the communications did a third party "breaker" join an ongoing conversation.

To summarize, local CB communications in the city studied are relatively evenly distributed throughout the week, with a high proportion of the communications taking place in the early evening hours. Of the calls that could be analyzed by the mode of transmission, most involved at least one mobile party (57 percent), but, perhaps more significantly, 90 percent included at least one fixed station, and a large proportion (44 percent) were between two fixed parties. In addition, most communications involved an attempt by one CB user to "call" or direct his communication to another specific CB user. Further, even though users who attempted to make contact with others on the home channel generally did not switch channels, their conversations did not usually include third parties.

Socializing on the Citizens Band: Maintaining the Community Bond

In 1958, when the FCC inaugurated Class D Citizens Band Service, the use of the new service for hobby purposes was specifically prohibited, because, the FCC reasoned, the amateur ("ham") service was already available for citizens who both use and discuss radio equipment as a hobby. Throughout the 1960s and early 1970s, the commission expressed increasing concern that the growing numbers of CB users were misusing Class D service, that is, spending a considerable amount of air time talking about their CB equipment—the quality of transmission, antenna heights, signal quality, etc.[24] In late 1975, the commission issued a modified set of regulations that, in effect, legitimated the use of CB for nonserious, hobby-oriented communications, commonly known as "chit-chat." One of the questions addressed by our research, then, was: What proportion of the total communications fall into this recently legitimized classification of use? In order to ascertain this, the mentions or conversation topics were divided into three categories based on the amount of information contained in each.[25]

Table 9.7 presents the means of these variables, expressed as percentages of the total. We found that 31.1 percent of the average communication was devoted to discussion about the locations of those conversing and/or arrangements for face-to-face contact or informal meetings. Other subjects that were mentioned frequently also related to socializing and establishing a community network of CB users. For example, mentions of other CB users and social events (the network phenomenon) averaged 22.2 percent, and 18.8 percent of the average communication was spent discussing equipment. Other subject categories, such as hobbies, weather, and police, occurred far less often; only 4.4 percent of the average conversation referred to hobbies, 2.2 percent involved weather, and 2.8 percent was devoted to requests for information, etc.

Another way of presenting the data is shown in Table 9.8. Here, the percentage of communications with one or more mentions are broken down by the categories listed. Since most conversations contained several categories of responses, the percents do not add up to 100. The table does show the relative importance of the various conversation subjects.

A definite conclusion can be drawn from the data in Tables 9.7 and 9.8: CB is used mainly as a means for establishing and maintaining a network of CB users. More than half of the typical CB communication is devoted to discussion of the parties' locations, plans for meetings, and references to other CB users. Approximately three-quarters of CB conversations included these kinds of mentions. On the other hand, only 13 percent of the CB communications contained even one request for information or direction, or mentioned the police; this is in direct contrast to the frequent discussion of the formal and informal aspects of CB activities in the media.

Table 9.7: *Profile of CB Communications by Content (Average Concentration of Topics by Category) (Stage Two)*

Category	Average Concentration
1. *Location, arrangement for face-to-face contact, informal meetings*	31.1%
2. *Mention of other CB users, social events, CB network*	22.2
3. *Discussion of equipment & technical aspects of CB*	18.8
4. *Mention of present activity or task*	11.3
5. *Discussion of hobbies, leisure*	4.4
6. *Discussion of weather*	2.2
7. *Mention of formal CB meetings & CB etiquette*	3.3
8. *Request for information, mention of police, other task-oriented communications*	2.8
9. *Miscellaneous & unclassified communications*	3.9
Total	100.0%
Number of communications	205

Table 9.8: *Percent of CB Conversations Containing One or More Mentions of Specified Categories*

Category	% of Conversations with One or More Mentions
1. *Location, arrangements for face-to-face contact, informal meetings*	82.9
2. *Mention of other CB users, social events, CB network*	70.2
3. *Discussion of equipment & technical aspects of CB*	68.3
4. *Mention of present activity or task*	44.4
5. *Discussion of hobbies, leisure*	25.4
6. *Discussion of weather*	16.1
7. *Mention of formal CB meetings & CB etiquette*	24.4
8. *Request for information, mention of police, other task-oriented communications*	13.2
9. *Miscellaneous & unclassified communications*	16.6
Number of communications	205

In Table 9.9 we have shifted the unit of analysis to the mention itself. The 205 conversations contain a total of 1,318 mentions, or an average of 6.4 mentions each. We reclassified the categories into three generic types. "Instrumental" refers to a utilitarian communication mechanism used to achieve purposeful ends that have little to do with the CB community; references to business, police, requests for assistance, advice on traffic conditions, etc., predominate—in other words, they contain any information that could be instrumental in benefiting the "real life" activities of the CB user and achieving some kind of gain. The "intrinsic" classification of messages contains much less information and is devoted to matters within the CB community: CB meetings, hobbies, the location of the participants, etc. Instrinsic mentions, therefore, could be used in the CB user's "real life" leisure activities, but are also directly oriented to developing and/or maintaining a network of social relations. The third classification, "expressive," includes gossip, messages containing little or no direction (e.g., "I am watching TV," "I am doing my homework"), and messages specifically related to CB equipment, other CB users, or the act of using CB itself.

Table 9.9 demonstrates that more than two-thirds of the mentions are classified as expressive and less than two percent of the mentions can be classified as instrumental. We have to conclude that an extremely small fraction of CB message content on the local channel is in any way concerned with the world beyond the CB itself and the flow of conversations over the CB.

The basic conclusions from these data are clear. CB is used to establish and maintain networks of CB users, much like the telephone is used as a social instrument. Although the standard telephone is used for a multiplicity of purposes, a substantial portion of domestic or nonbusiness use involves maintenance of a community or a network of social relations. Similarly, CB appears to be a way of making friends and of subsequently keeping up those freindships. It is a way for people to talk about their work, their possessions their hobbies and their lives.

CITIZENS BAND RADIO AND THE EMERGING MOBILE COMMUNICATING SOCIETY

We have tried to address several of the questions raised earlier regarding the mass popularization of a communication technology rooted in our mobile culture. Having established that in the city studied local use of Citizens Band radio is principally for intrinsic and expressive purposes, and that the CB subculture appears to have many attributes of a community, we will now turn our attention to some implications of these findings. The reader should be reminded again that these implications are derived from a study of limited scope, one not focusing on highway and emergency type patterns of use that predominate on bands other than the "home channel" monitored.

The loosening of bonds within a traditional spatial community results in an increasing number of people really not being tied to specific places. In addition, as a consequence of suburbanization, people are traveling greater distances to and from work. In terms of leisure time activities, the relative inexpensiveness of air travel, the packaged tour phenomenon, the recreational vehicle boom, and the upsurge of interest in camping, nature, and the outdoors have all contributed to the mobility of Americans, which separates people from specific places. This reduction of fixed "roots" has increased the disjunction between place and activity, social and spatial organization.[26]

But we have not conquered distance through travel alone, for telecommunications has come to play an equal, if not greater, role in the creation of a "global village." Telecommunications allows us to cover distance without having to travel; the telephone, radio, television, and satellite communications bring the world to our homes. At the same time, these technologies facilitate the interconnection of widespread business activities and organizations into networks.

Because of its technical and operational specifications and its patterns of use, Citizens Band radio appears to be qualitatively different from other mobile and nonmobile communication technologies (see Table 9.10). Unlike one-way paging

Table 9.9: *Number of Mentions by Type (Stage Two)*

Mention Type	Frequency	Percent
Instrumental	24	1.8
Intrinsic	386	29.3
Expressive	908	68.9
Total	1,318	100.0

Table 9.10: *Communication Technologies by Technical Configuration and Predominant Usage Pattern*

	Television	Radio	Citizens Band	Radio Dispatch	Paging	Fixed Telephone	900-MHz Cellular Systems
Number of parties who can participate in communications	Many	Many	Many	Few	Two	Two*	Two
Predominant use	Entertainment/passive listening	Entertainment/passive listening	Entertainment/passive listening/socializing/instrumental	Instrumental	Instrumental	Entertainment/socializing/instrumental	Entertainment/socializing/instrumental
Range of transmission	National with worldwide linkages	National with worldwide linkages	Five- to ten-mile radius	Fifteen- to twenty-mile radius	Twenty-mile radius	National and international	National [international set]
Mode of communication	One-way	One-way	Two-way	Two-way	One-way	Two-way	Two-way
Direction of communication	Top-down, hierarchical	Top-down, hierarchical	Horizontal	Primarily hierarchical; also multidirectional	Top-down,** horizontal	Hierarchical and multidirectional	Hierarchical and multidirectional
Possibility for developing new personal contacts	Closed	Closed	Open	Closed	Closed	Open, very limited	Open, very limited
Privacy	NA	NA	None	Limited	Limited	Private*	Private
Intrusiveness	Nonurgent	Nonurgent	Nonurgent	Urgent	Urgent	Urgent	Urgent
Availability/principal limitations	Virtually unlimited	Virtually unlimited	Possibly unlimited depending on FCC regulation	Limited by FCC regulation	Limited by FCC regulation	Virtually unlimited	Limited by high price and FCC regulation

*Except for party line.

**The primary factor that differentiates this technology from all others is the fact that access to the user's device can be limited simply by not disseminating the access code needed for contact.

devices (which are designed to facilitate scheduling and supervision of field personnel) and two-way dispatch radio (which serves the same functions but is limited to short, instrumental business messages), CB can encompass a wide variety of personal and social messages because of its technical form and the regulations governing its use. CB is characterized by the capacity to both send and receive messages in an open medium rooted in a community. The properties of the medium—two-way and nonprivate, open, nonurgent, nonintrusive—facilitate community formation by providing opportunities for people to listen to CB conversations and decide whether they want to get acquainted. Because of its anonymity, CB provides a riskless context within which people can form friendships. It is as if one were invisible at a cocktail party.

As presently constituted, the effective range (five to ten miles), power, and other technical specifications constrain CB to return at least a portion of the communications complex back to the local community, to the more physically proximate. CB represents a contraction from the scale and complexity of McLuhan's "global village" to the local community.

Citizens Band radio is both a broadcast and directed communication medium in the sense that it allows passive listening, like the radio, and open person-to-person conversations, like the telephone. The two-way feedback capacity of CB allows participants not merely to acknowledge a message, as in paging, but also to transmit subsequent reactions.

At the same time, Citizens Band radio increases the variety and diversity of communications and information available to people and thus offers distinct alternatives to mass communication technologies. Analysis of the content of CB communications in this study reveals that most conversations are intrinsic and expressive, with little "useful" information for the accomplishment of tasks. Rather, the typical CB communication is an end in itself. It is communication about communication, covering such topics as the comparative attributes of CB equipment, the quality of reception, the activities of fellow CB users, and arrangements for face-to-face meetings. In this way CB can be seen as a kind of wireless party line and, as such, is a medium through which communities can be established.

SOME IMPLICATIONS OF CB FOR THE FUTURE OF MOBILE COMMUNICATIONS, ESPECIALLY 900-MHz CELLULAR

Technologies do not develop in isolation from social and cultural forces. The forms that they take are dependent on the structure, values, and conditions of society. In the case of CB and some other mobile technologies, the increased pace, complexity, transcience, and anonymity of American life stand out as crucial social structural factors influencing their use and impacts. The familiar concept of "cultural lag" has its counterpart in "technological lag," or the development of technologies that both reflect and foster the expression of new social forms and cultural styles. Citizens Band radio appears to be an example of "technological lag," facilitating the desire of people to communicate while on the move, to extend their social bonds and form nonpropinquitous communities within our highly mobile culture.

In spite of the tremendous upsurge in personal uses of mobile communications, it should be realized that the diffusion of CB has extended it to only approximately 15 percent of the population. But the potential for utilizing CB or other more sophisticated communication technologies exists in every instance of mobility. If a higher proportion of this potential were realized, the character of society could be significantly altered. A number of factors suggest that more widespread personal use of mobile communications is likely, with important implications for the future of these technologies and American society.

As discussed earlier, Citizens Band radio differs in its technological configuration and patterns of use from other comparable communication technologies, such as radio, television, two-way mobile dispatch service, and paging systems. One can identify patterns of similarity and dissimilarity between the CB and each of these technologies.

Especially important, however, are some of the striking similarities between CB and certain kinds of telephone use, and these have special significance in evaluating the social implications of Citizens Band radio. Like the telephone, CB conversations often wander from subject to subject; the average CB conversation in our sample covered over six subject categories and lasted approximately 2.5 minutes. The average local home telephone conversation lasts 2.85 minutes.[27] Also, as with local telephone use, there is little effective time or cost restriction on CB communications, unlike that on all other mobile communication technologies. When users are consciously aware of *time* limitations on messages carried over a communications medium, they are more apt to "speak in a concise, to-the-point manner [as in business mobile communications]. Information, strict data exchange, becomes the prevailing mode and silence is no longer a virtue."[28] Our findings that CB communications have overwhelmingly expressive content, on the channel studied, suggest that CB communication promotes a far broader range of communication than purely informational and simplified message exchange.

Both the telephone and Citizens Band radio have two-way capacity and thus are personal, interactive technologies. Each one can also be seen as mitigating social isolation. Just as the telephone is used extensively to create spontaneous therapeutic or altruistic communities—especially in times of crisis or emergency—the CB performs an analogous function in the form of the widely publicized traveler and emergency assistance offered by individual users and organized REACT groups.[29] The telephone serves to link distant people and places;[30] Citizens Band radio serves the same purpose, though its physical range is less. And both reduce peoples' sense of isolation and boredom.

In several important respects, however, Citizens Band radio exhibits important departures from the technological configuration and usage patterns of the telephone. For example, the telephone has "high definition" in that it is characterized by individualism, urgency, intrusiveness, exclusiveness, and privacy in person-to-person communications.[31] These features will be present in the cellular telephone as well, because it will have all the technical characteristics of the fixed telephone, with the added crucial ingredient of mobility. Furthermore, the telephone creates "instant tension." It can stimulate a desire to pick up the receiver and talk to another party; it is difficult for most people to ignore its ringing. The cellular system telephone will have similar characteristics; in addition, because of its expected high price, peoples' use of it may be characterized by "use imperative"; that is, they may feel obligated to use their mobile telephone in order to justify its expense.

Citizens Band radio, on the other hand, tends to be communal, nonurgent, nonintrusive, and nonprivate. There is a strong element of volunteerism associated with this technology in the sense that one *chooses* to listen to it, or "sandbag." CB does not create "instant tension" because it does not ring. More crucial, perhaps, is the fact that CB is not a private communications medium like the telephone. Thus it does not appear that the proposed cellular system will satisfy the need being expressed by people using CB for an open, nonprivate, party line type of mobile service.

Except in one respect, the implication of the CB phenomenon for more sophisticated systems is not simple, straightforward, and unidimensional. The one exception is that the popularity of CB has demonstrated a widespread desire for communication on the move. However, more detailed and specific implications for the 900-MHz cellular system are less clear.

CB service will both compete with and stimulate deployment of the cellular system, and it is bound to remain significantly competitive because it provides a cheap and partial alternative to some cellular services. No matter how far we look ahead—twenty, thirty, or forty years—CB will continue to be an accessible, inexpensive, although not highly effective, communication system. However, the quality of CB communications may improve through technical advances and more supportive public regulatory policies. For the person who purchases a CB, there are important secondary benefits, such as participation in an entertainment hobby and a culture, that the cellular system will not provide because of cost and technological configuration. The cellular system is designed for directed private communications, while the CB fundamentally provides for a very large party line.

We can also expect that CB will stimulate demand for cellular services. Some people are using CB for instrumental and business purposes, and an even larger number are apparently employing it for the kinds of social functions that an ordinary telephone permits. It appears, then, that there might be a substantial market of people who would prefer a *mobile telephone* if they could afford it. Even though they may presently represent a small fraction of CB users, their total numbers are large. The history of technological and communications development in this country provides many examples of people steadily upgrading the quality of their equipment. It is reasonable to speculate that substantial numbers of people will enter the mobile market through CB and, partly because of dissatisfaction with the service, move up to a more complicated system.

The major public policy issue that has to be addressed is the need for development of appropriate public regulatory policies that recognize that different mobile communication technologies are not simply substitutes for each other. Public regulatory policies will have to be developed that provide the appropriate level of resources, such as spectrum and investment capital, to the various systems so that each of them can individually produce the maximum public benefit. The question is not whether we should have CB, paging systems, or the cellular system, but how we can attain the proper balance and mix of each.

This chapter completes the part of the volume that describes existing uses of mobile communications, and attempts to ascertain the developing trends. We have covered general patterns of use in organizations, specific uses in the transportation and public service delivery sectors, and the extraordinary growth in personal use manifested by the Citizens Band phenomenon. Now we turn to a series of broader issues that are not specific to any single sector. In the following pages we shall assess the scale of benefits expressed in economic terms of more widespread diffusion of mobile communication devices in our society. We shall explore whether broad dispersion of the devices might constitute a health hazard because of the radiation emitted. A number of legal issues will be discussed. Finally, we shall introduce an international dimension by considering developments in Japan. This analysis will provide insight into how another highly industrialized country is dealing with the problems of mobile communication, and will consider whether trade competition between the U.S. and Japan is likely to become a serious question in this area. These implications of expanded use will be covered in the next four chapters.

NOTES

1. Estimate provided by personal communication with members of Personal Radio Planning Group, Federal Communications Commission, Washington, D.C., May 1977.

2. Dean MacCannell, *The Tourist; A New Theory of the Leisure Class* (New York: Schocken Books, 1976), p. 7.

3. Carlos Valle Roberts, "Two-way Radio Communication Systems for Use by the General Public" (Master's thesis, University of Colorado, 1975), pp. 24-25.

4. Ibid., p. 7.

5. Many thousands of truckers were already using CB radios to facilitate emergency assistance, to avoid speed traps and weighing stations, and generally to keep in touch with each other to relieve the boredom on long, over-the-road trips. Most crucial, though, was the role that CB played in helping the truckers coordinate widespread strike activities and in avoiding speed traps being used to enforce the new 55 mph speed limit imposed as a result of the oil embargo. According to the president of the E. F. Johnson Company, a leading manufacturer of CB equipment, "The fuel shortage, the 55 mph speed limit and the trucker's strike were the events that lit the fuse" (Richard E. Horner quoted in Roberts, "Two-way Radio Communication Systems," p. 15). And, according to one journalist, "a lot of those who subsequently rushed to buy CB sets were attracted by what had already attracted many truckers—the glorious prospect of outwitting the Smokey Bears, the state

troopers with the forest ranger Sam Browne hats" (Michael Harwood, "America With Its Ears On," *The New York Times Magazine,* April 25, 1976, pp. 28, 60).

6. Harwood, "America With Its Ears On," p. 66.

7. Roberts, "Two-way Radio Communication Systems," pp. 8-9.

8. Ibid., pp. 11-12.

9. See Talcott Parsons, "The Social Structure of the Family," in *The Family: Its Functions and Destiny,* ed. Ruth N. Anshen (New York: Harper & Row, 1949).

10. See Maurice R. Stein, *The Eclipse of Community: An Interpretation of American Studies* (New York: Harper & Row, 1960).

11. Larry H. Long and Celia G. Boertlein, *The Geographical Mobility of Americans,* Current Population Reports Special Studies, Series P-23, No. 64, U.S. Bureau of the Census (Washington, D.C.: U.S. Government Printing Office, 1976), p. 20.

12. Ibid.

13. See William F. Whyte, *Street Corner Society* (Chicago: University of Chicago Press, 1955), and Warren G. Bennis and Philip E. Slater, *The Temporary Society* (New York: Harper & Row, 1968).

14. See Sam B. Warner, Jr., *Streetcar Suburbs: The Process of Growth in Boston, 1870-1900* (New York: Atheneum, 1969).

15. Melvin M. Webber, "Order in Diversity, Community Without Propinquity," in *City and Space: The Future of Urban Land,* ed. L. Wingo, Jr. (Baltimore: Johns Hopkins University Press, 1963), pp. 23-54.

16. See Herbert J. Gans, "Planning and Social Life: Friendship and Neighbor Relations in Suburban Communities," *Journal of the American Institute of Planners* 27:134-40; and William Michelson, *Man and His Urban Environment: A Sociological Approach* (Reading, Pa.: Addison-Wesley, 1970).

17. See Whyte, *Street Corner Society.*

18. Claire Selltiz, Marie Jahoda, Morton Deutsch, and Stuart W. Cook, *Research Methods in Social Relations* (New York: Holt, Rinehart & Winston, 1951), p. 335.

19. Ibid., p. 336.

20. In order to determine the hours during which communications would be monitored, quota sampling was employed. From stage one of the research, the frequency of directed, completed communications by hour and day was obtained. It was determined that, optimally, 220 conversations (or picks) should be monitored. The hour and day from which each of the communications was to be selected was determined by the following formula:

$$P_2(n) = \frac{t_1(n)}{T_1},$$

where $P_2(n)$ = probability that communications from hour "n" would be monitored during the stage 2 research,

$t_1(n)$ = number of directed, completed communications counted in hour "n" of stage 1,

T_1 = total number of directed, completed communications in stage 1 = $\sum_n t_1(n)$.

21. The figure 11,284 is derived by multiplying the number of sample communications (2,821) by four to adjust for the quarter-hour sampling procedure. The frequencies reported in Table 9.1 can be viewed as a valid sampling frame. It is reasonable to assume that the profile provided by this procedure (in terms of types of communications and distribution throughout the week) would be similar in other cities.

22. On the basis of the stage one results specified in Table 9.1, we derived the stage two sampling and recording scheme described earlier. In stage two we limited our attention to directed, completed communications, since these contained information on local patterns of CB use most helpful for our purposes. Also, in general, the incompleted and broadcast messages contained very little codable information.

23. It should be noted that the city studied did not have a complex network of interstate highways, and this fact may have skewed the results away from a valid representa-

tion of the mobile-to-mobile mode. However, there is no a priori reason why the distribution within the "no information" category should be erroneously skewed with respect to other categories.

24. Roberts, "Two-way Radio Communication Systems," pp. 7-8.

25. The content of CB communications was coded in the following manner: the coder divided the conversations into "mentions," a conversation unit on a particular topic. More than 100 specific topics were available for the coder to choose from. The topics were very specific, so that the coder could differentiate between, for example, a mention of hunting and of fishing. Some conversations only had one or two mentions, while others had twenty.

During analysis we combined relevant categories in order to provide a more understandable pattern, and then took the number of mentions in each "macro" category and divided it by the total number of mentions in each conversation. This yielded a proportion of mentions in a category type. By making the (perhaps unwarranted) assumption that each mention was approximately equal in length or importance, we derived our key dependent variables—the "concentration" of a category type in the communications.

26. Melvin M. Webber, "The Post-City Age," *Daedalus* 97:1095.

27. Telephone interview with an official of the New York Telephone Company, November 18, 1976.

28. David W. Conrath and Gordon B. Thompson, "Communications Technology: A Societal Perspective," *Journal of Communication* 23 (1973):53.

29. Suzanne Keller, "The Telephone in New (and Old) Communities," in *The Social Impact of the Telephone,* ed. Ithiel de Sola Pool (Cambridge, Mass.: MIT Press, 1977), p. 281.

30. Ibid., p. 15.

31. Marshall McLuhan, *Understanding Media* (New York: McGraw-Hill, 1964), pp. 233-40.

Part III
IMPLICATIONS
OF
EXPANDED USE

Chapter 10
QUANTIFICATION OF SOME BENEFITS OF MOBILE COMMUNICATIONS

Erwin A. Blackstone
Harold Ware

In this chapter we shall attempt to quantify some of the benefits of mobile communications that have been discussed in the volume. This attempt at quantitative analysis will be partial and speculative; many of the benefits are not readily expressible in quantitative terms and there are inadequate data to permit a more rigorous analysis. Also, some of the systems we shall discuss have yet to be deployed, so that information based on operating experience is not available.

THE GENERAL ECONOMIC SIGNIFICANCE OF MOBILE COMMUNICATIONS

Mobile communications used within a business firm or governmental unit can substitute for labor and capital;[1] thus, we can regard such communication systems as an "input" to the production process. A production unit employing improved communications can produce the same output with fewer other resources, or it can produce a greater output with the use of the same quantity of other resources. Examples of both situations can be found in earlier chapters.[2]

Our analysis will involve consideration of *private* and *social* costs and benefits. Private costs or benefits are those that the individual economic unit faces by choosing a particular course of action. In a *competitive* market economy such costs or benefits are transmitted via the price mechanism. If all of the costs and benefits are included in the economy's prices, the market can bring about an efficient allocation of resources. But the achievement of economic efficiency by a pricing mechanism is contingent upon several preconditions, one of which is the absence of externalities. For our purposes, an externality can be defined as either a benefit or a cost of an action that does not accrue to the person taking that action (often because external costs and benefits are not reflected in prices). *Social* costs and benefits include private costs and benefits as well as externalities. If externalities are present, optimal choices from the individual's point of view may no longer be optimal for society as a whole.

The numerous applications of mobile communications provide significant benefits through a unit's ability to respond more rapidly to a request and by potentially servicing the same number of customers with fewer vehicles. These are essentially private benefits in that they accrue primarily to those firms or other productive units employing the devices, with the consequence that resources are freed for use elsewhere. More widespread deployment of mobile communications will also involve social benefits and costs, which accrue to others besides the primary users. These benefits will be derived from the potential for a faster response time, especially in emergency situations such as accidents, crimes, fires, or illness. Increased use of mobile communications in the transportation sector could reduce energy consumption and resulting dependence upon foreign sources of supply; in addition, substituting communication for transportation is likely to be of growing importance.[3]

Mobile communications should have special significance within the large and expanding service sector of the economy where productivity gains have lagged behind those of the manufacturing sector.[4] Even small productivity increases in this sector could yield big net economic gains. As mobile communications become a more prevalent consumer item, the resulting costs and benefits could go beyond those that can be included in the concept of an "input" into the production process. The recent CB phenomenon indicates the many externalities that can arise if mobile communications become widely deployed; the externalities of the advanced systems under consideration may be substantial.

Several difficulties arise when analyzing the benefits of mobile communications. In addition to the frequent absence of reliable data, two other problems of a more conceptual character have presented themselves. First, communication is primarily a means to an end; in the language of economics, it is an "input" into other processes. Thus, the economic benefits are the result of its influence and leverage on these other processes. Our understanding of communications' economic effect is too primitive to precisely estimate the magnitude of this influence. We can discuss the scale, in economic terms, of those activities in the commercial and public sectors whose efficiency is likely to be increased through mobile communication, but the magnitude of the "leverage coefficient" must be a matter for speculation. Therefore, we shall present our data in such a way that readers may recompute the potential benefits based on their own estimates of the probable influence. Second, the benefits of mobile communications are associated with the *functions* performed, while the costs are specific to the technical *system* that is selected. In the end, private and public organizations— after studying the functions to be performed—will have to decide which system to purchase. This consideration, together with the FCC's decision to allocate spectrum in this area on the basis of systems, has led us ultimately to center the discussion of economic benefits around particular *systems.* This will facilitate the comparison of costs and benefits. Such a strategy for analyzing the benefits is not entirely straightforward, however, because various systems share common traits as well as possess specific properties. Obviously, the problem would be eliminated if each system performed a unique set of functions, but this is not the case.

In order to associate economic benefits with particular technological systems, it is important to know how suitable various systems are for providing specific functions and services. In discussing this topic, we shall recapitulate and summarize some of the relevant material contained in previous chapters.

THE SUITABILITY OF TECHNICAL SYSTEMS

ONE-WAY PAGING SYSTEMS.   Simple paging can be most efficiently accomplished with one-way devices. The devices have the advantage of being fully portable, but the simplicity of the technology involved limits use for other functions.

One-way paging systems are only partially suited for dynamic routing, and they are not suitable for police, fire, or ambulance service where prompt confirmation of receipt and, in some cases, explanation of the message are necessary. Effective taxi and dial-a-ride bus service also require verification. Thus, one-way systems would not be of great benefit to such commercial users, although they are sufficient for reducing the response time of many who have ready access to telephones.

While vehicle monitoring and emergency beacon service can be achieved with one-way systems, the systems would have to send signals rather than receive them; current one-way paging systems are not suitable for these functions. Finally, one-way paging systems have too limited a capability to be useful for general dispatch functions.

CONVENTIONAL AND TRUNKED SYSTEMS.   Despite the differing costs of conventional and trunked systems, we can consider these systems together in an analysis of benefits since they possess very similar capabilities. For example, conventional and trunked systems are well suited to dispatch service and also to paging, if a field operative is usually in a vehicle. Because of their size and weight, they are currently less convenient for use outside of a vehicle. With the availability of new spectrum and with technological advances leading to miniaturization, portable conventional and trunked devices may eventually be used more for paging, especially where short range is adequate and phones are not accessible.

Conventional trunked systems may be used for vehicle monitoring, but in their present form this can only be accomplished manually using verbal inquiry and response. Such a procedure has the disadvantage of tying up channels and reducing the number of units that can be served. Similarly, data transmission would be limited on these systems; trunked systems are better suited for this purpose, since several alternate channels exist for transmitting the data. The use of printing units that employ high-speed digital transmission would improve the suitability of both systems.

It should be stressed that transmission of long messages, such as those required for some forms of data transmission and for mobile telephone service, would reduce the number of users that can be served if waiting times and blocking probabilities are kept constant. The fact that current mobile telephone services provided by conventional and trunked systems are frequently characterized by relatively long waiting times to obtain a free channel, poor transmission quality, limited range, and lack of privacy influenced the FCC's decision to allow only cellular systems to offer mobile telephone service in the emerging 900-MHz band.[5]

If a noncellular (i.e., conventional or trunked) telephone system were authorized in the 900-MHz band, it would suffer from the following deficiencies: (1) lack of privacy—the system essentially involves party lines, though trunked systems are preferable to conventional systems in this regard (see Chapter 2); (2) short range—because of propagation characteristics at 900 MHz, a noncellular system would require numerous repeaters to cover a large market;[6] (3) limited access—since noncellular systems possess no automatic location capability, a call to a mobile unit would have to be prearranged if the unit were outside its normal service area; (4) inefficient spectrum use—this could indirectly result in either long waiting times or a high cost system with limited numbers of subscribers.

In short, conventional and trunked systems are best suited to dispatch uses involving short and frequent messages, such as dynamic routing. They can also be used in a limited manner for vehicle monitoring, data transmission, emergency beacon service, and for reporting breakdowns and accidents. However, these systems would be of little value for emergency beacon use in the event of hijacking, because they require verbal communication that would easily be detected by the perpetrators. Dynamic routing has long been accomplished with noncellular dispatch service, but where long range is desirable or spectrum congestion is a problem, the cellular system is a superior alternative.

THE CELLULAR SYSTEM.   Currently, all two-way systems are both over and under qualified for paging service. They are overqualified since paging involves little more than notifying a mobile person that a message exists. They are underqualified because they are not yet sufficiently portable to be kept with the recipient at all times. In addition, at 900 MHz, conventional and trunked systems are ill-suited to paging because they may not be interconnected with the wireline telephone system. On the other hand, the proposed, fully interconnected cellular system will ultimately provide an alternative superior to the current method of one-way paging plus wireline response.

The cellular system can be used for dynamic routing. AT&T has proposed a dispatch system based on cellular technology that is compatible with its mobile telephone system. The major difference between cellular dispatch and mobile telephone systems is that the former will be designed for shorter messages. Despite the consequent reduction in costs, however, cellular dispatch service may not be as economically attractive to many users as noncellular service.[7] Nevertheless, in markets where spectral congestion is a problem (such as urban and suburban markets where the bulk of the nation's commerce occurs and where the service sector has been experiencing rapid growth) or long range is desired, the cellular system's superior capabilities could outweigh the cost differential. Moreover, as our cost discussion in Chapter 4 indicates, a cellular system providing only dispatch may have the same or even lower average costs than trunked or conventional systems.

The nonemergency dispatch applications of the cellular system are not especially handicapped by its inability to provide efficient fleet call. In fact, in many commercial applications fleet call may actually be detrimental, since constantly listening to radio chatter can be irritating and distracting to the driver. Where fleet call is essential, however, such as in police and fire service, conventional or trunked systems are better suited. Although ambulance service may not require fleet call, as in other emergency services it is likely that a limited access system would be preferable to dependence on a system shared with the general public.

Besides the potentially unlimited range of the cellular system, an additional benefit of cellular dispatch systems for long-distance trucking may result from the possibility of vehicle monitoring. The extent of this use will depend on the type of locating system employed by cellular operators. A triangulation system would provide more precise location information than one based solely on signal strength. The benefits of opting for a more accurate system will be discussed later.

Also to be taken up later is the question of whether the cellular system might incorporate a system of emergency beacons. The location capability makes it possible to request emergency help by simply pressing a button on the mobile radio device, a method that is faster than verbal communication and eliminates the danger that a victim will be overheard by a criminal. Since beacon devices need to be activated only in the event of an emergency, the potential for abuse through surveillance is much less than with vehicle monitoring devices.

The cellular system's high spectral efficiency implies that it can be used for transmitting large amounts of data without significantly raising costs. Similar considerations also make this system the most suitable for mobile telephone service. In addition, improved privacy, longer range, ability to provide random access, and a very large subscriber capacity make the cellular system an attractive choice for mobile telephone service.

QUANTITATIVE ASPECTS OF SYSTEM BENEFITS

Having specified the merits of particular systems for performing specific functions, we can now compare the benefits in a more quantitative fashion.

PAGING.   The major benefits obtained from paging derive from its use for dynamic routing. If the person being paged is always close to a public telephone (e.g., within one-tenth mile) or has easy access to a private telephone, the effectiveness of a paging system can approach that of a two-way system. However, in the numerous cases where public telephones are not nearby and access to a private telephone is not easy, the benefits are substantially reduced. Thus, the benefits derivable from paging are only a fraction of the benefits from two-way systems—a fraction that will vary significantly with proximity to landline telephones. In order to quantify the benefits of paging systems, it would be necessary to consider a number of situations with telephone accessibility as a key parameter. We have not carried out such an analysis, because the main focus of our work is on two-way systems. While the benefits obtainable by paging are often smaller than those associated with two-way systems, the low cost of paging systems will render their use highly profitable in many circumstances.

CONVENTIONAL AND TRUNKED.   Conventional and trunked systems are probably most extensively used for dynamic routing; it has been estimated that land mobile radio communications used for this purpose can increase the productivity of some vehicle fleets by roughly 20 percent. For fleets engaged in fixed route and/or prescheduled operations, the efficiency gains would be less, though still significant.[8]

Since sizable savings are known to have been achieved, it might be expected that dispatch service would be widely used in the trucking industry. In fact, its use is significant but limited, because the costs of private systems are extremely sensitive to fleet size. As was pointed out in Chapter 6, approximately 50 percent of the larger Class I and Class II firms use mobile radio, while only five to eight percent of Class III firms—most of which only employ from three to five vehicles—use mobile radio. This situation is probably due to several obstacles faced by small firms, including the high cost per mobile for private systems; difficulty in obtaining an FCC license or in overcoming the transactions costs involved in setting up shared private systems; and relatively expensive common carrier service. Moreover, common carrier dispatch service has not been widely available in the bands below 900 MHz.

Despite the tendency for use of higher frequencies to be more costly, when the newly allocated spectrum becomes available the costs of radio common carrier service should come down for two reasons: excess demand caused by shortage of available channels should be reduced, and supply conditions should change to reflect the efficiencies of the new systems, as well as the economies of scale that may be realized as the industry develops. These factors should make it possible to provide the numerous small-scale truck operators with low cost common carrier service, and consequently to benefit society.

An examination of some relevant statistics indicates the large savings that could result. In 1972, there were approximately 20 million trucks in use; 84 percent, or almost 17 million, were in fleets of five vehicles or less. (This figure of 20 million trucks is the total number of trucks currently used for any purpose in the United States. The number of trucks associated with the "trucking industry," that is, common carriers, is much less, amounting to roughly a quarter of a million.[9] Thus the calculation of benefits in Chapter 6 refers to only a fraction of the potential users and is, moreover, concerned with private profits rather than total savings to society.) We estimate that of the 17 million trucks we are considering, just over half (55 percent)—or about 9.35 million (including those in manufacturing, service, utilities, etc., but excluding those in categories such as personal transportation)—would be able to use low cost common carrier dispatch service.[10]

The potential yearly savings that could be realized by the utilization of dispatch service can be estimated from the following product:

($/year) = ($\Delta$ in productivity) $\times$ (# of trucks) $\times$ (cost/truck/year).

Accepting 0.2 for the change in productivity, and assuming (conservatively) that $20,000 is the yearly cost of operating a truck, one can calculate that for each percentage point of the number of trucks in small fleets adopting the technology, the savings will be slightly less than $700 million per year. If 20 percent (3.4 million) of these trucks become radio equipped in the future, annual gross savings would be on the order of $13.6 billion. Savings *net* of the annual cost of operating mobile dispatch service for the 3.4 million trucks would be $12 billion. (The operating costs are $1.6 billion. See Chapter 4.) In the long run, assuming virtually universal deployment of mobile communication devices, the 55 percent penetration figure (9.35 million trucks) mentioned above might be achieved. In this case, the yearly gross saving will amount to $37 billion. While these figures are highly speculative, they are not unrealistic, though clearly it would take decades to reach the higher adoption levels that we have postulated and to realize the implied benefits.

If we consider appropriate figures for 1980 and beyond, the savings may be significantly larger, because the number of trucks will probably be greater and the operation costs will surely be higher. Given the potential net benefits in this application alone, mobile communications should be a strong contender for a place on any list of national priorities.

Dynamic routing can be used to reduce the response times of emergency vehicles. The benefits include decreasing property losses, saving lives, and/or reducing the severity of injuries. Under current FCC allocation schemes, emergency services are given priority; hence, the direct impact of the new 900-MHz systems may be limited to urbanized areas where spectrum congestion is a problem. Nevertheless, continued research and development directed toward improving the 900-MHz systems will undoubtedly result in improved dispatch services. This is especially true of trunked systems; since they offer superior spectral efficiency (vis-à-vis conventional systems), they will help to ensure that emergency services will continue to have high quality mobile communications. It has also been proposed that a single trunked system could provide efficient and less expensive service for the entire fleet of emergency and nonemergency public service vehicles within many municipalities, and perhaps simplify the coordination of the various services. (Similar conclusions hold for the taxi industry, if such firms could utilize a less costly common carrier system.)

Mobile communications for workers other than those involved in transportation and emergency service will also yield benefits. Let us consider the benefits for professional and other service sector workers who use vehicles. For example, there are approximately 360,000 active physicians in the United States. If each one experiences delays in locating a phone when responding to a page or calling his or her office at predetermined intervals, avoiding these delays with the use of two-way mobile radio could easily result in savings sufficient to justify the cost of proposed 900-MHz systems. Were the average saving only one hour per week per doctor, annual gross savings would be $850 million, assuming that the opportunity cost of a doctor's time is $50 per hour, probably a conservative estimate.

Because of the relatively high opportunity cost of a doctor's time, the savings to other professionals may not be as large. But the total saved by avoiding delay is potentially substantial, especially in light of the many persons involved. We estimate that there are about 13 million service workers (excluding physicians) who might be users of mobile communications (including administrators, foremen, medical personnel, technical workers, etc.).[11] If we assume that using mobile communications results in an accrued savings of only two hours per week per worker, and an opportunity cost of $15 per hour, the annual gross savings will be about $210 million for each percentage point of the number of potential service workers who adopt the technology. Thus, if 3.5 million out of the 13 million (i.e., 25 percent penetration) ultimately use this technology, annual gross savings would reach $5.25 billion.

Alternatively, a rough scale of the benefits from mobile communications can be estimated by examining possible changes in the productivity of its probable users. The service sector of the GNP amounts to more than $600 billion, so that each percentage point increase in productivity will save $6 billion. Consequently, if improved communications could increase the productivity of service workers by five percent (one-quarter the value estimated for parts of the trucking industry), gross annual savings would be on the order of $30 billion. Hence, although our detailed figures represent very crude estimates of the magnitude of the benefits, we must conclude that improving mobile communications is likely to be economically beneficial because it has leverage on very large sectors of the national economy. One would have to assume implausibly low figures for penetration and subsequent leverage to expect overall economic loss. Major savings seem attainable in one or two decades.

CELLULAR SYSTEMS.   The benefits mentioned above can be obtained from any two-way system, but their magnitude depends to some extent on the particular system and service employed. For instance, mobile telephone service should provide greater benefits than dispatch service, because it is more flexible and interconnected with the wireline systems, permitting direct interaction with a large number of people. As a result, more professional and technical workers are likely to use it. Since the cellular system has been chosen to provide more widespread mobile telephone service, it will have a central role in bringing about the potential benefits.

In an analysis of benefits, cellular technology's most important characteristic is its versatility. Cellular systems have the potential to perform, in superior fashion, nearly all of the tasks for which the other systems are employed; in addition, they can perform tasks for which the others cannot be used. Consequently, we shall assume that all of the benefits discussed earlier can be achieved with cellular systems, and that we need only analyze the *additional* benefits that can be obtained solely through the deployment of a cellular system.

One example is the cellular system's capability of providing widespread mobile telephone service (MTS). As discussed in other chapters, MTS is apt to have significant impact on our personal lives, depending on its degree of market penetration. Extensive utilization of MTS could result in greater freedom through increased mobility, facilitate social contact, lower the cost of living by dynamic routing, and increase people's sense of security by having better access to police and other emergency help. On the other hand, increased mobility could also produce feelings of alienation rather than freedom, and personal contact could be displaced by use of the telephone. Furthermore, the locational potential inherent in the cellular MTS could be misused for surveillance. Having a mobile telephone in the car could increase fuel consumption by making the vehicle more desirable to use. And, criminals could use the mobile telephone to avoid arrest. Such impacts are very difficult to discuss in quantitative terms.

The cellular system offers the possibility of greatly multiplying business and professional uses of mobile communications. Such uses will be facilitated and enhanced by the deployment of high-speed, digital mobile printing units that are already in the developmental stage. These units have the added advantage of increasing spectral efficiency by shortening the time required to transmit the information.

It can be argued that by making the information-gathering process of various bureaucratic agencies easier, land mobile communications will create serious problems. There is a potential threat to individual freedom posed by current computerized data gathering, storage, and dissemination systems; mobile communications, which supplements these systems by increasing the speed and number of inputs and outputs, could worsen this problem. A related danger is that we might collect and store more information than we can process in an efficient and beneficial manner. These are important "disbenefits" not readily amenable to quantification. Information "overload" may be a particularly difficult problem in some areas of public service delivery, especially police services.

A cellular system will be of the greatest benefit in applications and markets where consideration of range, spectral efficiency, and interconnection are critical. However, when the need for fleet call is great or the desire to promote competitive services is large, the desirability of utilizing a cellular system for dispatch service is reduced.

AUTOMATIC VEHICLE MONITORING AND
EMERGENCY BEACON SERVICE:
EXPLOITATION OF LOCATIONAL ABILITY

The locational capability inherent in cellular technology may generate a number of beneficial uses. For example, in discussing automatic vehicle monitoring (AVM) systems, Systems Applications, Inc. (SAI) states it is possible that "locational and other types of information needed to improve the efficiency of bus operations may be provided by multi-service, common-user, small-cell land mobile systems."[12] But the locational precision required for a cellular communication system will, in general, be much less than that necessary for effective automatic vehicle monitoring or beaconing, and achieving sufficient accuracy for these purposes will require substantially higher investments than those needed for a communication system. If over time the demand for cellular communication systems increases, however, cell sizes may shrink, thus permitting greater locational accuracy and perhaps reducing the costs of the additional components required for monitoring. The potential benefits of monitoring and beaconing are sufficiently large to warrant substantial attention, even though presently planned communication systems do not incorporate locational methods that are accurate enough for such uses. Moreover, some fraction of the benefits to be discussed are achievable with the cellular system as proposed.

If the benefits from automatic location prove to be much larger than those achievable with nonautomatic methods, society might be best served by deploying an automatic location system as quickly as possible, without waiting for the deployment of a cellular communication system. Subsequent development of the communication system might incorporate the monitoring technology. In some areas, though, it may be a wiser policy to build a combined location-communication system based on cellular technology. The cost of a combined system, which could share location sensors and monitoring equipment, could be lower than the cost of separate systems.

To gain some insight into these possibilities and policy issues, we shall analyze potential benefits that could be derived from automatic vehicle location.

Systems Applications, Inc. has performed extensive analysis on automatic vehicle monitoring.[13] Their work on urban bus service suggests that monitoring could lead to more regular service and reduce by approximately five percent the number of buses required for a given level of service. Significant direct reductions in total energy consumption would result; indirect reductions could also occur from increased utilization of an improved public transportation system. According to SAI, a total expenditure of about $55 million per year would provide a network of AVM systems capable of serving the 44 United States cities that have fleets of over 100 buses. These systems would yield "conservative" average net savings per bus of $500 and a total net savings of $15 million.[14] One can infer from the report that the total number of buses assumed in their calculations is 30,000, and that the cost of the AVM system is $1,818 per year per vehicle. This is substantially higher than the cost per year of MTS, which is

$720 based on an estimate of anticipated common carrier tariffs. Since complete mobile telephone service is not necessary for this use, $720 could be adequate to cover the additional equipment for monitoring, coordination, and display not required for standard cellular MTS. Thus, cellular systems may be able to provide the display and coordination equipment, as well as the normal MTS using the cellular system, for well below the $1,818 per year figure appropriate for an independent system.

Another benefit of locational capability may be the reduction of truck hijackings, a serious problem with significant economic consequences. The value of lost goods from hijacked trucks is not accurately known; one estimate places the value at $2 billion in 1972,[15] while a second puts it nearer $1 billion per year.[16] In addition, costs such as claims processing are said to exceed the value of lost merchandise, and insurance premiums are high.[17] It has been estimated that the average hijacking involves goods worth $47,000—far greater than the $4,500 associated with the average bank robbery. With vehicle monitoring, a company could see that one of its trucks is in trouble from information showing unreasonable deviation from prescribed routes or authorized stops. Countermeasures involving the police or security personnel could then be initiated.

Hijacking losses from trucks alone could amount to $3 billion by the 1980s if present trends continue. Automatic vehicle monitoring and/or automatic beaconing (to be discussed below) would save $30 million for each percentage point reduction in losses. If these technologies were to reduce losses by one-third, the direct savings to society would be $1 billion. Such a decrease is not implausible, since a reduction as high as 80 percent has been achieved using a combination of methods based on monitoring and beaconing.[18] The savings in claims processing and insurance premiums would also be large.

The benefits of AVM can be applied to any fleet of vehicles. We have noted that cellular dispatch systems would probably not be employed by emergency organizations. Nevertheless, they could potentially make use of location facilities that are incorporated in cellular systems.[19]

Another system described by SAI that could utilize the facilities of a cellular network is an automatic beacon (AB) system designed to cut response times in emergency situations. It employs a small beacon that when activated can be located by a grid of sensors; fixed equipment then relays the location information to monitors in emergency service agencies. Similar benefits can be achieved with any mobile system capable of transmitting a distress call. Emergency beacon services can be based on two modes—verbal and automatic; semiautomatic "hybrid" devices involving manual activation are also possible.

Response to vehicular accidents provides a good example of possible benefits. (The reader is reminded that in Chapter 8 the relation between reduced response time and *overall* safety service outcome is shown to be complex and uncertain. In the ensuing discussion, we shall restrict ourselves to those particular situations where rapid response is most likely to be beneficial.) In recent years, the number of fatalities associated with automobile accidents has been roughly 50,000 per year.[20] Additionally, SAI estimated that 1.2 million people would suffer at least moderate injuries due to traffic accidents in 1976.[21] Critical injuries involving heart stoppage and loss of breathing must be treated within five minutes of the accident for the victim to have any chance of survival. A large percentage of serious injuries such as hemorrhaging, shock, or extensive burns must be treated within about 40 minutes for the victim to have much chance of survival.[22] These types of injury comprise approximately 3.4 percent of casualties from automobile accidents, or about 40,800 injuries annually. Currently, the median delay for receiving medical attention in urban areas is 24.4 minutes. This total response time includes a median delay of 8.4 minutes before emergency authorities are notified of an accident.[23] With widespread beaconing or two-way mobile communications, the notification delay could probably be substantially reduced, perhaps to one or two minutes. This measure alone would reduce total response time by nearly 30 percent. It must be stressed that significantly lowering the median notification delay would require that a substantial percentage of all vehicles have these devices. Thus, our analysis concerns the long-term benefits ultimately realizable on a national scale with widespread deployment of the technology. (In rural areas, the median notification is probably longer than 8.4 minutes, so an even greater potential benefit from two-way mobile communications exists.) SAI suggests that both fatalities and serious disabilities could be reduced by 18 percent.[24] Other analysts cited by SAI would attribute as much as a 40 percent reduction in fatalities to shortened response times in certain cases. While it is impossible to predict the benefits precisely, it is clear that death and serious injury could be prevented in a certain percentage of cases. Assuming that fatalities and disabilities could be reduced by 18 percent implies that 9,000 lives could be saved and 216,000 (or 18 percent of the 1.2 million who suffer at least moderate injury) disabilities prevented.

Obviously, one cannot express the full value of saving a human life in purely economic terms; the most fundamental effects of lives lost by accident are not quantifiable on any scale, least of all financial. However, for our purposes it is instructive as we assess benefits to compute some economic consequences based on the present discounted value of the income or production lost due to the premature curtailment of an individual's earning capacity. Two hundred thousand dollars is a reasonable figure to use for this purpose; similar and higher magnitudes have been proposed by others.[25] If we associate $200,000 with each fatality and half that figure with each serious disability, then $23.4 billion becomes the direct economic benefit that may be possible through improved response to accidents by beaconing.

Similar benefits can be obtained by shortening the response time in other medical emergencies; heart attacks, the leading cause of premature death, provide a good example. There are approximately one million heart attack victims in the United States annually.[26] Of the 650,000 who die from these attacks, 350,000 die before reaching a hospital. Although another 300,000 die in the hospital, intensive treatment has reduced the mortality rate for those who reach the hospital by 50 percent.[27] As a result, the major recommendation made to improve the treatment of heart attack victims has been to have highly trained technicians and specially equipped mobile units.

The Advanced Coronary Treatment Foundation has estimated that prompt treatment might save as many as 100,000 of the 350,000 persons who now die from heart attacks before reaching a hospital. Others estimate that perhaps 35,000 to 70,000 (or 10 to 20 percent) could be saved.[28] Widespread mobile communications will lead to faster reporting of heart attacks in situations where a wired telephone is not quickly available; unfortunately, it is not clear how many such cases exist. Out of the total number of 35 to 100 thousand lives that could be saved, we shall assume that 10 percent could be saved with beaconing from mobile communications. Even using this very conservative figure of 3,500, the annual economic benefit based on curtailed lifetime earnings amounts to $700 million. Response to other medical emergencies may also be improved and the level of disability may be reduced, but it should be remembered that in all medical emergencies there is the potential problem that decreasing the response time may lead to increasing the number of those who are legally alive but are incapable of life in any real sense. This consideration will lower the net benefits.

A final area in which automatic beaconing will benefit society is in shortening police response times. These response times are composed of delays in detection, reporting, dispatch, and transit.[29] The monitoring of emergency vehicles reduces transit delay, and monitoring vehicles involved in the emergency reduces the detection and reporting delays. Yet, emergency beaconing may cut the detection and reporting delays even further, since the victim would often be aware of his predicament earlier than someone monitoring his status. Some evidence to substantiate this conclusion may be derived from the fact that in cases where a victim activated an alarm, the arrest rate in Los Angeles in 1966 was 60 percent, while the arrest rate for robberies in the United States in 1966 was only about 30 percent.[30]

Improved mobile communications should have an impact on certain crimes, such as robbery, vandalism, arson, drunken driving, homicides, and assaults. The cost of these crimes in 1974 has been estimated as follows: robbery, burglary, theft, shoplifting—$3.0 billion; vandalism, arson—$1.3 billion; homicides, assaults (loss of earnings, medical costs)—$3.0 billion; and drunken driving (wage loss, medical costs, costs to victims, property damage)—$6.5 billion.[31] Thus, the total cost of these crimes was $13.8 billion. A conservative assumption, considering the arrest statistics for on-site alarms, that 10 percent of the losses from these crimes may ultimately be avoided by widespread utilization of mobile communications for beaconing would place the annual savings at approximately $1.4 billion. The above figures do not take into account the costs of law enforcement, which totaled $20.6 billion in 1974.[32] Although it can be argued that law enforcement costs will increase because more criminals will be apprehended and adjudicated, the greater likelihood of apprehension and incarceration may well have a deterrent effect that could reduce the cost of administering the criminal justice system. Measured in economic terms, the net effect of these two factors is uncertain; but, when measured in broader, more human terms, it is likely to be beneficial.

The potential benefits from automatic beaconing and vehicle monitoring depend on the accuracy with which such systems can locate a mobile unit. As mentioned earlier, it may be technically possible, but more costly, to modify a cellular system to monitor vehicles to within the necessary tolerance for automatic beaconing and monitoring. A public policy issue centers on whether the net benefits of these functions warrant the anticipated extra cost of more precise location capability. As we have noted earlier, some fraction of the benefits we have discussed in this section will be achieved through widespread mobile communications without the incorporation of precise locational methods and automatic emergency signaling.

SAI estimates that a widely deployed beacon system would entail relatively low costs. A manually operated beacon could be produced, at retail, for $10 per unit; a more sophisticated beacon for vehicles, requiring accelerometers for automatic turn-on, might cost anywhere from $50 in early operation to $25 in large-scale operations.[33] The fixed facilities of a system covering a 20 X 20 mile area are estimated to cost between $750,000 and $1,000,000 for a sensor network, and $150,000 for central processing equipment.[34] The $150,000 figure should be largely independent of the size of the service area, while the cost of a sensor network should vary in proportion to the service area. Thus, for purposes of comparison with the other systems that we have analyzed, whose service area was assumed to be 1,300 square miles, we can scale up the above figures to approximately $2,500,000 to $3,250,000 for the sensor network and $200,000 for the central processor.

Using these figures, it can be seen that the extra cost required to incorporate beacon service into a cellular communication system would be some fraction of $2,700,000 to $3,450,000. Even if we assume the added cost is $2,000,000, or more than half of the higher figure, the impact on the cost per user of improving the cellular system's locational capability to the accuracy necessary for use in AVM and AB systems may not be very great. Specifically, for the 50,000 subscriber dispatch-mobile telephone cellular system, the cost of the extra fixed equipment would be $40 per subscriber—a figure roughly eight percent more than the $485 average fixed cost per user of the cellular communication system.[35] The additional cost of the mobile unit would be an even smaller portion of the total cost per subscriber. Therefore, the development of a cellular system with the location ability necessary for AVM and AB services should receive serious consideration.

The benefits that can be associated with radio locational ability are certainly large. Yet, it is unclear whether these possible benefits will stimulate the large demands for mobile radio services required to realize the potential, because of the "public good" dimension of the benefits in question. In the cases of automatic beaconing and automatic vehicle monitoring, the benefits accruing to a firm that paid for accurate locational equipment would be too small to justify the expenditure unless it could charge those benefiting for the service. Even if a pricing mechanism could be established, an underallocation of resources would occur, since the external effects on society associated with deterrence and reduction of crime would not be taken into account by those who paid for the service.

Moreover, the capital investment for the communication system alone is quite large, so firms may have to absorb substantial losses for a long period while promotional pricing is used to obtain enough subscribers to drive down average costs. Adding the cost of the extra locational equipment would mean either that firms would have to convince customers that the uncertain benefits associated with these novel applications of mobile technology would justify paying higher rates or that greater losses would have to be sustained due to longer periods of promotional pricing. Here again the external benefits might imply that an insufficient amount of resources would be allocated for these services. It is therefore possible that government action may be needed if these benefits are to be translated into an effective demand for mobile communications.

Individuals' desire to increase personal security may contribute to the demand for mobile communication. But in this case also there are significant externalities that are unlikely to be considered by private decision makers, and sufficient resources may not flow into the industry. One implication is that land mobile radio should be subsidized by the government. However, it is possible to interpret the lack of spectrum pricing as a subsidy that may in fact abrogate the desirability of additional government support. Furthermore, without a detailed evaluation of all externalities, including radiation hazards, public nuisance, safety hazards, and privacy problems, it is impossible to make a final determination about the desirability or magnitude of the subsidy.

Even this partial accounting of the economic benefits that could result from the deployment of more sophisticated mobile communication technology suggests the benefits will exceed the systems' direct costs discussed in Chapter 4. In addition, there are many benefits associated with private and organizational uses discussed in earlier chapters that cannot be readily quantified. At the same time, the deployment of these devices will be accompanied by a range of indirect costs and disbenefits—some of which have been briefly mentioned here—and most of these "negative externalities" are also very difficult to quantify. These shall be discussed in the next three chapters, placed in the broader context of potential health hazards caused by radiation, legal issues, and problems of international trade.

NOTES

1. See, for example, M. J. Fogarty, "Closing the Mobility Gap," *Electronics Weekly,* October 9, 1974, pp. 21-22.

2. See also U.S. Advisory Committee for the Land Mobile Radio Services, *Report* (Washington, D.C.: Federal Communications Commission, 1967), pp. 399-409, cited in Harvey Levin, *The Invisible Resource: Use and Regulation of the Radio Spectrum* (Baltimore: Johns Hopkins University Press, 1971), pp. 133-34.

3. M. J. Fogarty, "Closing the Mobility Gap," pp. 21-22.

4. Victor R. Fuchs, *Production and Productivity Gains in the Service Industries* (New York: National Bureau of Economic Research, 1969), pp. 9-10.

5. "Memorandum Opinion and Order," Docket 18262, *Federal Communications Commission Reports,* 2nd series, vol. 51, March 1975, p. 945.

6. In some cases this limitation can be overcome by using remote *receivers* (as opposed to repeaters) to amplify mobile-to-base transmission, but this less costly alternative would not solve mobile-to-mobile range problems. In addition, high gain antennas offer a temporary solution, because the 900-MHz band is not as saturated with "noise" as are lower bands.

7. See Chapter 4.

8. See Chapter 4 and M. J. Fogarty, "Closing the Mobility Gap," p. 21.

9. Based on the average fleet size per Classes I, II, and III firms, as published in Interstate Commerce Commission, "90th Annual Report" (Washington, D.C.: U.S. Government Printing Office, June 1976).

10. U.S. Department of Commerce, Bureau of the Census, "Truck Inventory and Use Survey," *1972 Census of Transportation* (Washington, D.C.: U.S. Government Printing Office), vol. 2, p. 8.

11. U.S. Department of Commerce, Bureau of the Census, *Historical Statistics of the States* (Washington, D.C.: U.S. Government Printing Office, 1975), pp. 140-45.

12. U.S. Department of Commerce, "Land Mobile Communications and Public Policy," prepared by Systems Applications, Inc. (Springfield, Va.: National Technical Information Service, 1972), vol. 1, p. 167.

13. Ibid., p. 163.

14. Ibid.

15. "Crime: A Crushing Burden for Shippers," *Nation's Business,* December 1972, p. 67.

16. U.S. Department of Transportation, Office of Transportation Security, "A Cooperative Approach to Cargo Security in the Trucking Industry," prepared by Executive Services (Washington, D.C.: Department of Transportation, 1973). The smaller figure cited here apparently deals only with the common carrier trucking industry.

17. "Crime: A Crushing Burden," p. 67. For example, in 1970, when the direct cost of cargo losses for shippers (including trucks, trains, and planes) was $1.47 billion, the indirect costs were as high as $7 billion.

18. Ibid.

19. Herbert H. Isaacs, "A Study of Communications, Crimes, and Arrests in a Metropolitan Police Department," in President's Commission on Law Enforcement and the Administration of Justice, *Task Force Report: Science and Technology,* prepared by the Institute for Defense Analyses (Washington, D.C.: U.S. Government Printing Office, 1967), pp. 100-101.

20. Prior to the nationwide 55 mph speed limit, there were more than 50,000 such deaths per year. The reduction in highway speeds has brought the number of fatalities down, however; in 1976, there were 47,100 deaths due to automobile accidents. *New York Times,* March 6, 1977, p. 26.

21. Department of Commerce, "Land Mobile Communications," p. 224.

22. Ibid., p. 220.

23. Ibid., p. 221.

24. Ibid.

25. Bryan C. Conley, "The Value of Human Life in the Demand for Safety," *American Economic Review,* March 1976, p. 54.

26. William F. Renner, "Emergency Medical Service," *Journal of the American Medical Association,* October 14, 1974, p. 253.

27. Ibid., p. 253.

28. Ibid.

29. Department of Commerce, "Land Mobile Communications," p. 201.

30. Arrest rate for Los Angeles is from Isaacs, "A Study of Communications," p. 99, and the U.S. robbery rate is from the Federal Bureau of Investigation, *Uniform Crime Reports for the United States* (Washington, D.C.: U.S. Government Printing Office, 1966), p. 15.

31. "Crime: A High Price Tag that Everybody Pays," *U.S. News and World Report,* December 16, 1974, p. 32.

32. Ibid.

33. Department of Commerce, "Land Mobile Communications," p. 218.

34. Ibid.

35. Bell Laboratories, "High Capacity Mobile Telephone Technical Report," Unpublished document, December 1971, section 4.

Chapter 11
RADIATION EMISSIONS FROM NEW MOBILE COMMUNICATION SYSTEMS

Alfred M. Lee
Beth Baldwin
Jeffrey Frey

It is necessary to examine whether the widespread deployment and use of 900-MHz systems pose problems for health and safety that may require product modification, legislation, or safety rules for their control. One important concern—the potential for new driving hazards—has already been discussed in Chapter 7. A second and more controversial concern is the possibility that deleterious biological effects could be caused by exposure to the radiation emitted from mobile devices. In this chapter we shall analyze the potential radiation hazards.

NONIONIZING RADIATION HAZARDS

All living organisms are continuously bombarded by various kinds of radiation. Mobile communication systems will emit significant quantities of *nonionizing* radiation. This type of radiation, unlike emissions from nuclear armaments or reactors, does not cause electrically charged ions to be formed; its characteristic frequencies lie below those of X rays (see Figure 11.1). Exposure to such radiation may have two possible effects upon biological entities. Biological reactions will occur when a large amount of energy is absorbed by an organism. Joule heating can increase the kinetic or internal vibrational energy of the constituents of a biological entity; if unchecked, such absorption will lead to damage of a thermal nature.[1] Generally, energy fluxes in excess of a hundred milliwatts (mW) per square centimeter are required for thermal damage to human organisms. A second, more controversial biological effect of high-frequency radiation[2] occurs at lower power levels. These more subtle biological reactions materialize at power levels that are too low to cause significant direct

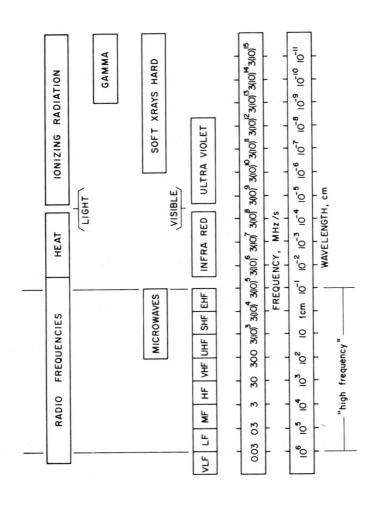

Figure 11.1: *The electromagnetic spectrum*

SOURCE: Adapted from W. W. Mumford, "Some Technical Aspects of Microwave Radiation Hazards," **Proceedings of the IRE 49** (February 1961):427.

heating, and their existence has been subject to considerable dispute. The proponents claim that these reactions are caused by modifying the potential energy of the system's constituents, which give rise to alterations in the electrical or chemical balance.[3] We shall refer to these effects as athermal; the term "athermal" is used instead of "nonthermal" to emphasize the current uncertainty about the cause of low radiation intensity effects.

The possibility of athermal mechanisms has produced major disagreement concerning exposure thresholds of nonionizing radiation (i.e., power density and exposure duration) that will produce harmful biological effects. Abnormal and deleterious effects may be expected if exposure differs qualitatively and/or quantitatively from that which an organism usually encounters in its environment. In order to appraise the possibility of biological effects from operating mobile communication systems, we will first summarize the current state of knowledge concerning nonionizing radiation. Next, we shall discuss current levels of high-frequency radiation in our environment. Finally, we will attempt to ascertain whether widespread mobile communication systems will produce a significant increase in that level and assess whether deleterious biological effects can be expected in light of our present knowledge.

RESEARCH ON BIOLOGICAL EFFECTS OF
NONIONIZING RADIATION

Prior to the middle of this century, very little systematic research in this country had been concerned with the biological effects of nonionizing radiation. It was not until 1956 that a major, joint Army-Navy-Air Force program to determine the bioeffects of this ubiquitous form of radiation was begun in the United States. Findings of this tri-service program were reviewed annually throughout the four years of research.[4] The most significant conclusion was that "convincing evidence for nonthermal effects has not been demonstrated to date." However, several reviewers of the program noted that the protocols were designed, and the research conducted, with a preconceived idea that all effects were of a thermal nature. It has been argued that athermal effects could not have been observed in a majority of cases because the experiments were designed to test only high exposure levels.

Since the tri-service program was conducted, a theme underlying U.S. safety regulations in this area has been that the principal hazards from high-frequency radiation result from thermal reactions. In contrast, research over the last 40 years in the Soviet Union has led to the belief that subtle athermal effects do occur as a result of exposure to power densities significantly below the 10 mW/cm^2 safety guideline recommended by several U.S. agencies (e.g., the Occupational Safety and Health Administration and the Department of Defense). To avoid these low-level effects, Russian and Eastern European exposure safety standards have been set at levels 1,000 times more stringent than their U.S. and Western European counterparts.

American research workers continue to show interest in this field, particularly in the Soviet work. Several factors may account for this. Equipment emitting high-frequency radiation has proliferated in the past thirty years. For example, numerous uses have been found for microwave devices in the manufacturing sector, ranging from drying potato chips to process control. Microwave devices beam long-distance telephone conversations and monitor aircraft flight patterns; they open doors and serve as intrusion alarms. These are only a few current applications and more uses have been proposed for the future.[5] Approximately one-half of the U.S. population now "risks" low-level electromagnetic radiation exposure from microwave ovens while at home or in commercial establishments.[6] According to the Bureau of Radiological Health, the growth of radio-frequency and microwave transmitters is accelerating.[7] The increased power output of this growing number of sources has further stimulated interest in nonionizing radiation research.

In 1968, the Office of Telecommunications Policy (OTP) undertook a study of the bioeffects of high-frequency radiation, reviewing the literature and research projects in the U.S. and abroad. OTP concluded that little was known about the impact of high-frequency radiation upon humans except at high power densities, where adverse and even lethal effects due to heating were known to occur. As a result of this review, the Electromagnetic Radiation Management Advisory Council (ERMAC) was established, composed of nine extragovernmental experts from engineering, physics, biology, and medicine. Its purpose was to assist the director of OTP in assessing both the possible health hazards of radiation and the adequacy of present and future control of high-frequency radiation arising from communication activities.[8]

In December 1971—after a comprehensive review of current information, existing programs, and potential problems—ERMAC proposed a five-year interagency program, coordinated by OTP, aimed at improving the factual basis for future policy in this area. The "primary emphasis of this Program is to determine whether there are effects of exposures at relatively low power density levels, particularly over extended periods of time, and the effects of different electromagnetic conditions."[9] Research is being conducted in "areas where earlier experimentation yielded results suggesting effects, particularly at moderately low power density levels (e.g., on the order of 1-15 mW/cm^2)."[10] Though the research program is still in progress, OTP summarized some preliminary observations in their Fourth Report issued in June 1976:

> Biological effects have been observed and reported at scientific meetings and in the literature at levels around and even below 10 mW/cm^2 (e.g., 1-10 mW/cm^2); findings which evoke considerable interest because of current U.S. safety standard levels: These observations include changes in learned behavior or task performance, electromagnetic radiation perception, apparent alterations in some electrophysically observable nervous system activity, and possible changes in growth and development, some metabolic indicators, and the blood and the hematopoietic system. Such

reports frequently cause apprehension, can be misinterpreted and could lead to premature conclusions as to whether health hazards exist at these levels. Systematic study must be vigorously carried out to confirm such observations and, if confirmed, to determine the conditions under which they occur.[11]

Published American data on human response is almost exclusively derived from clinical studies of occupationally exposed workers—exposures that occurred at various frequencies and are characteristically of low intensity. A recent review of epidemiological studies mentioned three American studies surveying the general health of occupationally exposed radar workers.[12] All three projects were undertaken prior to 1955 and were funded by the Defense Department; they found no evidence of microwave induced pathology. Clinical research conducted outside of the U.S. has had different conclusions, however.

In a majority of Soviet and Eastern European studies of occupational personnel, functional disorders were observed within clinical norms or tolerances; they were reversible if subjects were removed from the exposure environment. The recorded symptoms most commonly associated with prolonged low-level exposure included headache, increased tiredness, dullness, partial loss of memory, decreased sexual ability, irritability, sleepiness and insomnia, emotional instability, sweating, and hypotension. Parasympathetic disturbances of the nervous system included sinus arrhythmias and a tendency toward bradycardia. The most commonly reported objective physiological changes were altered EEGs, increased thresholds for sensory perception, and lengthened latent periods in conditioned reflex reactions. In the circulatory system, alterations were reported in protein fractions, ions, histamine content, hormone and enzyme levels, and immunity factors. Additionally, shifts in leukocyte indices were the most commonly reported change. Such shifts are a common response to various physiological stressors.[13]

No definitive conclusions can be drawn from the conflicting clinical reports. Nevertheless, some scientists feel that the evidence used to set the very low Eastern bloc exposure standards "deserves serious evaluation by U.S. researchers."[14] Collaborative efforts are now in progress to reconcile differences in research results.

Two major concerns of nonhuman experimentation have focused on the possibility of radiation induced damage to the ocular organ and also alterations in human genetic development. Yet conclusions drawn from the results of this research are often ambiguous with respect to bioeffects from low-level, high-frequency radiation exposure. Some researchers attribute teratogenic effects to nonionizing radiation.[15] In one study, VHF radiation has been shown to adversely affect embryonic development.[16] Rugh et al. exposed mice fetuses to 123 mW/cm^2 on the eighth day of gestation and induced gross anomalies; there appeared to be a dependency between dose and effect without evidence of a threshold phenomenon.[17] The opposite conclusion was reached when McRee et al. exposed quail eggs during the first five days of incubation to 14 mW/cm^2 of 2,450-MHz radiation and observed no gross malformations.[18]

In 1965 Sigler et al. reported a higher incidence of Down's syndrome among children whose fathers were previously exposed to radar.[19] They reported that 8.7 percent of the fathers of mongoloid children had had occupational contact with radar, while 3.3 percent of the control fathers of normal children had been thus exposed. However, this small-scale study has questionable statistical significance. The experimental evidence of biological effects of 900-MHz radiation at low intensity is scant and inconclusive regarding human genetics and development, but research is continuing.[20]

Reports of microwave induced cataracts are probably the most publicized of the biological effects from this type of radiation. The lens of the eye is avascular and located at least two millimeters or more from any blood supply; hence, it is ineffective as a heat dissipating structure. In some experimental animals, notably the rabbit, microwave heating at levels of about 100 mW/cm^2 causes protein denaturation and a clouding of the lens, or lens opacity; this reaction is commonly termed a cataract. There are indications, accepted by several investigators, that a cumulative effect of exposures at low intensities may induce cataract formation after a certain latency period.[21] Tengroth et al. report the frequency of significant lens opacities in young radar workers exposed to low-level microwave radiation to be greater than expected.[22]

In a study of the middle ear, Lebovitz points out that vestibular effects caused by microwave radiation could stimulate sensations of angular acceleration or disturbed equilibrium, and hence feelings of nausea.[23] There are also controversial reports of death and nausea in monkeys exposed to electrocardiogram type modulated, low-level, microwave radiation.[24] Again, these results are interesting but inconclusive. Clearly, more research will be necessary to develop definitive answers to unresolved questions of athermal bioeffects of high-frequency radiation.

SAFETY STANDARDS FOR HIGH-FREQUENCY RADIATION EXPOSURE

Determining maximum nonhazardous exposure levels of high-frequency radiation is a complex process, dependent upon experimental design factors that include: size of irradiated body, duration of exposure, type of radiation (i.e., pulsed or continuous), and surrounding environmental conditions. Because experimental results have been inconsistent and have not agreed with clinical observations, a wide variety of safety standards have been proposed over the last 20 years.

Based upon his clinical observations, Hirsch reported in 1952 that a 100 mW/cm^2 exposure level was hazardous. One year later Bell Telephone Laboratories suggested that 0.1 mW/cm^2 would be a safe level. In 1954 General Electric recommended a 1.0 mW/cm^2 average limit. Shortly thereafter Schwan and Li determined that 10 mW/cm^2 of radiation over an hour period could be a

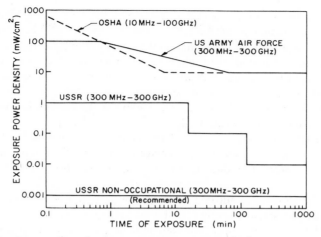

Figure 11.2: *Some selected microwave exposure standards*

SOURCE: Adapted from U.S. Environmental Protection Agency, "An Analysis of Broadcast Radiation Levels in Hawaii," prepared for the Office of Radiation Programs by Richard A. Tell (Washington, D.C.: U.S. Environmental Protection Agency, August 1975), p. 26.

reasonable safety limit, and by 1959 the military and several industrial organizations had adopted the 10 mW/cm^2 level.[25] Yet, in 1972 the Occupational Safety and Health Administration (OSHA) adopted a maximum permissible exposure of 10 mW/cm^2 averaged over any six-minute period.

The standards continue to evolve. Figure 11.2 illustrates several of the organizational exposure standards effective in 1975, with Soviet standards included for comparison. Notice that these selected standards vary with respect to duration of exposure and emission frequency. The Soviet nonoccupational recommended standard is 1 μW/cm^2 over any time interval, but the maximum radiation level permitted in occupational situations is time dependent, ranging from 10 μW/cm^2 up to 1 mW/cm^2. The U.S. Army and Air Force have an exposure standard of 10 mW/cm^2 for periods of more than one hour; exposure levels may be proportionally higher for shorter periods until the 100 mW/cm^2 maximum is exceeded. Nonmilitary American personnel have occupational standards that place no upper limit on allowable level as long as the required shorter exposure periods are observed. For exposure periods of six minutes or more, the American occupational standard is 10 mW/cm^2.

CURRENT LEVELS OF NONIONIZING RADIATION IN THE ENVIRONMENT

Electronic equipment causes ambient levels of *nonionizing* radiation to exceed that of the natural background by several orders of magnitude in some frequency ranges, a situation that differs from the case of *ionizing* radiation. Normally, the emission of ionizing radiation, even from nuclear industrial processes, is at a level far lower than that of the natural background. Those who claim that these emissions are safe frequently point out the small magnitude of this addition to the natural ambient level. Clearly, such an argument cannot be made for the emission of nonionizing radiation.

Almost the entire U.S. population is exposed to low-level, high-frequency radiation from communication devices.[26] A smaller group is continuously exposed to levels that the Soviets feel are occupationally unsafe.[27] To assess the current level of radiation in the environment, we shall rely upon several government-sponsored studies. However, a caveat is worth noting before continuing.

While the devices monitored are numerous and varied, ranging from high-power radar and satellite sources employing very intense and focused beams to a large number of lower power broadcast sources, past environmental studies have excluded particular emitters. Classified military sources of high-frequency radiation are routinely omitted from agency studies (except perhaps DoD studies), despite indications that strategic areas such as Washington, D.C., may have ambient levels two orders of magnitude larger when such sources are taken into account.[28] Further, standard Environmental Protection Agency (EPA) and Bureau of Radiological Health (BRH) environmental monitoring practices exclude many emitters, including mobile radio devices, amateur radio stations, microwave oven installations, and medical diathermy machines, which do not contribute significantly to "far field" radiation, but which can produce relatively intense local electromagnetic radiation power densities.[29] Therefore, the actual ambient level may be somewhat higher than reported. We shall make sample calculations of emissions from typical sources and also cite environmental monitoring measurements reported by the EPA and the BRH.

An estimate of the order of magnitude of the power density (PD) in W/m^2 produced directly by a high-frequency transmitter (i.e., neglecting reflections) can be calculated from the relationship:

$$PD = \frac{P_{rad}}{4\pi R^2}, \tag{1}$$

where P_{rad} is the effective radiated power (in watts) and R is the distance from the source (in meters). We shall use this equation in subsequent calculations.

Powerful signals emitted by communication satellite stations, radio and television broadcasting stations, and two-way radio services have recently been surveyed. The Electromagnetic Compatibility Analysis Center identified the most powerful unclassified, domestic, nonpulsed sources.[30] For the twenty most powerful civilian emitters, the maximum average value of the effective isotropic radiated power ranges from 5 GW to 31.6 GW.[31] Since the original search was conducted in August 1972, an experimental source in California with 3.2 TW of average effective isotropic radiated power has commenced operation. Emissions from these transmitters produce rather large power densities over limited areas when the maximum power is utilized. Obviously, in a satellite communication system, the highly focused beam is normally directed upward. Thus, for these systems a beam's side lobe radiation can be considered the principal concern with respect to irradiation of humans; such side lobe power densities are likely to be 40 to 50 decibels (dB) lower than the main beam measurements.[32] But the extraordinary power levels of these emitters and their potential for emitting hazardous radiation can be illustrated by noting that if there were such a thing as a 3 TW point-source emitter, its radiated power density at full power, based on equation (1), would be on the order of 10 mW/cm^2 at a distance of more than 50 km!

Civilian radar installations that are used predominantly for air traffic control also contribute significant amounts of pulsed microwave radiation to the environment. For example, ambient readings made at Washington's National Airport by the BRH reached 8 μW/cm^2, a power density level very close to the Polish occupational safety standard of 10 μW/cm^2.[33] It is clear that these very powerful radar and communication satellite sources significantly contribute to the ambient level of radiation in their localities.

Although only a small segment of the American population is exposed to pulse microwaves from radar, virtually all are subject to the medium, high, very high, and ultrahigh frequency electromagnetic radiation of radio and television broadcasting. The biological significance of combined high-frequency radiation from several stations is unclear, since it is not known whether the composite effects are synergistic. Nevertheless, some meaningful statements can be made about emissions from single broadcast sources.

The radiated power of radio and television stations is limited by Federal Communications Commission (FCC) regulation. Figure 11.3 is a plot of power density, produced in the main beam, from single sources broadcasting at maximum power as a function of distance from the antenna. Exposure levels in the

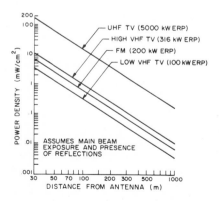

Figure 11.3: *Maximum power density from FM and TV stations*

SOURCE: Richard A. Tell and Dave E. Janes, "Broadcast Radiation: A Second Look," in **Biological Effects of Electromagnetic Waves,** eds. C. C. Johnson and M. L. Shore (Washington, D.C.: U.S. Government Printing Office, December 1976), vol. II, figure 6.

immediate vicinity of the antenna (e.g., at the base or on the tower) can be very high. Other estimates of power density from broadcast sources at varying distances can be obtained by using equation (1).

Numerous broadcast sources concentrated at one location significantly affect density levels of the surrounding area. For example, there are 27 FM and television emitters located at Mt. Wilson, California, serving the Los Angeles metropolitan area. Measurements in the vicinity of the local post office showed that radiation levels were as high as 4.8 mW/cm^2. Typical levels ranged from 0.2 to 2 mW/cm^2 outside the building (0.06 to 0.6 mW/cm^2 inside), while readings taken on the tower were much higher.[34] In addition, measurements taken on rooftops of tall urban buildings, such as in Houston, have shown a power density of 4 to 6 mW/cm^2.[35] though ambient levels are not as large.

Ambient radiation density readings have been made at several sites over a broad frequency range. A study of the Las Vegas area reported a maximum observed level of 0.8 μW/cm^2. Another study found readings of 0.4 μW/cm^2 in Washington, D.C.[36] Measurements made in Boston indicate ambient levels below 2 μW/cm^2, commonly between 0.1 and 0.5 μW/cm^2.[37] In general, "environmental levels and exposure of the general public appear to be quite low (i.e., lower than 10 mW/cm^2, and for the most part generally lower than 1 mW/cm^2)."[38]

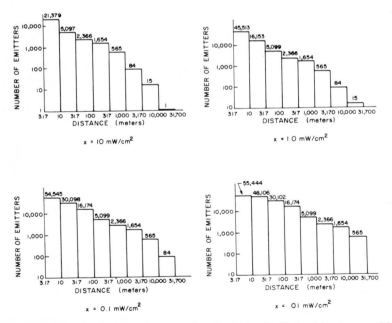

Figure 11.4: *Cumulative distribution of emitters in the United States capable of producing a power density equal to or greater than x mW/cm^2, as a function of distance*

SOURCE: W. D. Rowe, D. E. Janes, and R. A. Tell, "An Assessment of Adverse Health Effects of Telecommunications Technology" (Paper presented at the National Telecommunications Conference, Atlanta, Ga., November 1973), figures 1–4.

The localized power densities emitted from more than 30,000 sources operating across the nation[39] will continue to increase as the number of stations continues to grow. Figure 11.4 illustrates the cumulative distribution of domestic emitters that are capable of producing various power densities as a function of distance. It has been estimated that if a safety standard of 10 μW/cm^2 were adopted in the U.S., over 5,100 FCC authorized emitters would be creating "unsafe" exposure environments within distances of half a mile from the transmitter.[40]

This suggests that some civilian emitters produce relatively large localized power densities besides contributing to the general ambient level. The operation of radar equipment has increased the ambient radiation densities at some airports to levels that are one order of magnitude greater than that at which the Soviets identify functional abnormalities. Numerous radio and television broadcast sources also produce radiation densities of a highly localized nature. The increasing proliferation of electromagnetic devices has caused present radiation levels in certain major U.S. urban areas to be a billion times greater than the highest level that occurs naturally, during sunspot activity. In some special locations, power densities from already existing equipment may be biologically significant.

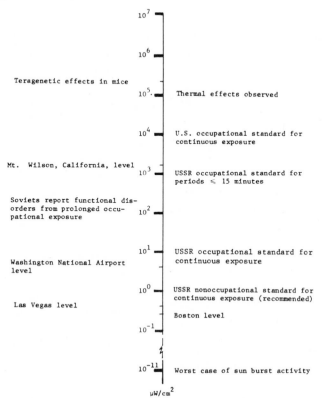

Figure 11.5: *Comparison of various safety standards and ambient radiation levels*

Figure 11.5 summarizes information on various safety standards, ambient radiation densities, and levels at which biological effects have been observed.

RADIATION EMISSIONS FROM MOBILE COMMUNICATION SYSTEMS

A medium or high deployment of 900-MHz mobile communication systems could have two potential effects: all persons within an area serviced by such systems will be exposed to an increased ambient radiation level; in addition, some members of that population, particularly the users of hand-held portable units, might be directly exposed to larger amounts of radiation during transmissions. Radiation from portable units is probably of greater concern, because the users of properly constructed vehicular units will be largely shielded from the outside antenna by the metal structure of the vehicle (barring transmitter leakage). We shall estimate the radiation levels likely to result from both mobile and base equipment of 900-MHz mobile communication systems.

Several simplifying assumptions can be made to facilitate our calculations. If we assume that the antenna of a portable unit will be a simple thin, straight wire, its radiation pattern can be roughly approximated by that of a quarter-wave monopole antenna. In the far field of such an antenna,[41] the power radiated can be expressed as:

$$P_M = \frac{Z_o I_m^2}{8\pi^2 R^2} \; \frac{\cos^2(\frac{\pi}{2}\cos\theta)}{\sin^2\theta} \tag{2}$$

where P_M = power density from a mobile unit (W/m^2),
Z_o = impedance of free space (377 ohms),
R = distance from the antenna,
θ = angle between antenna axis and ground plane ($^\circ$),
I_m = maximum antenna current flow (amp).[42]

Further calculation shows that the total power radiated is equal to

$$0.609 Z_o I_m^2 / 4\pi. \text{[43]}$$

Consequently, the numerical value of I_m^2 (in amp^2) is equal to 0.055 times the total radiated power in watts. If the mobile unit is operated with the antenna perpendicular to the ground (i.e., $\theta = 90°$) and the above condition is assumed, equation (2) can be simplified:[44]

$$P_M = [0.26/R^2] P_{rad}. \tag{3}$$

Equation (3) has been used to construct Figure 11.6, which shows approximate power densities in the far field of a quarter-wavelength long portable unit antenna at 850 MHz for several radiated power values, as a function of distance.

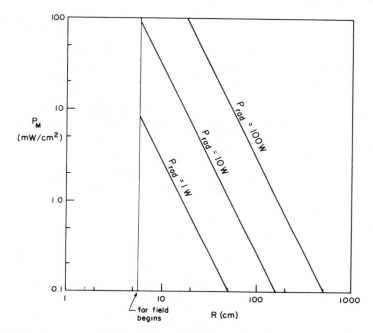

Figure 11.6: *Power densities from 850-MHz portable units as a function of distance from antenna*

Notice that these values decrease very rapidly as the distance from the antenna increases.

Before estimating the emissions from mobile communication systems, it is necessary to develop one more equation. D. O. Reudink has proposed the following procedure for calculating radiation power densities from base station transmitters:[45]

> In order to predict the median power received by a mobile unit from a base station antenna in a basic urban environment, we may use the following equation (all quantities in decibels):

$$P_p = P_o - A_m(f,d) + H_b(h_b,d) + H_m(h_m,f), \qquad (4)$$

where

P_p represents the values of the predicted received power;
P_o is the power received for free-space transmission;
$A_m(f,d)$ is the median attenuation relative to free space in an urban area where the effective base station antenna height, H_m, is 3 meters;
$H_b(h_b,d)$ is the base station height-gain factor expressed in decibels relative to a 200-meter high, base station antenna in an urban area;
$H_m(h_m,f)$ is the vehicle station height-gain factor expressed in decibels relative to a 3-meter high, vehicle station antenna in an urban area.[45]

Correction factors have also been developed to allow for propagation in various types of environment and terrain.[46] To simplify the subsequent calculations, we shall assume a flat urban environment in which all of these factors are zero.

The preceding base station power equation can now be readily applied to our analysis. We shall assume a high deployment urban situation with corner illuminated, hexagonal, tertiary offset cells of one kilometer raddi, with 200-meter base antenna heights to determine an upper bound power density estimate. In this case the median free space attenuation, A_m, will be 20 dB at a distance of one kilometer from a 900-MHz base station.[47] The base station height gain, H_b, will be zero for a 200-meter base antenna height.[48] Finally we assume a six-foot tall user, which will yield a mobile station height gain, H_m, of -2 dB.[49] Therefore, applying these assumptions to equation (4), we find that,

$$P_p = P_o - 20 - 2. \qquad (5)$$

This expression indicates that the free space power density, P_o, is to be multiplied by 0.0063 to obtain corrected estimates for the power density received. If we approximate the transmitting antenna as a point source—see equation (1)—the corrected base station power density equation with implied assumptions is:

$$P_p = 0.0063(\frac{1}{4\pi R^2})P_{rad}. \qquad (6)$$

Using the maximum allowable P_{rad} of one kW, we find that each base station will at most contribute 5×10^{-8} mW/cm^2 at the cell center. The power density contributed by each base station will be very small in this high usage situation.

Estimating the power radiated from hand-held portables is more complicated, requiring an additional assumption. Let us assume that a large number of persons are uniformly distributed within the central region of our 1 km radius hexagonal cell approximated by a rectangle, as shown in Figure 11.7. This number will be sufficient to fully load N channels. Then, the power density throughout the cell, independent of the contribution by portables in surrounding cells, can be approximated (ignoring reflections) by summing individual density values so that,

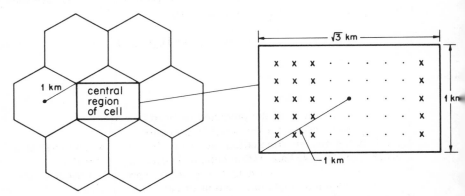

Figure 11.7: *Projected central region of a hexagonal cell*

$$PD_{port} = \sum_{i=1}^{N} [0.26/R_i^2] P_{rad}. \tag{7}$$

For instance, if 100 operating ten-watt transmitters were evenly distributed throughout our 1 km by $\sqrt{3}$ km box, the power density contributed by the portables would be less than 0.3 $\mu W/cm^2$. If the 30 MHz of reserve spectrum were allocated to cellular, in addition to the 40 MHz already granted, more than 150 channels would be available per cell. If all of these channels were in use with operating portables uniformly spaced in the same area, the power density would approach 0.4 $\mu W/cm^2$. These calculated power density levels for both the medium and high user density cases would not be substantially higher even if we were to add the contribution of radiation from portables in immediately surrounding cells.

While it is clear that the portable units could contribute much more radiation to the ambient level than the base stations, we cannot know the actual power density without measurement. Radiation measurements throughout each cell are apt not to be uniform because of local reflections from structures and terrain, and because of unevenly dispersed users. Therefore, our calculations represent only order of magnitude estimates in these medium and high deployment cases.

Irrespective of deployment level, the users of the mobile communications equipment will be exposed to greater power densities over short periods of time. Again, assuming the conditions specified to arrive at equation (3), an approximate value can be estimated. We find that a subject holding a ten-watt portable unit is likely to receive on the order of 2 to 3 mW/cm^2 at a distance of 30 cm from the antenna while transmitting. For current mobile equipment, EPA officials have found the values can exceed 10 mW/cm^2 in small regions very close to walkie-talkie antennas.[50]

By comparing our estimates of radiation densities produced by 900-MHz mobile communication systems to the current level of electromagnetic radiation in the environment, several observations can be made. In an extremely high use situation within a congested urban market, a new system could contribute something less than 1 $\mu W/cm^2$ of radiation in particular parts of cells. This density is roughly equivalent to the emissions from a strong FM radio station at a distance of 2,500 meters. While we cannot determine exactly how much the ambient level will rise in response to a high deployment of mobile communication technology, we can conclude that the new level will be five orders of magnitude less than intensities that cause thermal damage (i.e., 100 mW/cm^2) and well below the safety standard of 10 mW/cm^2. Moreover, this *average* calculated ambient level is below the much more stringent, nonoccupational standard recommended by Soviet authorities. In lower deployment scenarios, radiation levels will be even less.

Additionally, users of portable, hand-held equipment might be exposed to significant amounts of radiation, independent of deployment levels. Though the very localized density may be on the order of a few milliwatts per square centimeter, certain body organs in close proximity to the emitter may nevertheless be affected. The essentially avascular middle ear may be affected by microwave power densities as low as 2 mW/cm^2.[51] However, such dangers might be avoided by modifying the receiver headset so that it is isolated from the unit's antenna. Such product safety measures, should they become necessary, fall under the jurisdiction of the Consumer Product Safety Commission.

Finally, safety procedures will be needed to protect operating and maintenance personnel who are occupationally exposed to radiation hazards. Technicians working on operating antenna towers could already be exposed to power densities above the U.S. safety guidelines.[52] Massive deployment of mobile communication systems could only worsen the problem if no strict safety rules are established.

It is difficult to ascertain whether radiation from new mobile communication systems will produce deleterious biological effects, since controversy still remains over safe exposure thresholds. R. A. Tell of the EPA suggests one interpretation. "At the present stage of evaluation, exposure levels over 10 mW/cm^2 are considered as potentially hazardous and should be reduced."[53] He continues:

> Exposure levels between 1 and 10 mW/cm^2 are considered as significant and such exposure situations should be documented in the event that research indicates that a limiting criteria should be established somewhere in this range. Exposure levels below 1 mW/cm^2 are rather prominent in the environment and at the present time there is no scientific evidence of them being a hazard.[54]

Thus, we conclude that the power radiated by 900-MHz mobile communication systems will not create field intensities that are large enough to be a substantial hazard. Yet, given the major gaps that exist in our understanding of biological effects of nonionizing radiation—and because of unresolved questions associated with athermal reactions—we cannot rule out the possibility that some potential hazards may exist. Environmental monitoring studies and biological research on low-level nonionizing radiation exposure should continue to improve our understanding of the nature of the risks involved. Given the uncertainties and the fact that local radiation levels may depart substantially from average ambient levels, it seems prudent to incorporate some monitoring devices into new systems as they are deployed. Information gained by this means may clarify whether in the long run more extensive monitoring is desirable and may even establish whether any hazard is likely.

Radiation hazards involve only one category of effects that could cause legal problems. Consequently, we shall return to this topic in the next chapter when we consider a wide range of legal issues that could arise if mobile communication devices proliferate in our society on a scale far greater than exists today.

NOTES

1. Stephen F. Cleary, ed., *Biological Effects and Health Implications of Microwave Radiation* (Symposium Proceedings, Richmond, Virginia, September 1969), p. 7.

2. We shall use the term "high frequency" to include those bands below 100 GHz designated as: low frequency, middle frequency, high frequency, very high frequency, ultrahigh frequency, superhigh frequency, and extremely high frequency. It can also be defined as a subset of the nonionizing portion of the spectrum that includes microwaves but excludes heat and visible light. (See Figure 11.1.)

3. Jeffrey Frey and Raymond Bowers, "The Impact of Solid State Microwave Devices: A Preliminary Technology Assessment," *Advances in Electronics and Electron Physics* 38 (1975):187.

4. Paul E. Tyler, ed., "Biological Effects of Nonionizing Radiation," *Annals of the New York Academy of Sciences* 247 (February 1975):7.

5. Frey and Bowers, "The Impact of Solid State," p. 178.

6. U.S. Congress, Senate, Committee on Commerce, *Hearings on Radiation Control for Health and Safety Act of 1968,* 90th Cong., 1st sess., p. 312.

7. Ibid., p. 228.

8. Richard A. Tell, "Broadcast Radiation, How Safe is Safe?", *IEEE Spectrum* 9, no. 8 (August 1972):45.

9. Executive Office of the President, Office of Telecommunications Policy, "Program for Control of Electromagnetic Pollution of the Environment: The Assessment of Biological Hazards of Nonionizing Electromagnetic Radiation" (Washington, D.C.: Office of Telecommunications Policy, June 1976), p. 1.

10. Ibid., p. 3.

11. Ibid., p. 6.

12. Senate, Committee on Commerce, *Hearings on Radiation Control,* p. 154.

13. Cleary, *Biological Effects,* p. 150.

14. Ibid., p. 95.

15. Tyler, "Biological Effects," p. 367.

16. Ibid.

17. World Health Organization, *Biological Effects and Health Hazards of Microwave Radiation,* proceedings of an international symposium, Warsaw, Poland, October 1973 (Warsaw: Polish Medical Publishers, 1974), p. 106.

18. Tyler, "Biological Effects," p. 377.

19. Senate, Committee on Commerce, *Hearings on Radiation Control,* p. 151.

20. Office of Telecommunications Policy, "Program for Control," p. 5.

21. Cleary, *Biological Effects,* p. 80.

22. World Health Organization, *Biological Effects,* pp. 304-5.

23. Tyler, "Biological Effects," p. 182.

24. "Monkey Deaths Denied in RF Bio Tests at Saunders," *Microwaves,* April 1969, p. 18.

25. W. W. Mumford, "Some Technical Aspects of Microwave Radiation Hazards," *Proceedings of the IRE* 49 (February 1961):430-37.

26. Senate, Committee on Commerce, *Hearings on Radiation Control,* p. 312.

27. Ibid., p. 330.

28. Bureau of Radiological Health, Product Testing and Evaluation Branch, *Radio Frequency and Microwave Radiation Levels Resulting from Man-Made Sources in the Washington, D.C. Area* (Washington, D.C.: Bureau of Radiological Health, November 1971), p. 32.

29. Ibid., p. 5.

30. U.S. Environmental Protection Agency, "An Evaluation of Selected Satellite Communication Systems as Sources of Environmental Radiation," Technical Report prepared for the Office of Radiation Programs by Norbert N. Hankin (Washington, D.C.: U.S. Environmental Protection Agency, December 1974), p. 5.

31. Effective isotropic radiated power is a rating based on the hypothetical total power that a focused source would emit if its signal were radiated equally in all directions. Because the radiation emitted by such a source is highly focused, the actual radiated power is very much smaller than these rating figures.

32. Personal communication with D. E. Janes, Environmental Protection Agency, Silver Spring, Md., December 1976.

33. S. Smith and D. Brown, "Nonionizing Radiation Levels in the Washington, D.C. Area," *IEEE Transactions on Electromagnetic Compatability* EMC-15 (February 1973):5.

34. Office of Telecommunications Policy, "Program for Control," p. 30.

35. R. A. Tell and D. E. Janes, "Broadcast Radiation: A Second Look," in *Biological Effects of Electromagnetic Waves,* ed. C. C. Johnson and M. L. Shore (Washington, D.C.: U.S. Government Printing Office, December 1976), vol. II.

36. Ibid.

37. Office of Telecommunications Policy, "Program for Control," p. 28.

38. Ibid., p. 4.

39. U.S. Department of Commerce, *Statistical Abstract of the United States: 1975* (Washington, D.C.: U.S. Government Printing Office, 1975), p. 517.

40. Senate, Committee on Commerce, *Hearings on Radiation Control,* p. 332.

41. The far field is a distance larger than $\lambda/2\pi$; in the case of 850-MHz radiation, for instance, $\lambda/2\pi$ is 5.6 cm.

42. E. C. Jordan and K. G. Balmain, *Electromagnetic Waves and Radiating Systems* (Englewood Cliffs, N.J.: Prentice-Hall, Inc., 1968), p. 329.

43. Ibid., p. 332.

44. This calculation is only a rough approximation of the complicated behavior of a hand-held antenna at some distance above the ground plane; the actual density will be influenced by local reflections and near-field effects.

45. D. O. Reudink, "Large Scale Variations of the Average Signal," in *Microwave Mobile Communications,* ed. W. C. Jakes, Jr. (New York: John Wiley and Sons, 1974), p. 124.

46. Ibid., p. 125.

47. Ibid., p. 102.

48. Ibid., p. 103.

49. Ibid., p. 104.

50. D.E. Janes et al., "Radio-Frequency Radiation Levels in Urban Areas" (Paper presented at USNC/URSI meeting, Amherst, Mass., October 10-15, 1976).

51. Tyler, "Biological Effects," p. 190.

52. U.S. Environmental Protection Agency, "A Measurement of RF Field Intensities in the Immediate Vicinity of an FM Broadcast Station Antenna," Technical Note prepared by R. A. Tell (Washington, D.C.: U.S. Environmental Protection Agency, January 8, 1976), p. 7.

53. U.S. Environmental Protection Agency, "An Analysis of Broadcast Radiation Levels in Hawaii," prepared for the Office of Radiation Programs by R. A. Tell (Washington, D.C.: U.S. Environmental Protection Agency, August 1975), p. 27.

54. Ibid.

Chapter 12
NEW MEANS OF LAND MOBILE COMMUNICATIONS
Some Legal Implications

Kurt L. Hanslowe

INTRODUCTION

There is tension between technology assessment and the analytical technique of the law. The former is speculative and seeks to be prophetically oriented toward general problems of the future. The latter builds heavily on concrete cases and tends to be responsive rather than predictive. This makes legal analysis in the context of a technology assessment a hazardous enterprise. There is the risk of futility in, say, drafting a statute to meet an anticipated problem that may never arise. And as to some problems with potential legal overtones, reflection may lead to a negative conclusion. To some problems, no answer from the law seems necessary; in others no answer is to be found there. The following pages contain instances of such abortive efforts.

A chapter on legal implications will, of necessity, focus not only on problems already familiar to the law. It will also contain discussion of the impact of the assessed technology on the legal system itself. Hence the discussion below contains sections dealing with implications for law enforcement and the matter of access to counsel. Withal, we have sought to focus attention on those situations suggested in previous chapters where the legal implications were both plainly visible and qualitatively most significant. Thus, we have explored legal implications involving the provision of medical or emergency care through the agency of the new technology. Aspects of accidents and safety are also important.

New communication technologies have the potential to facilitate and to deter criminal activity. We have therefore thought it desirable to speculate about potential use by criminals and attempted to analyze new patterns of crime that may result from such use. While we will not be able to come to any definite conclusions, we also wish to at least touch upon the question of the advantages and disadvantages of relying on electronic devices as a means of discouraging street crimes, especially personal assaults.

No technology assessment of new electronic means of communication can be complete without examining the implications for privacy. In this case it is necessary to think about not only the question of surveillance of individuals, but also the question of the invasion of solitude and the more general "right to be left alone."

In an even more speculative vein, we have considered some legal implications of a society that becomes increasingly mobile and more transient in its living arrangements through the combination of new modes of physical transportation and improved communication systems. It should be recognized that, in the long run, these developments could have substantial effect on the role of residence and domicile as legal concepts.

ACCIDENTS AND SAFETY

The safety implications of a new technology are paramount in any social and legal assessment. Safety problems attendant to the use of the mobile communication devices are: radiation damage. resulting from concentrated electromagnetic radiation at close range (see Chapter 11), and the greater risk of automobile accidents when the device is used in a moving vehicle (see Chapter 7).

Problems of physical safety are, of course, not unfamiliar to the law. Indeed, when courts undertake the assessment of damages for harm caused by the negligent construction of a device or the abatement of a polluting nuisance, they are involved in a primitive kind of technology assessment. Yet the legal liability mechanism, when applied to new technologies, is often inadequate in its present form. The existence of hazards and injuries may not be known for years. Injuries are often of a nonspecific nature, for which it becomes difficult to show causation adequate for "legal" purposes. In addition, even if a causal relationship *is* shown, under a test of negligence, the proponent of the technology may be shown to be nonnegligent and therefore not liable.

Corrective measures are hard to come by. Courts act retrospectively and responsively, and legislative action usually does not come until a sizable social or economic problem exists. Private litigation may also be too episodic and unpredictable to exert a timely and continuous influence.[1] Courts are usually lacking in expertise.[2] The typical lawsuit results in an award of damages to the plaintiff or an order directing the defendant to alleviate the problem.[3] At best, civil liability as a method of internalizing the secondary costs of a particular technology may be used to protect individuals from private damage, but is insufficient to afford prospective protection against technological mistakes.

Let us assume that a plaintiff alleges that prolonged exposure to microwave radiation from a mobile communication device has caused a severe impairment of vision. The law of torts is largely based on sudden injuries from single sources. Yet many deleterious effects of modern technology are not caused in this manner. Case law focuses on the unique circumstances of the parties. There is a balancing of the utilities of the defendant's conduct against the hazards that it creates.[4] In such an analysis, expensive large-scale technology has an inherent advantage over the damage done to the individual. Another problem is that the plaintiff must usually bear the burden of litigation, and he will in most instances have less technical expertise than the promoter of the technology.

One avenue of relief for the plaintiff would be to shift the burden of proof to the defendant, with the plaintiff being merely required to make out a prima facie case. The related problem, of having more than one defendant, must also be solved. There may be numerous purveyors of microwave radiation. The plaintiff, however, will often be unable to identify which of the potential defendants caused his injury. Several cases[5] and the *Restatement (Second) of Torts* § 433(B) (3) place the burden on multiple defendants, so that each must prove the absence of wrongful conduct on his own part.

Another approach would be to apply a theory of strict products liability.[6] The manufacturer is liable for a defective product regardless of negligence or fault on his part. Dispute has, however, arisen concerning what is a defective product. Section 402(A) of the *Restatement (Second) of Torts* defines a product for which the manufacturer is strictly liable as a "product in a defective condition unreasonably dangerous to the user or consumer." But what is a defect? What is unreasonably dangerous? Defect ordinarily refers to the situation that exists when materials or workmanship of particular articles fall below levels of quality justifiably expected by their purchasers—for example, the failure of an altimeter to register the correct altitude. Comment (i) to the *Restatement (Second) of Torts* § 402A (1965) states that "many products cannot possibly be made entirely safe for all consumption." As Milton Katz points out,[7] much depends on the meaning of the word *cannot*. Should "cannot" be measured by the current operational capacity of an industry, by the newest scientific and technical knowledge already available but not yet generally applied, or by the potentialities of new research? Courts have been willing to set the standard on the basis of the available technology known to experts rather than the technology currently used by manufacturers.[8]

The doctrines of "proximate cause" and "scope of the risk" are applied in cases involving huge damages to numerous plaintiffs, so as to limit defendants' liability by using the criterion of "foreseeability." In strict products liability the proximate cause doctrine interconnects with the questions of who is the ultimate user or consumer, and whether and how far beyond such users strict liability should run. If our plaintiff was not a user of the mobile device himself, one would have to argue for an extension of strict liability.[9] We may therefore conclude that utilization of the strict products liability doctrine, along with its expansion to cover third parties, may greatly increase a plaintiff's chances of recovery. If this approach is not followed, shifting the burden of proof to the defendant, as the party with the greater resources and technical expertise, would also serve to alleviate the disadvantaged position of such a plaintiff.

In addition to possible radiation hazards, heed will need to be paid to the risk of automobile accidents due to use of communication devices in a moving vehicle. It is obvious that any car driver who is conducting a conversation by such means is necessarily diverting some of his attention from his driving. The extent of that diversion—and thus the risk of accident—depends on several factors: the complexity and emotional level of the conversation; whether the user attempts to refer to visual aids; the physical location of the mouthpiece and earphone; and the physical location of the dialing mechanism. Yet while all of these factors will affect safety, only the latter two can effectively be controlled by law, short of prohibiting using the device while the vehicle is in motion. Connecticut has, in fact, since 1949, required civilian motorists to pull off to the side of the road when using a mobile telephone.[10] But doing this will also create problems. Motorists are unlikely to obey such a law; and even if they do, their stationary vehicles can themselves become traffic hazards. Moreover, such a law would negate much of the convenience of the mobile telephone and would thereby make its ownership less attractive.

Hence, reducing the likelihood of accidents will have to depend on placing the device in a sensible position and on requiring additional driving aids.[11]

MEDICO-LEGAL IMPLICATIONS

Mobile communication devices such as the mobile telephone will enhance medical service in several ways. (1) A physician will be able to consult and treat patients while commuting to his office. This will effectively expand the physician's working day and increase the number of patients he can treat. The physician would, of course, be able to consult with specialists at all levels, while either or both are mobile. (2) Ambulances with sophisticated monitoring equipment could permit emergency ward physicians to read the vital signs of an incoming patient, and direct a time-saving course of treatment to be initiated immediately upon the patient's admission. (3) Citizens with mobile telephones will be able to call for medical aid from the scene of an accident, just as they could report suspicious activity to the police. Moreover, they could administer limited aid to the victim under the direction of a physician in the hospital until an ambulance arrived. (4) Miniature monitors could be implanted in or attached to a patient so that he could pursue his normal activities while still being under medical supervision.

The legal problems arising from the above concern medical malpractice, and they already exist to some degree. Physicians presently consult over the telephone, and deal with their patients that way as well. When a physician uses his phone, however, his entire attention is normally focused on the discussion and, seated at his desk, he is able to examine charts and write down notes. But if he were driving, he would not be able to devote such complete attention and would obviously not be able to examine visual material. Thus he could discuss only those cases that would not need graphic aids, and his failure to resort to such aids when he reasonably should have done so might result in a new form of liability.

The question of establishing some negligence standard for physicians is the subject of much discussion because of the staggering rise in malpractice insurance rates and, indeed, the inability of some physicians to get insurance at all.[12] Whatever standard is used in the future, it is logical to assume that it will apply to negligent acts performed while using a mobile telephone. An argument can be made, though, for applying a relatively low standard for such acts, lest physicians adamantly refuse to use the device with the consequence that many of the medical gains prophesied for the device will be neutralized.

The problem is compounded by increased resort to paraprofessional medical personnel made possible by use of mobile communications. At present only some 30 states have statutes outlining the legal restrictions placed on a physician's assistant, and they vary significantly in what they permit and prohibit.

The law of agency may also pose problems. A physician will likely stand in a relationship of principal to any agent following his instructions by phone and will thus be liable for any actions that his agent has authority to take.[13] The physician may, in addition, be considered a master in the legal sense, that is, one who has the right to control in detail the performance of a service by another. If this is so, the physician will also be held liable for the negligent and tortious acts of the agent, even if done without authorization.[14]

A passerby who reports a medical emergency and then administers aid to the victim pursuant to a physician's telephoned directions might also be liable for tort damages. A person normally has no duty to aid someone in distress unless he tortiously caused the situation, but once the rescuer has begun to help, he has a duty to act reasonably and not exacerbate the injury. Indeed, if something goes wrong, he could be subject to a malpractice suit for practicing medicine without a license. In order to encourage passerby to aid their fellows, perhaps a "good Samaritan" exception to the licensing statute and a lower standard of care would be in order for volunteers who act under the directions of a doctor at the other end of a mobile communication device. Such an approach might also be desirable in connection with paraprofessionals acting under a physician's direction via the communication device.

ACCESS TO COUNSEL

The Sixth Amendment assures persons accused of a crime the right to have the assistance of counsel in making a defense. The question has arisen *when* in the criminal process this right to counsel attaches. In the past, lawyers were often prevented from seeing their client until time of trial.[15] The Supreme Court, in a historic series of cases, has enlarged the constitutional right to counsel.[16] In 1963, the Court, in *Escobedo* v. *Illinois,*[17] held that a criminal suspect was entitled to legal counsel during the course of an accusatory police interrogation.

Should devices such as a mobile telephone achieve a high degree of market penetration, it is not improbable that persons who are arrested will have one. If not, the patrol car may be so equipped. In either case, the accused, if permitted, could make instant contact with counsel. It is not clear how much police interrogation of an accusatory nature occurs *immediately* upon arrest. If the suspect were willing to talk to the arresting officers immediately upon being informed of his rights, that evidence might be highly informative, especially when following closely upon the alleged commission of the crime. But the sooner contact is made with counsel, the sooner the suspect will be told to remain silent. Thus, resort to the mobile telephone to contact counsel could significantly affect the police's investigatory task. Analysis comparable to that which has occurred in connection with the "stop and frisk" problem may be required.[18] In that situation, the constitutional protection of persons stopped by the police against unreasonable searches has had to be placed within the context of the police's proper concern with safety.

IMPLICATIONS FOR LAW ENFORCEMENT

A potentially important role of mobile communication devices is to enable the ordinary citizen to contact law enforcement authorities more quickly than before, a subject that is discussed in more detail in Chapter 8. If the fully portable telephone becomes a reality, the ease and rapidity with which citizens can report crimes will be transformed. Even though fixed telephones are commonplace, it can still take many minutes to gain access to one for reporting. The new ease of reporting may lead some to propose measures that increase the obligation of citizens to report crimes and even penalize failure to do so.

Hence, an important question relevant to our assessment is what the law should be respecting a citizen's *duty* to report observed crime. This involves consideration of the ancient offense of misprision of felony, which has been a common law or statutory crime for more than 700 years, and has been known by its present name since 1577.[19]

Misprision of felony is the concealment and/or nondisclosure of the known felony of another. The elements needed to constitute the infraction are: the felony of another; *actual knowledge* of that felony on the part of the misprisior; nondisclosure of the crime; and/or a positive act to conceal it by the misprisior. American jurisdictions that recognize the offense have usually required both concealment and nondisclosure. Acting with an "evil intent" toward the administration of justice is another component that must be found in a few jurisdictions.

At the federal level, misprision has been a felony (not a misdemeanor) in statutory form virtually unchanged since 1790. Section 4 of 18 U.S.C. (1970) currently provides:

Whoever, having knowledge of the actual commission of a felony cognizible by a court of the United States, conceals and does not as soon as possible make known the same to some judge or other person in civil or military authority under the United States, shall be fined not more than $500.00 or imprisoned not more than three years, or both.

The first case citing this section did not arise until 1922,[20] and the statute has since appeared before the courts fewer than two dozen times. But a series of prosecutions between 1966 and 1971 indicates that the crime still has vitality in at least four of the nation's eleven Circuit Courts of Appeals.[21]

Several factors have served to limit the use of the federal misprision law. First and foremost, it is well established that the conjunctive nature of the statutory phrase "conceals *and* does not . . . make known" creates two separate elements to the federal crime, each of which must be fulfilled for conviction. "Under it some affirmative act toward the concealment of the felony is necessary. Mere silence after the commission of the crime is not sufficient."[22]

A second limitation on federal misprision is the Fifth Amendment right against self-incrimination. The Ninth Circuit, in *United States* v. *Pigott,* has held that a conviction for misprision cannot stand where an analysis of the facts indicates that the defendant was simultaneously involved in the crime at the moment when his duty to notify could have arisen.[23] The Seventh Circuit recently rejected the result of *Pigott* when it decided *United States* v. *Daddano,*[24] but several factors make it likely that this conflict among the circuits will be resolved in favor of the Ninth Circuit.[25]

A third restriction of federal misprision is in the area of traditional privileges. A person is probably privileged from prosecution for misprision in instances where his report would reveal a confidence obtained in a spousal, attorney-client, physician-patient, or other protected relationship.

Insofar as mobile communications are concerned, two common themes run through all federal misprision indictments, even the unsuccessful ones. They have all involved concealments and actions so blatant and egregious as to place the alleged misprisior in jeopardy of prosecution as a principal or accessory under separate federal statutes.[26] Also, every defendant had a relationship with the felon of some degree of intimacy and length of time; the defendants have never been mere casual bystanders. These two factors create a threshold for prosecution well beyond what is contemplated from a telephone equipped driver or pedestrian who just happens upon a felony. Thus, misprision of federal felony as presently constituted would not impose any duty on users of mobile telephones.

Misprision of felony has a varying status in the several states, but it probably does not exist as a crime in the overwhelming majority of them. Where it does lie, prosecutions are few indeed, and defendants are entitled to the same privileges against conviction that are available in the federal courts.

It is appropriate to examine why misprision has been so severely restricted in application, and where found at all, why there has been universal reluctance to employ it.

Chief Justice Marshall long ago observed that:

> It may be the duty of a citizen to accuse every offender and to proclaim every offense which comes to his knowledge; but the law which would punish him in every case for not performing this duty is too harsh for man.[27]

The concern for fairness evident in this statement has been preeminent whenever misprision has been limited in scope or eliminated altogether. American jurisprudence generally illegalizes only positive actions, not nonactions. Hence, several courts have said that punishing a person who has done nothing to further a crime save accidentally appearing during its commission is "wholly unsuited to American criminal law."[28]

Another fear concerning misprision of felony is that it:

> would do violence to the unspoken principle of the criminal law that "as far as possible privacy should be respected." There is "as strong reluctance on the part of judges and legislators to sanction invasions of privacy in the detection of crime." [Citations omitted.][29]

Of course, it is not altogether clear that misprision does, in fact, inhibit privacy. But in support of the contention, one may turn to some of the greatest moralists in Western tradition, such as Johnson and Thoreau who thought they had no obligation to report every observed wrong.[30]

The mobile telephone will permit subscribers almost effortlessly to report crimes. Whether those users should be required to do so depends on whether this new ease of communication outweighs the arguments heretofore advanced against misprision. Some of the arbitrariness of misprision can be eliminated by restricting the felonies that must be reported to a statutory list of only the most basic crimes, or to offenses so serious that a reasonable man would consider it his duty to inform the police, or perhaps both. But however grievous the crime that is not reported, misprision would be punishing the witness for essentially a nonaction, and whether that is still "too harsh for man" is a major policy question.

If misprision is reinvigorated, there may well be problems in judicial integrity and administration. Any number of drivers may pass by the scene of a crime, and it will be difficult to track all of them down. Yet, if only one witness among many is prosecuted, the indictment is vulnerable to a charge of selective prosecution.

And aside from misprision's alleged inhibitory effect on the right of privacy, there are many other aspects of mobile communications that will inevitably have that obnoxious result. The law will doubtless be preoccupied with protecting people from new forms of electronic surveillance and bugging; it would be ironic indeed if the law imposed on users a duty that (arguably) invades their privacy from a different direction.

Of course, whether a reporting duty *should* accompany the use of mobile equipment is a far different question from whether such an obligation *will*, in fact, be imposed. The modern trend is unmistakably away from prosecuting for nondisclosure. This has occurred despite the concurrent growth of the landline telephone, a system that already makes reporting crime quite easy. Thus it might be concluded that misprision is not common now, and will likely become even less common in the future, regardless of developments in the communications field.

Withal, a reporting ability that is practically effortless may encourage reporting, regardless of legal duty, by those other than the direct victims of crime or accident. Use of CB for such purposes is becoming common (see Chapters 8 and 9). The evidence shows that people will report even without a legal duty to do so, and the mobile telephone could make reporting more rapid and easier than using CB radio. Rather than sending out a general call, the message will be directed to some emergency police reporting center by the operator or by dialing the local emergency number. Implementation of a mobile counterpart to a 911 emergency number system could facilitate easy and timely reporting.

A potential problem may come from overutilization by the populace. If the communication devices become common, police and other emergency agencies might be flooded with such a volume of calls that complex procedures might be needed to cope with them. Such problems are discussed in more detail in Chapter 8.

CRIMINAL ACTIVITY

Conclusions concerning the possible implications of mobile communication devices on criminal activity appear at several points in this chapter, when monitoring of citizens and use in the parole system are discussed. The aim here is to predict in broad terms some possible effects of the widespread deployment of such devices.

USE BY CRIMINALS.   Mobile telephones are likely to be helpful to conspirators and other criminals acting within a group in coordinating their activities. Use of mobile telephones by organized crime is apt to parallel the coordination uses in legitimate business. The mobile telephone is unlikely, however, to have much effect on common crimes. For instance, it seems improbable that it would have more than a merely marginal impact on the efficiency of a lookout aiding a bank robbery.

PATTERN OF CRIME.   But to postulate that criminals may not have great need for the mobile telephone is not to say that there will be no effect on the pattern of crime. Crimes may become more brutal in order to prevent a quick call for help. This might be especially true of street crimes where the assailant is interested in isolating and incapacitating the victim. As an alternative to brutality, crimes may involve theft of or damage to the unit to accomplish the same purpose. Thus, even if the victim has little of value to give his mugger, he may lose his portable phone. This loss will probably be in the form of destruction of the unit, since the assailant will want to avoid the locational capabilities of the device and the phone will be of little use to him if it must remain turned off. If market penetration of the unit was only moderate (let us assume for reasons of cost), those without phones may be increasingly victimized as being susceptible, lower risk targets for crime. Under such an assumption, crime would focus more heavily on the poor than it already does. The question arises, in any event, whether theft or malicious destruction of mobile telephone units should be discretely criminalized. Provision will need to be made for the reporting of lost or stolen telephones, similar to the current practice in connection with credit cards, so as to prevent costly misuse of the unit.[31] Access to and theft of computer data would not appear to be greatly affected by deployment of mobile telephones, since access codes and passwords would still be necessary.

EVIDENTIARY IMPLICATIONS

Hearsay evidence, evidence either oral or written that is not based on the personal knowledge of the witness, is generally inadmissible in a court of law, although this rule is subject to many exceptions. All United States jurisdictions, for example, have an exception to the rule against hearsay that allows for the business records of a company to be admitted as evidence.[32] The New York statute[33] is representative and is closely paralleled by Rule 803(6) of the new Federal Rules of Evidence. The former provides:

> Any writing or record, whether in the form of an entry in a book or otherwise, made as a memorandum or record of any act, transaction, occurrence or event, shall be admissible in evidence in proof of that act, transaction, occurrence or event, if the judge finds that it was made in the regular course of any business and that it was the regular course of such business to make it, at the time of the act, transaction, occurrence or event, or within a reasonable time thereafter. All other circumstances of the making of the memorandum or record, including lack of personal knowledge by the maker, may be proved to affect its weight, but they shall not affect its admissibility. The term business includes a business, profession, occupation and calling of every kind.

Under this section telephone company records may be introduced, showing information on the number of calls, to whom made and when, etc. Records produced by a mobile telephone system may provide additional information showing the cell locations of receiving and transmitting units.[34] The locational capabilities, however, are likely to come under attack when such records are attempted to be introduced as evidence, because the accuracy of the system to provide this information will only be recognized by a court if it is generally accepted by the scientists concerned with these matters.[35] Its similarities to accepted radar evidence will strengthen the case for admission.

PRIVACY

When asked to define the ultimate objective of telephone communications, Dr. Frank B. Jewett, former President of Bell Telephone Laboratories, is reported to have given the following answer: "When you can take out of your pocket a device the size of a watch, dial the number of a particular individual who may be anywhere in the world, and if he did not answer, you knew he was dead; that is the ultimate objective."[36]

We are devoting extensive attention here to the subject of privacy because any enhanced electronic communication capability—beyond the added efficiency, convenience, and enlargement of personality it furnishes—unavoidably creates risks for the privacy of the user. Any electronic communication device can be "bugged" and is as much a means of intrusion upon a user's repose as it is a convenience. Furthermore, insofar as the cellular system provides the means of even approximate tracking of persons (say, within a one- or two-mile radius), a significant human control mechanism will have become potentially available.

"Privacy" is a complex concept with psychological, social, and legal elements.[37] The above quotation reveals one of its critical dimensions. The notion that if one does not answer his telephone, one is dead, implies constant availability to anyone else. And the possibility that all telephone conversations can be monitored contains the seed of the elimination of all confidentiality. The combination of the two potentially creates a substantial threat to the integrity of the individual human personality.

A legal analysis of the concept of privacy should begin with an enumeration of interests calling for protection. These can be stated as follows: integrity of the individual personality (e.g., freedom from unwarranted, nonconsensual behavior modification); physical protection of the person (e.g., freedom from physical torture); protection of private property and space (e.g., freedom from trespass and unreasonable searches and seizures); protection from access to and dissemination of private information (e.g., freedom from unwanted and unwarranted publicity); protection of private communication (e.g., freedom from eavesdropping).[38] Historically, the law has more readily compensated for tangible loss than for injury to less concrete interests. Until relatively recently, the "right to be let alone"[39] was regarded as a bar only to unwanted physical interference.

In 1890 the *Harvard Law Review* published the Warren and Brandeis article "The Right to Privacy."[40] In calling for the recognition and protection of an individual's claim to privacy, the authors pointed to "recent inventions and business methods" that threatened "the sacred precincts of private and domestic life."[41] The article explores the bounds of the right and potential remedies for its deprivation. In conclusion, the authors observed:

> The common law has already recognized a man's home as his castle, impregnable, often, even to its own officers engaged in the execution of its commands. Shall the courts thus close the front entrance to constituted authority, and open wide the back door to idle or prurient curiosity?[42]

Response to this question was forthcoming. In 1903, reacting to a hint from the courts,[43] the New York legislature enacted the first privacy statute.[44] By the early years of this century a right to privacy was recognized by statute in New York, Oklahoma, Utah, and Virginia, and by case law in nearly all U.S. jurisdictions.[45]

In 1960, William Prosser examined the then existing authority and discerned "four distinct kinds of invasion of four different interests of the plaintiff which are tied together by the common name, but otherwise have almost nothing in common. . . . "[46] He identified the four torts as: (1) intrusion upon the plaintiff's seclusion or solitude or into his private affairs; (2) public disclosure of embarrassing private facts about the plaintiff; (3) publicity that places the plaintiff in a false light in the public eye; (4) appropriation for the defendant's advantage of the plaintiff's name or likeness. Prosser warned that development of the fledgling right to privacy was proceeding without sufficient reflection and, if left unchecked, threatened other substantial interests, such as the freedom of communication protected by the First Amendment.[47] Although Prosser's catalog is the leading work, there have been other recent efforts to identify the nature of the interest protected under the designation "the right to privacy"; especially noteworthy are the writings of Parker, Westin, and Cantor.[48]

A constitutional right to privacy was explicitly identified in the 1965 Supreme Court decision *Griswold* v. *Connecticut.*[49] The decision includes discussion of a "zone of privacy protected by several fundamental constitutional guarantees,"[50] embodied in the First, Third, Fourth, Fifth, and Ninth Amendments, the Fourteenth Amendment due process clause as a safeguard of fundamental rights, the right to marital privacy implicit in the Ninth Amendment,[51] and the concept of ordered liberty as a source of the protected right.[52] In enumerating the specific amendments from which the right to privacy flowed, Mr. Justice Douglas pointed to the areas in which the interest in privacy had earlier been discussed. The case law is substantial. The intersection of the right to privacy and the First Amendment is complex and troubling.[53]

A similar conflict may arise when the right to privacy clashes with effective governmental administration. The 1972 decision *Laird* v. *Tatum*,[54] involving military surveillance and storage of information concerning civilian political activity, is especially pertinent. Because the Supreme Court refused to hear the case, however, finding a lack of standing, a clear statement from the Court on the limits of an individual's right to control the accumulation of data about him was not forthcoming. In contrast, the physical security—of persons and property—that the Fourth Amendment protects has received considerable recognition from the Court. Described as the core of constitutional privacy,[55] the amendment's scope has been frequently analyzed.

The parameters of the constitutional constraint on search and seizure were developed in a famous line of cases beginning in 1928. These are especially relevant to potential abuses of mobile communication technology. In *Olmstead* v. *U.S.*,[56] the use of a wiretap by legal authorities was held not to constitute a Fourth Amendment breach. Emphasizing the absence of a tangible physical invasion, the Court observed:

The reasonable view is that one who installs in his house a telephone instrument with connecting wires intends to project his voice to those quite outside, and that the wires beyond his house and messages while passing over them are not within the protection of the fourth amendment.[57]

Apparently the justices viewed the consent given by the telephone customer as all-encompassing and unqualified; access once given could not be limited. But a vigorous dissent was registered by the coauthor of "The Right to Privacy," now Mr. Justice Brandeis. He discerned the future threat of a growing technology's impact on the Fourth Amendment interests and counseled the Court to draw the line now:

The progress of science in furnishing the Government with means of espionage is not likely to stop with wiretapping. Ways may some day be developed by which the Government . . . will be enabled to expose to a jury the most intimate occurrences of the home.[58]

Nine years later the legality of wiretapping was again before the Supreme Court. In the interim, however, the Federal Communications Act of 1934[59] had been enacted. Section 605 of this statute forbade wiretaps and on this basis, rather than that of the Fourth Amendment, evidence obtained by such illegal means was excluded in *Nardone* v. *U.S.*[60]

In *Goldman* v. *U.S.*,[61] the use of a detectaphone, a listening device attached to an adjacent wall, passed constitutional muster. Again, as in *Olmstead,* the Court stressed the absence of a physical invasion. The distinction was drawn to and perhaps beyond its logical limits in *Silverman* v. *U.S.*,[62] when a "spike mike"—whose operation required minute penetration of the wall of the "bugged" room—was found constitutionally impermissible on the Fourth Amendment grounds. The justices pointed to the "actual intrusion into a constitutionally protected area."[63]

The "property approach"[64] to Fourth Amendment liberty was halted in the 1967 decision *Katz* v. *U.S.*[65] Evidence obtained through eavesdropping on a telephone booth was held inadmissible. Rejecting the earlier reasoning, the Court said, "the Fourth Amendment protects people, not places."[66] Although the booth was glass walled and any passerby could gaze upon the speaker, the "bug" was nonetheless illegal. The Court indicated that in this instance the Fourth Amendment would protect what the accused "seeks to preserve as private," since "to read the Constitution more narrowly is to ignore the vital role that the public telephone has come to play in private communication."[67]

In a recent discussion of the constitutional limits of warrantless eavesdropping, Mr. Justice White further examined the *Katz* expectancy theory:

Our problem is not what the privacy expectations of particular defendants in particular situations may be or the extent to which they may in fact have relied on the discretion of their companions [but] what expectations of privacy are constitutionally "justifiable"—what expectations the Fourth Amendment will protect in the absence of a warrant.[68]

The safeguarded interest in privacy narrows then, from one of expectancy to one of *justifiable* expectancy.

New developments in technology have not escaped the attention of the Supreme Court. In striking down the New York eavesdropping statute in *Berger* v. *New York,*[69] it was said, "the law, though jealous of individual privacy, has not kept pace with . . . advances in scientific knowledge."[70] In a similar vein, Mr. Chief Justice Warren, concurring in *Lopez* v. *U.S.,*[71] warned "the fantastic advances in the field of electronic communication constitute a great danger to the privacy of the individual."[72]

All this provides cold comfort for those who would seek to combat the intrusions made possible by modern technology. In the words of one commentator:

> Having created the right to protect the most familiar and established zones of intimacy, the Court has precluded its application to many new and foreign situations where other privacy interests equally deserving of constitutional protection are threatened. Beyond one's body, mate, children, family and abode, governmental intrusion goes unchecked. If the particular activity does not have some physical, tangible aspect, and has not been revered for centuries, then most courts fail to see its relevance to individual privacy.[73]

Coupled with the Court's holding in *Laird* v. *Tatum,*[74] the restrictive treatment of the post-*Griswold* right to privacy suggests that judicial protection of this interest is likely to be limited. Since the right to privacy is recognized as essentially related to the home, it protects "places" not "people." In any event, it is more closely concerned with tangible intrusion upon traditional interests.

What weapons are available, then, to combat the considerable threat modern technology poses to personal privacy interests? The legal arsenal seems ill-stocked indeed. The private tort causes of action are limited. Primarily designed to provide economic compensation, they may fail to recompense less tangible harm. It is questionable whether freedom from electronic intrusion or data gathering would be recognized as a protectible interest.[75]

The constitutional right to privacy suffers from severe constraints. Except for the *Katz* case, the Court has largely imposed a restrictive definition focusing on matters of home life, sexuality, and childbirth. Although this could change, the constitutionally necessitated state action requirement is a far greater bar to recovery against usurpations practiced by private industry. Senator Proxmire suggested that any business that affects "fundamental aspects of life in a modern society"[76] should be subject to procedural due process demands despite the absence of governmental administration. The concept of state action, however, has never been extended so far.

With the foregoing discussion as background, we may now turn to some specific problems of privacy, as affected by the technology of mobile communications.

MONITORING OF CITIZENS.   A mobile telephone based on the cellular principle raises not only the possibility of the interception of voice communication. It also has the potential of at least partial "tracking" of persons, to the extent that it becomes feasible to store and retrieve call-locational records about where a person has been. The capability of checking the approximate present location of an individual might also become available. As the earlier discussion has suggested, it is doubtful that a zone of purely "informational" privacy exists within the realm of constitutionally or other protected rights. Recent cases, while affording the opportunity for decision on grounds related to informational privacy, have not been decided on that basis.[77] The computer compounds the problem by making it possible to gain access to far-ranging data on specific individuals.[78] A computer-mobile telephone hookup might conceivably supply large quantities of information within seconds.

The federal statutes most directly concerned with the problem are Section 605 of the Federal Communications Act, and Title III of the Omnibus Crime Control and Safe Streets Act of 1968. These leave the situation with respect to electronic eavesdropping as follows. (1) Section 605 prohibits "interception and divulgence of any wire or radio communication."[79] This has not, in practice, prevented mere *interception* of communications by law enforcement officers. (2) Title III of the Omnibus Crime Control and Safe Streets Act prohibits the *warrantless* interception of all wire and oral communications. A protected "oral communication" is one uttered under circumstances justifying an expectation of privacy. Title III applies to law enforcement officers as well as private citizens.[80] (3) A wire communication is one "made in whole or in part through the use of facilities for the transmission of communications by the aid of wire, cable, or other like connection between the point of origin and the point of reception. . . . "[81] This provision apparently does not apply to *purely* wireless communications. (4) A mobile telephone system would seem to involve wire communications by virtue of the wireline connections between cellular units and the central processing unit. And even if the "wireless" part of the transmission is not protected from interception by Title III,[82] presumably it would be protected if the tap is made on a "wired" part of the transmission in one of the processing units. Finally, while the "wireless" portion of the transmission is protected by Section 605, the latter does not effectively prevent law officers from its interception without divulgence.

Problems of coping with the tracking potential of the cellular system may also arise. Will the locating signal be secured by the Federal Communications Act? This signal is presumably "unimpregnated" with normal communication information, yet monitoring it could identify the rough location of the transceiver. Is this a sufficiently specific type of radio broadcast to involve Section 605's prohibition on interception and use, or will the very purpose of the location device (to identify and locate someone in a public area) vitiate any claim to privacy?

One possible approach might be to develop the concept of personal information as a property right. An individual would retain some control over the information about himself in circulation, if such personal information were treated as a property right—with all the restraints on interference by public or private interests and due process guarantees that the law of property has been so skillful in devising.[83] A similar theory is necessary since the mobile telephone may yield more than mere locational data. A mobile telephone may be used as a bug. The owner of the unit may call a number and then proceed to initiate a face-to-face conversation that will be overheard by whomever he has called. The law has treated this sort of participant monitoring as simply being one form of a person's privilege to repeat what is said to him,[84] yet such a betrayal of the nonconsenting conversant and the potentialities of widespread monitoring pose an additional threat to privacy.

These and similar problems may be alleviated by the incorporation of an on-off switch within the mobile unit. This feature would minimize possible abuses while the unit was off. But the essence of the present phone system is quick communication without notice or prearrangement. If the mobile telephone is to compete on this level, the mobile units would always have to be able to receive an incoming call or at least a signal that someone is attempting to call; this implies the capability of being tracked. Nor can the possibility of tampering with the on-off switch be ignored.

USE IN THE PAROLE SYSTEM.   There have been increasing efforts in recent years to resort to modern technology as means of dealing with complex law enforcement problems. Use in the parole system is one example. The most enthusiastic backer of the use of electronic devices for parolee monitoring is R. K. Schwitzgebel.[85] He postulates that the frequent and continuing monitoring of parolees might allow for the release of those who might otherwise remain in jail. Incentives or legal constraints might be experimented with to encourage voluntary adoption of the system, which could work to the parolee's benefit by eliminating false accusations against him. The monitoring system could be used to single out behavior specifically related to certain offenses (e.g., location within certain "crime areas"). Arguments against such use of mobile telephones or simpler monitoring devices on the basis of privacy intrusion are met with the claim that in the parolee's case, incarceration itself is the greatest invasion of privacy. Incarceration is also more expensive to society than parole; prison is five times more costly than parole as a method of control over people.[86] In addition, a paroled individual may work and pay taxes, thus contributing to the economy. Also the recurrence of crime is lower with parolees than with mandatory or direct dischargees.[87] Hence the cost of crime and further imprisonment may be lowered by an effective parole system.

There are, however, many legal obstacles to immediate use of the system, even with volunteers. "A waiver of constitutional rights in return for a governmental benefit may be invalid if the threat of withholding the benefit in question renders the waiver in some appropriate sense 'involuntary'."[88] Waiver must be voluntary, knowing, and specific.[89] Furthermore, the doctrine of unconstitutional conditions holds that the power to withhold a privilege entirely does not imply the power to grant it only if the recipient renounces a constitutional right.[90]

An interesting note in the *Harvard Law Review* discussed other possible legal constraints,[91] and mentioning a few of them will show the complexity of the problem. Tracking intended to control religious or political behavior would violate the First Amendment, for example. Tracking could be considered an impairment of the parolee's freedom of movement, association, and speech. Electronic monitoring that continued within the home would constitute a Fourth Amendment search. The privilege against self-incrimination would also be compromised. And tracking might impede parole therapy, since excessive surveillance destroys self-confidence and trust.[92]

More fundamental objections include the possible use of tracking as a conditioning device for humans. "Conditioning is seen as more of a threat to our freedom of choice than the use of incentives or disincentives."[93] But even if we disregard the conditioning aspect, tracking gives the tracker excessive power. The tracker might "follow" large numbers of people continuously, easily, and relatively cheaply. The privacy of those not being tracked may also be affected. Various mobile devices may be utilized for limited locational monitoring, and, with slight adaptation, for more intensive total monitoring. Obviously the former lacks many of the latter's offensive features and would be more easily accomplished.

Schwitzgebel has proposed some ground rules to alleviate these difficulties. First, the system should only be used with confirmed recidivists on parole (not probation). Second, the information should be limited as much as possible to conduct relating to prior offenses. Last, long-term therapeutic effectiveness should be demonstrated as a prerequisite for the use of such a system.[94] Limited data concerning tests with such a system show that, of 14 individuals who accepted the option of monitoring, seven changed their minds within five days, four continued for ten days, two for 35 days, and only one went for 167 days.[95] Adapting to the system is possible but probably will not occur without legal incentives.

TELEPHONE SOLICITATION. If a substanial number of people are constantly in the presence of a telephone, be it mobile or stationary, there is a likelihood that advertising over the telephone will increase. Salesmen and others who currently use the telephone only at night for soliciting may well pursue this endeavor full-time if their targets can always be reached.

To many, telephone solicitation is already a nuisance and an invasion of privacy, but it has not been subjected to any controls except the blanket Federal Communications Act and state prohibitions against harassing, obscene, and anonymous calls.[96] Any attempts to regulate or prohibit uninvited telephone solicitation might run into First Amendment problems; but in view of several Supreme Court cases, a strong argument could be made that such regulation is warranted and not unconstitutional.

In *Valentine* v. *Christensen,*[97] the Court held that "purely commercial advertising" is not protected by the First Amendment. The difficulty in proscribing "purely commercial" solicitation is in defining just what it encompasses. Calls on behalf of political or religious organizations are presumably protected under the First Amendment. But what about telephone pollsters such as Gallup and Harris? Is their solicitation a protected act of journalism or purely commercial speech?

Assuming that a law is drafted with sufficient strictness to overcome the above problem, however, it should pass constitutional muster as a reasonable and rational regulation of the use of telephones. In *Breard* v. *City of Alexandria*[98] the Court sustained a "Green River" ordinance banning the practice of summoning occupants of a residence to the door, without their prior consent, for the purpose of soliciting subscriptions for nationally known magazines.

IMPACT ON JOURNALISM.   New mobile communication technologies will provide a more flexible means of reportorial communication than the equipment currently used by radio and television stations. No massive hookups will be needed for on-the-spot live coverage of events that occur in peculiar locations or are mobile. This capability, though, should not create any legal problems that do not already exist regarding live coverage. Any time there is a live broadcast, there is the possibility that obscenities or slanders will be uttered by the individuals or groups who are being covered, in violation of the Federal Communications Act and local libel law. The mere fact that live coverage would be made easier with the new devices at present is not apt to create legal problems of a novel character.

NUISANCE AS A REMEDY FOR THE RIGHT TO BE LEFT ALONE. Nuisance is related to the broader discussion of privacy. What legal course can be pursued if a person objects to the constant ringing, beeping, or buzzing of mobile devices and the ensuing conversation in public places (e.g., in the concert hall, library, etc.)?

Noise is difficult to deal with because at many levels it is largely subjective. Noise essentially is sound not wanted by those who hear it.[99] Furthermore, private nuisance traditionally is a civil wrong based on a disturbance of the rights to enjoy the use of land by conduct occurring outside that land.[100] It is this aspect of proprietary interest that will limit the usefulness of nuisance doctrine in the hypothetical cases described. *Any* interest in property has, however, been held sufficient to support a nuisance action. "Public nuisance" is a catchall criminal offense, consisting of interference with the rights of the community at large. Normally, the state holds the remedy for public nuisance. But, an aggrieved individual capable of showing special damages (i.e., damages different in kind from those of the community at large) may take private action for public nuisance.

In awarding damages for nuisance, the *Restatement*'s approach that each person must put up with some inconvenience and annoyance, and its formula for weighing the gravity of harm to the plaintiff against the utility of the conduct of the defendant, is often used by the courts.[101] The utility of the defendant's conduct is deemed to be one of the social values that the law views as one of its purposes.[102]

"Reasonable" use is often marched out as a standard, although it is by no means self-defining. Society must have factories and airports somewhere. Among the noises that, if they do not cause substantial discomfort, residents in a large industrial city may have to tolerate are those that accompany and are incident to the reasonable recreation of a crowded population.[103] Of course, a more exacting standard is applied when the "noise" emanates from public entertainment enterprises than would be imposed for more vital economic services.

It is hard to classify use of mobile devices as completely in one or the other category, that is, public entertainment or vital economic service. One who purchases property in an area already devoted to noisy activities "comes to the nuisance" and takes his property subject to some sort of easement. Courts are understandably reluctant to grant injunctions if the effect would harm a business being properly conducted, without there being at least some negligence in design and/or operation of an improper location. The unintentional by-product of noise from modern living is hard to combat.

An added obstacle for a plaintiff challenging the noise levels of mobile units was erected long ago in *Richards* v. *Washington Terminal Co.*[104] No right of action exists under federal law for private property owners for noise caused by an entity functioning under authority of government, if the activity is sanctioned by law and properly designed and operated. Claims against public utilities may not be enforced if they are performing acts authorized by the government. Such would appear to be the case with the phone company's introduction and use of mobile telephones. Noise alone does not appear to be a Fifth Amendment taking of property. Several well-known cases involving airports[105] have been similarly concerned with this problem.

Turning from private nuisance litigation to the public arena is of little consolation. New York law makes one guilty of disorderly conduct if one creates unreasonable noise.[106] Intent is, however, a necessary element of the offense, and mere private annoyance will not support a disorderly conduct charge. A state statute dealing with the disruption or disturbance of religious services may possibly be used in appropriate circumstances.[107]

A look at another relevant New York statute concerning criminal nuisance is not too helpful either.[108] Public nuisance formerly required little, if any, intent. The revised statute on criminal (i.e., public) nuisance requires greater culpability. A person must knowingly or recklessly create or maintain a condition that endangers the safety or health of a considerable number of persons. The evidence needed to sustain this is not likely to be compiled by the government, which must bring such criminal nuisance actions against individuals or the communications company.

Although, as we have seen, there are possible constitutional arguments to be made concerning privacy, the average person will seemingly be saddled with the burden of noise from mobile communication devices. In any event, nuisance theory, as presently designed, does not afford much help.

RESIDENCE AND DOMICILE

The new technologies we are analyzing will facilitate greater mobility, even with respect to living arrangements. In conjunction with mobile homes, transient living arrangements may become more common. Obviously, the introduction of a new communication technology will not, in and of itself, affect the traditional means of determining the residence and domicile of individuals. Residence is defined as living in a particular locality; a person may have more than one residence, and physical presence suffices to establish it.[109] Domicile is much less transient than its connotations. One's domicile is the place of one's permanent home,[110] not a temporary abode, and is sometimes called a person's *legal* residence. Though a person may have more than one residence at the same time, he can have only one domicile.

Phone company records, showing the origin of calls, may become an additional element to be considered in establishing residence. Unless the cell size were quite small, this would still probably not pinpoint a physical residence exactly, without other proof. At best it might show residence within the cell—domicile, of course, being much harder to prove by resort only to telephone records.

There is another way in which such considerations may come to bear. If an increasingly mobile society were to include great numbers of "vehicular homes," that is, mobile units that served as domiciles or primary residences, the initial assignment of a phone number might be a factor in determining the vehicle's home base. This point of origin might then be considered as a domicile for the vehicle's inhabitants. Without a concept of "home base," its inhabitants might have a series of successive residences without ever establishing a domicile. As already indicated, the phone company records could be used to make a case for residency at a particular time and place. But a person lacking a domicile or permanent, "legal" residence may encounter difficulty respecting such matters as eligibility to vote and other participation in political or legal processes in which permanence of residence plays some part.

NOTES

Citations of standard legal references are given in the conventional legal format. A list of common abbreviations used is given below. Further elaboration of this format can be found in *A Uniform System of Citation,* 12th ed. (Cambridge, Mass.: The Harvard Law Review Association, 1976).

Legal Reference	*Abbreviation*
Code of Federal Regulations	C.F.R.
Federal Reporter	F.
Federal Reporter, Second Series	F. 2d
United States Code	U.S.C.
United States Supreme Court Reports	U.S.

1. Laurence H. Tribe, *Channeling Technology Through Law* (Chicago: Bracton Press, 1973), p. 85.

2. Learned Hand, "Historical and Practical Considerations Regarding Expert Testimony," *Harvard Law Review* 15 (1901):40.

3. Barry M. Portnoy, "The Role of Courts in Technology Assessment," *Cornell Law Review* 55 (1970):861; see also Philip Bereano, "Courts as Institutions for Assessing Technology," in *Scientists in the Legal System,* ed. William A. Thomas (Ann Arbor, Mich.: Ann Arbor Science Publishers, 1974), p. 73.

4. See, e.g., *U.S.* v. *Carroll Towing Co.,* 159 F. 2d 169 (2nd Cir. 1947).

5. E.g., *Summers* v. *Tice,* 199 P. 2d 1 (Cal. 1948); *Ybarra* v. *Spangard,* 154 P. 2d 687 (Cal. 1944).

6. *Seely* v. *White Motor Co.,* 403 P. 2d 145 (Cal. 1965); *Goldberg* v. *Kollsman Instrument Corp.,* 191 N.E. 2d 81 (N.Y. 1963); *Greenman* v. *Yuba Power Products, Inc.,* 377 P. 2d 897 (Cal. 1963); *Henningsen* v. *Bloomfield Motors, Inc.,* 161 A. 2d 69 (N.J. 1960).

7. Milton Katz, "The Function of Tort Liability in Technology Assessment," *University of Cincinnati Law Review* 38 (1969):587.

8. See, e.g., *March Wood Products Co.* v. *Babcock Wilcox Co.,* 207 Wis. 209 (1932).

9. The *Restatement* adopts a neutral stance on this point. America Law Institute, *Restatement of the Law, Second: Torts* (St. Paul, Minn.: American Law Institute Publishers, 1965), § 402, comment (o).

10. C.G.S.A. § 14-259 (1958).

11. Examples include: wide-view mirrors to compensate for the loss of audio cues signaling vehicles approaching and overtaking the mobile technology user; steering knobs on the steering wheel for easier one-handed control of the car when the other hand is holding a handpiece; locating the device within easy arm's reach of the driver but just outside normal vision while driving, perhaps where the radio is normally placed today; requiring the device to be mounted securely so that it does not disengage and become a missile should the car have an accident.

12. Joel J. Reich, *Telemedicine: The Assessment of an Evolving Health Care Technology* (St. Louis, Mo.: Center for Development Technology, Washington University, 1974).

13. Harry G. Henn, *Agency, Partnership and Other Unincorporated Business Enterprises* (St. Paul, Minn.: West Publishing Co., 1972), p. 104.

14. Ibid., pp. 132-33.

15. See, e.g., *Crooker* v. *California,* 357 U.S. 433, 441 (1958).

16. *Powell* v. *Alabama,* 287 U.S. 45 (1932); *Johnson* v. *Zerbst,* 304 U.S. 458 (1938); *Lisenba* v. *California,* 314 U.S. 219 (1941); *Chandler* v. *Fretag,* 348 U.S. 3 (1954); *Moore* v. *Michigan,* 355 U.S. 155 (1957).

17. 378 U.S. 478 (1964).

18. See *Terry* v. *Ohio,* 392 U.S. 1 (1968).

19. Concern over the Watergate scandals has evoked renewed interest in this shadowy transgression, as is indicated by the publication of two law review articles from which much of the following discussion has been condensed. Royal G. Shannonhouse, III, "Misprision of a Federal Felony: Dangerous Relic or Scourge of Malfeasance?", *University of Baltimore Law Review* 4 (1974):59; Michael S. Stone, "Misprision of Felony: A Reappraisal," *Emory Law Journal* 23 (1974):1095.

20. *Presont* v. *United States,* 281 F. 131 (6th Cir. 1922).

21. *United States* v. *Pigott,* 453 F. 2d 419 (9th Cir. 1971); *United States* v. *Daddano,* 432 F. 2d 1119 (7th Cir. 1970); *Sullivan* v. *United States,* 411 F. 2d 556 (10th Cir. 1969); *United States* v. *Norman,* 391 F. 2d 212 (6th Cir. 1968); *United States* v. *King,* 402 F. 2d 694 (9th Cir. 1968); *Lancey* v. *United States,* 356 F. 2d 407 (9th Cir. 1966), *cert. denied,* 385 U.S. 922 (1966).

22. *United States* v. *Farror,* 38 F. 2d 515, 517 (D. Mass. 1930), *aff'd* 281 U.S. 624 (1930); see also *Sullivan* v. *United States,* 411 F. 2d 556 (10th Cir. 1969); *Bratton* v. *United States,* 73 F. 2d 795 (10th Cir. 1934).

23. *United States* v. *Pigott,* 453 F. 2d 419 (9th Cir. 1971).

24. 432 F. 2d 1119 (7th Cir. 1970).

25. Shannonhouse, "Misprision of a Federal Felony," p. 71; Stone, "Misprision." pp. 1105-6.

26. 73 F. 2d 795, 797 (10th Cir. 1934).

27. *Marbury* v. *Brooks,* 20 U.S. (7 Wheat.) 556, 575 (1822).

28. E.g., *People* v. *Lefkowitz,* 293 N.W. 642, 643 (Mich. 1940).

29. *United States* v. *Worcester,* 190 F. Supp. 548, 566 (D. Mass. 1960).

30. See also Zechariah Chafee, *The Blessings of Liberty* (Westport, Conn.: Greenwood Press, Inc., 1973), pp. 180-82; A.S.L. Farquharson, ed. and trans., *The Meditations of Marcus Aurelius* (Oxford: Clarendon Press, 1968), pp. 39, 167, 168, 354; E. M. Forster, *Two Cheers for Democracy* (New York: Harcourt, Brace Jovanovich, Inc., 1962), pp. 68-69; Morgan, *The Silkworm and the Loom or the Liberty to Mind One's Own Business, Liberties of the Mind* (London, 1951), pp. 226-29, 231; Sir Frederick Pollack, ed., *Table Talk of John Seldon* (London: Quaritch, 1927), pp. 139-40. The foregoing are cited by Judge Wyzanski in *United States* v. *Worcester,* ibid.

31. Consumer Credit Protection Act, 15 U.S.C. §§ 1602(o), 1643 (1970). These sections limit the liability of cardholders for unauthorized use of credit cards to a maximum of fifty dollars. If the company issuing the card is notified of its loss or theft prior to the unauthorized use, the cardholder is not liable for any amount.

32. E.g., Federal Rules of Evidence, Rule 803(6); Code of Ala. tit. 7, § 415; Ga. Code Ann. § 38-711; N.Y.C.P.L.R. § 4518(a); Vernon's Ann. Civ. Stat., art. 3737(c) (Texas).

33. N.Y.C.P.L.R. § 4518(a).

34. Records of location may be generated cheaply, but their storage and retrieval may be prohibitive. Also, local calls within the same exchange need not be located by computer for billing purposes.

35. Irving Younger, "Computers and the Law of Evidence," *New York Law Journal,* February 25, 1975, pp. 1-5.

36. David Talley, "A Prognosis of Mobile Telephone Communications," *IRE Transactions on Vehicular Communications* VC-11 (August 1962):27-39.

37. Sidney M. Jourard, "Some Psychological Aspects of Privacy," *Law and Contemporary Problems* 31 (1966):307-8; Barry Schwartz, "The Social Psychology of Privacy," *American Journal of Sociology* 73 (1968):741; Glenn Negley, "Philosophical Views on the Value of Privacy," *Law and Contemporary Problems* 31 (1966):319, 321.

38. For further discussion, see Harry Ralvin, "Privacy in the Year 2000," *Daedalus* 96 (1967):876.

39. Thomas McIntyre Cooley, *Law of Torts,* 2nd ed. (Chicago: Callaghan and Co., 1888).

40. Samuel D. Warren and Louis D. Brandeis, "The Right to Privacy," *Harvard Law Review* 4 (1890):193.

41. Ibid., p. 195.

42. Ibid., p. 220.

43. The plaintiff in *Roberson* v. *Rochester Folding Box Co.,* 64 N.E. 442 (N.Y. 1902) was denied relief for want of legislative interest in protection of the right to privacy.

44. N.Y. Civ. Rights Law §§ 50-51 (McKinney 1948), originally enacted at ch. 132 §§ 1, 2 (1903) N.Y. Laws.

45. William L. Prosser, *Handbook of the Law of Torts,* 4th ed. (St. Paul, Minn.: West Publishing Co., 1971), pp. 802-18.

46. William L. Prosser, "Privacy," *California Law Review* 48 (1960):383, 389.

47. Ibid., pp. 422-23.

48. Richard B. Parker, "A Definition of Privacy," *Rutgers Law Review* 27 (1974):275; Alan F. Westin, *Privacy and Freedom,* 1st ed. (New York: Atheneum, 1967); Norman L. Cantor, "A Patient's Decision to Decline Life-Saving Medical Treatment: Bodily Integrity versus the Preservation of Life," *Rutgers Law Review* 26 (1973):228.

49. 381 U.S. 479 (1965). The case is discussed in "Symposium—Comments on Griswold Case and the Right of Privacy," *Michigan Law Review* 64 (1965):197-283.

50. Ibid., p. 485.

51. Ibid., pp. 494-96 (concurring opinion, of Goldberg, J.).

52. Ibid., p. 500 (concurring opinion, of Harlan, J.).

53. See *NAACP* v. *Alabama,* 357 U.S. 449 (1958). Recent articles examining this topic include: Edward J. Bloustein, "The First Amendment and Privacy: The Supreme Court Justice and the Philosopher," *Rutgers Law Review* 28 (1974):41; Don R. Pember and Dwight L. Teeter, "Privacy and the Press Since *Time, Inc.* v. *Hill," Washington Law Review* 50 (1974):57. Major cases in this area are *New York Times Co.* v. *Sullivan,* 376 U.S. 254, 279-80 (1964) (requiring a public official to show "actual malice" to recover damages for a defamatory statement); *Time, Inc.* v. *Hill,* 385 U.S. 374 (1967); *Rosenbloom* v. *Metromedia,* 403 U.S. 29 (1971); *Gertz* v. *Welch,* 418 U.S. 323 (1974); *Cantrell* v. *Forest City Publishing Co.,* 419 U.S. 245 (1974).

54. 408 U.S. 1 (1972).

55. R. H. Clark, "Constitutional Sources of the Penumbral Right to Privacy," *Villanova Law Review* 19 (1974):833, 856.

56. 277 U.S. 438 (1928).

57. Ibid., p. 466.

58. Ibid., p. 474.

59. 47 U.S.C. § 151 et seq.

60. 302 U.S. 379 (1937).

61. 316 U.S. 129 (1942).

62. 365 U.S. 505 (1961).

63. Ibid., p. 512.

64. Jerold H. Israel and Wayne R. Lafave, *Criminal Procedure in a Nutshell,* 1st ed. (St. Paul, Minn.: West Publishing Co., 1971), p. 183.

65. 389 U.S. 347 (1967).

66. Ibid., p. 351.

67. Ibid., p. 352.

68. *White* v. *U.S.,* 401 U.S. 745, 751-52 (1971).

69. 388 U.S. 41 (1967).

70. Ibid., p. 49.

71. 373 U.S. 427 (1963).

72. Ibid., p. 441.

73. "The Constitutional Right of Privacy: An Examination," *Northwestern University Law Review* 69 (1974):263, 282-83.

74. 408 U.S. 1 (1972); see note 54 and text accompanying.

75. Arthur R. Miller, *The Assault on Privacy: Computers, Data Banks and Dossiers* (Ann Arbor: University of Michigan Press, 1971), pp. 210-38.

76. William Proxmire, "Consumer Credit: Privacy the Collateral," *Trail* 11 (January-February 1975):34-36.

77. *California Bankers Association* v. *Schultz,* 416 U.S. 21 (1974); *Cox Broadcasting Corp.* v. *Cohn,* 420 U.S. 469 (1975).

78. Arthur R. Miller, "Personal Privacy in the Computer Age: The Challenge of a New Technology in an Information-Oriented Society," *Michigan Law Review* 67 (1965):1091; and Keith Mossman, "A New Dimension of Privacy," *American Bar Association Journal* 61 (July 1975):829.

79. Federal Communications Act of 1934, 47 U.S.C. § 605 (1962).

80. 18 U.S.C. §§ 2510 et seq. (1970).

81. 18 U.S.C. § 2510(1) (1970).

82. See *United States* v. *Hall,* 488 F. 2d 193 (9th Cir. 1973).

83. But see Miller, "Personal Privacy," p. 1223.

84. Kent Greenwalt, "The Consent Problem in Wiretapping and Eavesdropping: Surreptitious Monitoring with the Consent of a Participant in a Conversation," *Columbia Law Review* 68 (1968):189.

85. Ralph K. Schwitzgebel, "Issues in the Use of an Electronic Rehabilitation System with Chronic Recidivists," *Law and Society Review* 3 (1969):597.

86. George Torodash, "Parole Must Not be Abolished," *New York State Bar Journal* 47 (June 1975):292-94.

87. Hoffman, "Mandatory Release," *Criminology* 11 (February 1974):541, 547.

88. Tribe, *Channeling Technology,* p. 365.

89. *Miranda* v. *Arizona,* 384 U.S. 436 (1966).

90. See generally, "Unconstitutional Conditions," *Harvard Law Review* 73 (1960): 1595; and "The Parole System," *University of Pennsylvania Law Review* 120 (1971):282-377.

91. Note, "Anthropotelemetry: Dr. Schwitzgebel's Machine," *Harvard Law Review* 80 (1966):403.

92. Charles Fried, *An Anatomy of Values: Problems of Personal and Social Choice* (Cambridge, Mass.: Harvard University Press, 1970), pp. 143, 148.

93. Note, "Anthropotelemetry," p. 414.

94. Schwitzgebel, "Issues in the Use," p. 610.

95. Ibid., p. 600. Note that the "system" used was not one completely parallel to mobile telephone. The "experiment system" involved the use of pagers, telephones, and companion observers to create surveillance situations.

96. 47 U.S.C. § 223 (1970).

97. 316 U.S. 52 (1942).

98. 341 U.S. 622, 644 (1951).

99. George A. Spater, "Noise and the Law," *Michigan Law Review* 63 (1965):1373.

100. Prosser, *Handbook,* Chap. 15.

101. American Law Institute, *Restatement,* § 826.

102. Prosser, *Handbook,* Chap. 15.

103. William H. Lloyd, "Noise as a Nuisance," *University of Pennsylvania Law Review* 82 (1934):567.

104. *Richards* v. *Washington Terminal Co.,* 233 U.S. 546 (1914).

105. *United States* v. *Causby,* 328 U.S. 256 (1946).

106. New York Penal Law § 240.20 (1967).

107. New York Penal Law § 240.21 (1967).

108. New York Penal Law § 240.45 (1967).

109. Henry Campbell Black, *Black's Law Dictionary,* 4th ed. rev. (St. Paul, Minn.: West Publishing Co., 1968).

110. Ibid., p. 572.

Chapter 13
LAND MOBILE COMMUNICATIONS IN JAPAN
Technical Developments and Issues of International Trade

INTRODUCTION

The development of advanced mobile communication systems is proceeding in a number of countries. The once formidable gap in the technological capabilities of the U.S. on the one hand and Europe and Japan on the other has been closing, especially in electronics and telecommunications. Japan has been particularly successful in areas of advanced technology; it is therefore not surprising that Japan is devoting considerable effort to the development of a sophisticated mobile communications capability. In the context of our study, this fact led us to consider two very different types of questions. First, what are the characteristics of the system being developed in Japan as compared with those in the U.S., and second, what are the possible implications for the economic and political relations between the U.S. and Japan that may result from trade competition.

From the end of World War II until at least the early 1960s, two super-powers—the United States and the Soviet Union—held positions of military and political predominance, and their actions had the widest and most profound influence on world affairs. The development of economically strong and ideologically compatible allies in then-devastated Europe and Japan was central to U.S. strategy. To achieve this, the U.S. government subordinated its foreign economic policy to broader political and strategic ends. Thus, foreign economic aid, favorable tariffs for allies, incentives for private investment overseas, aid for foreign currencies, exportation of U.S. technology on comparatively favorable terms, and a host of other "economic" policies were enlisted to serve broader U.S. aims.

So long as the U.S. economy remained dominant and the major international goal of the U.S. continued to be the containment of the Soviet Union and its allies, the policy of utilizing economics as a tool of international politics went relatively unchallenged. However, neither the international situation nor U.S. goals remained the same. Japan and most of Western Europe achieved economic success. The Sino-Soviet split and the manifestation of national autonomy among the states of Eastern Europe belied the unity and central direction of communism. At the same time, emerging social problems and discontents within the U.S., worsening of the U.S. balance of payments, decreasing expectation of nuclear confrontation, declining international strength of the dollar, and other factors combined to break down the conditions that had previously sustained U.S. foreign policy. Bipolarity and military confrontation could no longer be taken as the norm in international relations.

With such changes, economics increasingly regained its own intrinsic significance, no longer to be subordinated to broader international politics. Military and political allies frequently found themselves with highly competitive economic interests, particularly at the microlevel of specific products and firms. Foreign economic policies among advanced industrial societies, which once involved the subordination of economic considerations to broader political commonalities, soon came to reflect the political differences that emerged from these competing interests.

Despite close political and military ties between the U.S. and Japan, a number of economic issues began to weaken these bonds. The 1971 textile controversy, the unannounced devaluation of the dollar, and the unannounced alteration in U.S.-China policy led to the sharp deterioration in relations between the two countries.[1] Also of significance were problems over Japanese regulation concerning importation of capital and technology, the soybean crisis, and U.S. insistence on "voluntary restrictions" against the exportation of a variety of Japanese goods to the U.S.[2]

These issues provide striking examples of the way in which microlevel economic problems can influence the much broader course of relations between two countries. In the case of textiles, for example, Japanese exports to the U.S. accounted for less than two percent of total U.S. consumption; yet the problem was treated by the Nixon administration as critical for domestic political reasons, and two decades of relations were jeopardized over efforts to restrict Japanese exports to the U.S.[3] Soybeans would seem to be of interest to few beyond their growers and speculators in agricultural futures, but Japan depends on imports for 97 percent of its heavy soybean consumption,[4] and a U.S. refusal to deliver specified quantities to Japan created a sizable political problem for both governments. Similarly, grapefruit, semiconductors, televisions, computers, and auto emission control systems—each with varying degrees of significance to the total economies of both countries—have generated substantial political tensions. As economic confrontations have become more central to relations among industrialized democracies, on a bilateral level numerous minor economic problems have clearly acquired a significance that goes far beyond their direct economic importance.

It is from this perspective that we wish to examine the potential for developments in the mobile communications field. What is the likelihood that they will generate significant political tensions between Japan and the United States? To assess such probabilitieṣ, it is essential to examine the developments taking place in Japanese mobile communications, and to consider the extent of technological lead or lag, and the existence of mutual market possibilities.

JAPANESE TRANSPORTATION AND TELECOMMUNICATIONS

In transportation and telecommunications, the two fields that provide the basis for mobile communications, Japan is not nearly as advanced as the U.S., and U.S. prominence in the number and range of its transportation and telecommunications systems is so overwhelming that it is unlikely to be overtaken by any other country for quite some time. Nevertheless, Japanese progress in both areas is significant, and in the development of an auto telephone, Japan is now comparable to, and possibly ready to surpass, most Western European nations.

The Japanese transportation system is extremely well developed and has been advancing rapidly. Its shipping industry is the largest in the world; its railroad network has a coverage per square kilometer that is one of the world's highest. In 1973, the tonnage of goods shipped within the country was 3.7 times higher than in 1960,[5] and the number of "person-kilometers" traveled in 1974 was nearly triple the 1960 figure.[6] Most central to the development of the automobile telephone, however, is the increasing number of autos and the distances traveled in them.

The number of automobiles registered in Japan (about 28 million) is second only to the U.S., which has nearly 100 million more, but is well ahead of third-place West Germany (18 million), France (17 million), England (16 million), and Italy (15 million).[7] For land mobile communications it is especially significant that a relatively high 35 percent of Japanese vehicles are used for commercial purposes, with no other country having over 20 percent.[8] Japan has more than 10 million commercial vehicles, behind only the U.S. with 22 million—a gap much smaller than that between the total numbers of automobiles. Moreover, within the last decade there has been a fivefold increase in the total number of vehicles and a ninefold increase in passenger cars, and there is no reason to suspect that growth has peaked; indeed, though the oil crisis has shocked Japan at least as much as any other industrial nation, rapid expansion is still expected to continue in this area. Official predictions are that by 1985 the number of "person-kilometers" driven in Japan will be double the 1970 figure.[9]

Telephone service, meanwhile, has also expanded rapidly, from 746,000 phones in operation in 1945, to 11 million in 1965, to nearly 43 million in 1975. Japan has a switching system as advanced as that in the U.S., and virtually all of the telephones in use are dial phones. There are 28.8 subscriber lines per 100 persons in Japan, which ranks a comparatively low twelfth in the world, just ahead of West Germany (26.8) and Belgium (23.8). But here again, while Japan is well behind the world-leading U.S. penetration rate of 62.8 per 100 persons,[10] it is rather close to Holland (29.9), Finland (30.5), England (31.4), Norway (32.0), Australia (34.0), and Denmark (37.9). Furthermore, Japan has the second largest absolute number of phones in the world, 1.7 times more than England and 1.8 times more than West Germany. Nippon Telegraph and Telephone (NTT), in its fifth five-year plan that began in FY 1973, is undertaking a major construction investment that projects a total of 52 million telephone lines in the country by 1977–1978. If this figure is reached as expected, it would put Japan's per capita telephone ratio close to present U.S. figures.[11]

Expansion in both these fields has occurred within the broader context of the world's fastest growing economy; demands for a communication capability while on the move was a logical outgrowth and there has been a consequent expansion in all areas of Japan's mobile communications field.

JAPANESE MOBILE COMMUNICATIONS

Much of the technology used in present Japanese land mobile radio communications[12] grew out of systems originally developed for the country's extensive fishing fleet and harbor traffic. Japan introduced its first primitive marine mobile system in 1923. Fishing and transport vessels remain the prime users, all vessels over 20 tons being required by law to maintain a mobile communications capability.[13] As of March 1975, there were some 50,575 marine mobile stations in Japan, a figure second only to the U.S.[14]

On land, private companies have provided many mobile communication systems, but it was not until 1948 that the first significant mobile radio communication system was introduced into the public sector—the police radio system, operating in the 30-MHz band.[15] Other emergency uses, most notably fire and flood protection, have since been added, and a large proportion of the present land mobile units in Japan are dedicated to such purposes. There is also a growing use of mobile communication systems by private construction companies, taxi dispatching systems, newspaper services, and a host of other businesses (see Table 13.1). By March 1975, there were over 300,000 mobile units in operation, necessitating the utilization of portions of the 40-, 60-, and 150-MHz bands, as well as the original 30-MHz band. At present the most common forms of mobile communications are either simple paging devices or devices involving comparatively limited communication from one fixed station to one or more mobile units. A high capacity cellular mobile telephone system is now being planned, but in order to see its possible significance a brief discussion of a few alternative and competing forms of mobile communications that bear directly on the mobile telephone will be useful.

Table 13.1: *Distribution of Land Mobile Radio Stations by Industry (March 1975)*

Taxi service	141,972
National railroads	33,516
Police	30,891
Fire and emergency	22,695
Public communications	20,680
Electric power	14,010
Manufacturing and sales	8,903
Civil engineering and construction	5,139
Flood defense and road service	4,956
Local government	3,553
Gas service	3,292
Other	39,445
Total	329,052

SOURCE: The Electronics Association of Japan, **Electronics in Japan '75** (Tokyo: ELAJ, 1975).

Pocket Bell Paging System

The simplest of the mobile communication systems offered by NTT is a one-way paging system, known as "Pocket Bell."[16] An individual subscriber carries with him a lightweight, compact unit having a discrete call number that can be dialed from an ordinary telephone.

The system was first introduced in 1968. The unit presently in use consists of a solid-state receiver weighing about 150 grams and slightly smaller than a pack of cigarettes (97 mm × 50 mm × 18 mm). It operates in the 150-MHz band; phase modulation is employed and selective calling for this service utilizes multifrequency signaling. To identify a particular unit, two audio range frequencies, whose minimum separation is 15 Hz, are transmitted simultaneously, followed by a second pair of frequencies. These signals are sent by base stations having 250-watt transmitters.

Generally operable within a 20-km radius, the units' transmission facilities are adequate to ensure coverage throughout an entire city such as Tokyo. The 150-MHz band allows for excellent building penetration with the result that contact efficiency is rated at 97 percent.

The pocket bell system was introduced commercially in Tokyo in 1968 and was subsequently deployed in other cities. Its use expanded very rapidly, soon exceeding the number of U.S. subscribers for such systems. By early 1977, nearly 640,000 subscribers were being served in 42 cities, and demand for the pagers was running about 100,000 units ahead of supply. As can be seen from Table 13.2, subscribers are concentrated in the areas of sales, construction, service, and manufacturing, though the unit clearly provides an important service for other sectors that may be less significant numerically.

Table 13.2: *Pocket Bell Paging System Distribution of Present Users and Applicants*

Category	Present Users		Applicants Added to Present Users	
	Number	% of Total	Number	% of Total
Sales	149,729	37.8	185,057	36.5
Construction	77,613	19.6	99,092	19.6
Service	65,672	16.6	86,486	17.1
Manufacture	42,178	10.7	55,587	11.0
Transport	12,772	3.2	17,087	3.4
Printing and publishing	11,031	2.8	13,601	2.7
Self-employed	9,501	2.4	15,092	3.0
Doctors	7,901	2.0	9,199	1.8
Finance	7,766	2.0	9,119	1.8
Trade	4,530	1.1	6,156	1.2
Government	2,487	0.6	2,964	0.6
Organizations	2,240	0.6	4,142	0.8
Newspapers	1,386	0.4	1,576	0.3
Broadcasting	631	0.2	770	0.2
Total	395,437	100.0	505,928	100.0

SOURCE: Figures supplied by NTT as of March 1975.

Subscribers pay initial deposits of ¥ 20,000 (ca. $67) and a monthly use charge of ¥ 2,000 (ca. $6.70) for a single pager.

The leasing and maintenance of pocket bell units are provided by fourteen private companies under NTT franchise; five companies are franchised to produce the units in conformity with NTT specifications. Through this and the leasing system, NTT is virtually guaranteed the needed supply without becoming overly dependent on any single manufacturer.

One key weakness of the present pager, and an obvious disadvantage vis-à-vis any telephone, including cellular, is the inability to determine the identity of one's caller. For most professional use this is not a major drawback, since one's office is the usual caller. However, long-range plans at NTT call for a "Super Bell" system with a base station memory capacity that would allow the subscriber to call a prearranged number whenever he is paged and receive automatically the telephone number of the party paging him.[17] Yet at the moment, the high costs of developing the needed memory circuitry are working against its early deployment. In the interim, NTT has introduced a paging system that will emit distinct tones, allowing discrimination between two different callers and thereby providing greater identification capabilities than at present. This new system will also be able to accommodate three times as many subscribers per channel as the present system, through higher speed transmission, and will need battery recharging only monthly.[18]

One of the peculiar features of the Japanese communication system that helps sustain the pocket bell paging system is Japan's widespread and easily accessible public telephone system. With 11.8 public phones per 1,000 population (and equally significant to jaded users of U.S. public telephones, with virtually none out of order), Japan's public telephone service is in quantity and quality unmatched in the world. This excellent telephone service, when combined with pagers, is in some respects a lower cost challenger to cellular.

Cordless Telephone

Somewhat more complex than the pocket bell is the cordless telephone. This device involves a radio telephone in which a pair of small UHF or VHF radio units eliminate the wired indoor wall hookup of a subscriber's unit so that the telephone can be moved freely inside a limited enclosed area.[19] Such phones are fully capable of all the functions of a wired phone. Up to 128 units can be accommodated by a single system, making it particularly useful in convention halls, department stores, business offices, and so forth. No equivalent service is offered by U.S. telephone companies, although the equipment to do this may be purchased from private interconnection manufacturers for $300-400.

NTT began field testing the system in Tokyo during 1972. It proved to be particularly useful as a means of reducing the time and labor required for installing and removing temporary telephones. The system is currently also in test use within NTT headquarters and two large trade centers; in 1976 more extensive commercial testing was begun in a variety of cities.

The rental cost for an individual unit will be approximately ¥ 3,000 ($10) per month, though, of course, multiple units are required for the system to operate with maximum cost efficiency.

Equipment for the cordless system consists of the necessary number of telephone sets, a base station unit and antenna, and multiplexer. The individual telephone set has the same case as the standard touch-tone telephone, but inside it also houses a radio transmitter and receiver, a control circuit, a rechargeable nickel cadmium battery, a whip antenna, and a dial signal oscillator. To reduce power consumption, an electromagnetic transceiver is used in place of a carbon microphone. The battery is a DC 2.4 V with a capacity of 3.5 AH; it weighs 330 grams, can be used for about ten days and then be recharged. Speech quality is designed to approximate that of the ordinary telephone, with sound articulation above 80 percent, AEN below 49 dB, and the s/n ratio above 40 dB.

The system's service area is limited to approximately 100 square meters. Because of the low transmitting power of both the fixed base unit and the individual telephones, and since field strength is adjusted to less than 15 MV/m at a point one hundred meters from the radio facilities, no radio license is needed to operate the system. Though in principle it is conceivable that private manufacturers will offer such systems to customers without an authorized franchise from NTT, much as they are doing within the U.S., in reality the

Public Telecommunications Law requires all such units to meet technical approval from NTT.

While the cordless telephone appears to have many of the attributes of the mobile telephone service, it does so only in limited and specified areas. No widespread usage of the cordless telephone is possible due to capacity restrictions, and mobility is restricted to its specified area. Nonetheless, NTT anticipates a demand of 500,000 units by 1987.

The Mobile Telephone and Transport Services

Development of mobile telephones began with the work on a marine telephone, which was introduced in 1953 for use in Kobe, Osaka, and Tokyo. This system, which operates in the 150-MHz band, has since been expanded and now services all coastal areas of Japan, including Okinawa, to a distance of 50 km from shore. By 1974 there were 6,600 ship telephone stations averaging over 24,000 calls in a typical day,[20] and further expansion of the service is planned.

In addition to shipboard telephones, Japan has a system of railroad telephones on its long-distance trains;[21] this system was introduced in the late 1950s and operates in the 400-MHz band. On the super express trains now running from Tokyo to the southern island of Kyushu, base stations with an effective transmitting range of about 20 km are established at intervals along the tracks, with extra boosters to handle transmission through tunnels. Eight channels are utilized for the service. Phones are located in two of the sixteen cars, and a paging service makes it possible to locate any passenger for an incoming call. The quality of transmission and reception is high, and costs to the user are low; the service is currently provided in over two hundred train runs daily.

Both ship and train mobile telephone services are significant, but they are used by a comparatively limited and select portion of the population. Far broader in potential impact and usage is the mobile telephone system that operates from automobiles.[22] Plans were formulated by the NTT secretariat for the development of such a service in 1961. Prototype testing began in 1964, and the Ministry of Posts and Telecommunications, which has responsibility for radio frequency management, began planning for spectrum allocation in the 400-MHz band. In 1967 research was completed by NTT, and a system similar to the Bell system's Improved Mobile Telephone Service (IMTS) was developed. Still considered experimental, this service has been mostly for emergency use (though those with access frequently use it for less urgent calls), and only a few units are actually used for nonemergency purposes.

The Cellular System

Since 1967 major consideration has been given to a completely different type of mobile telephone system that would have sufficient capacity to accommodate the growing public and private demand, but at the same time would not require an inordinate expansion into already limited or crowded portions of the radio spectrum.[23] The result has been the development of a cellular system in the 800-MHz range with very high service capacity. The design of the system is virtually identical to that proposed by AT&T to the FCC in 1971. It is to be fully automatic and private; however, somewhat smaller cells have been proposed by NTT for urban areas (3–5 km vs. 7 km), and the channel spacing is to 25 kHz. This system is compared to the proposed U.S. cellular system and current U.S. IMTS in Table 13.3.

In spring 1976, the Japan Radio Technical Council (Denpa Gijutsu Shingikai) of the Ministry of Posts and Telecommunications approved use of the 800-MHz band for land mobile, and NTT hopes 50 MHz of spectrum will be allocated for this use (25 for transmission and 25 for reception).[24] This allocation would allow an eventual capacity of one million mobile units, NTT's anticipated demand nationwide; surveys undertaken by NTT project a medium-term demand of 400,000 (about 50,000 in Tokyo) with 100,000 units in use by 1987. Service will begin in Tokyo, then expand to other major cities until the entire country is serviced.

Table 13.3: *Operational Features of Current IMTS, Planned Cellular MTS, and Japanese Cellular MTS Compared*

Operational Features	United States	United States	Japan
System	IMTS	HCMTS (cellular)	Cellular
Mode of operation	Automatic	Automatic	Automatic
Two-way direct dial automatic connect	Yes	Yes	Yes
Special land-to-mobile paging channels	No	Yes	Yes
Channel selection	Automatic	Automatic	Automatic
Marked idle free channels	Yes	Paging/access	Paging/access
Selective call addressing	Yes	Yes	Yes
Call privacy	Yes	Yes	Yes
Automatic identification	Yes	Yes	Yes
Call timing and billing	Automatic	Automatic	Automatic
Automatic mobile location and call hand-off between cells or zones	No	Yes	Yes
Mobile equipment price	$2,100–2,500	$1,000	----
Mobile service and equipment monthly rental fee	$100–120	$50–60	----

SOURCE: R. L. Legace and H. L. Pastan, "Public Mobile Telephone—A Comparative Analysis of Systems Worldwide," **Proceedings of the 26th Conference of the IEEE Vehicular Technology Group (New York**: Institute of Electrical and Electronics Engineers, 1976), p. 23, Table III.

For at least the short-term future, this cellular system will be utilized primarily to provide Japan with an automobile telephone system, including both private and public service. Ultimately, though, it furnishes the possibility for a fully portable telephone, allowing both incoming and outgoing service comparable to that of the existing fixed telephone system. Research is currently underway on such a system, but there are varying expectations regarding the problems to be faced and the time it may take for commercial availability.

Hand-Held Portable Telephone

Attention has been generated in the United States by discussion of Motorola's Dyna-Tac system, which is expected to provide a telephone capable of being carried in one's hand or attache case, and performing all the functions of a desk top telephone.[25] In Japan NTT foresees a number of problems that must be solved before the widespread commercial introduction of such a system could take place. Frequency use efficiency, miniaturization of the transceiver, power source for the transceiver, exchange control technology, zone identification, and billing are among the largest problems.

In reality, however, these problems can be reduced to those of miniaturization and cost: How small can a unit be without reducing its power so much that a very large number of fixed stations are required that in turn would lead to prohibitive costs? Research on such problems is just beginning in the Electrical Communications Laboratories of NTT[26] and it is unlikely that field tests on propagation capabilities, etc., would be held before 1978–1979; widespread usage of cellular hand-held telephones is not expected for several years after that. Nevertheless, a prototype has been created[27] and NTT feels it will be possible to introduce a commercial version of such a system sometime around 1980.

THE MOBILE COMMUNICATIONS MARKET IN JAPAN

It is important to stress that across the entire spectrum of mobile communications Japanese technological development is quite high. The pocket bell system is comparable in size, efficiency, and cost with American systems, for example; far more Japanese mobile equipment—at least CB equipment—is used in the U.S. than vice versa; and, in the more relevant area of the cellular system for mobile automobile telephones, technical progress in Japan is on a par with that of the U.S., and both are well ahead of any other nation. Indeed, due to regulatory issues in the U.S., NTT might begin commercial service considerably ahead of any U.S. company, in March 1979. Also, certain cellular research done by Japan was relied upon in Bell's presentations concerning FCC Docket 18262.[28] On the other hand, Bell, with its long-term commercial auto telephone service experience, may have a slight edge on Japan. In general, though, neither country should be much ahead or behind the other in cellular technology.

This technical sophistication has been an important factor in making the Japanese domestic market for mobile communications equipment the second largest in the world, well over $100 million. This is significantly behind the U.S. with its nearly $800 million market, but it still represents about ten percent of the world market. Table 13.4 indicates the sales level of Japan's land mobile radio equipment.[29]

While mobile equipment is only a small portion of the total communications market, $100 million in sales is substantial. More impressive than the present value of the market, however, is the rapidity of its expansion: the number of mobile stations in 1970 was just short of 160,000, yet by March 1975 this figure had more than doubled to over 300,000. Although pagers were not introduced until 1968, by 1977 there were about 640,000 being used in Japan, more than in any other country in the world, including the U.S., which introduced paging ten years before Japan and which has twice the national population. Deployment of the cellular system promises to contribute further to Japanese market expansion, particularly when it reaches a relatively developed and comparatively widespread status in the mid-1980s.

The degree to which NTT has committed funds and prestige to developing a cellular system, as well as the extent of the research it has completed, suggests that NTT is convinced that there is, or at least there can be created, a vast demand for mobile telephone service throughout Japan. Certainly, the present backlog of requests for pocket bell systems and cordless phones, and the long-standing demand for automobile telephones point in the direction of an expanding mobile communications market. Furthermore, the comparatively large number of commercial vehicles in Japan would appear to provide a far better market for mobile autotelephone systems than would an equal number of private vehicles.

One question of importance to our assessment is just how extensive this potential market is and, more significantly from a political standpoint, whose it will become. Projecting long-range demand for a particular product, especially

Table 13.4: *Sales Value of Mobile Equipment in Japan (U.S.$ × 1,000)*

	1972	1973	1974	1975
Land mobile radio	44,669	60,657	64,808	56,904*
Paging	7,964	13,592	12,906*	14,489
Other mobile	17,578	23,970	23,845*	29,604
Subtotal	70,211	98,219	101,559	100,997*
Total Communications Equipment	1,658,304	2,328,461	2,138,668	1,868,731*

*Figures all given in face value, not constant $; apparent drop due to rate of exchange shifts.

one not yet widely used commercially, is an extremely hazardous and uncertain venture. So much depends on the general growth of a nation's economy, the price level for the product, competition from similar products or technologies, governmental regulation or support, advertising for both the product and its competition, development of possible user industries, the psychological climate of the potential buying public, as well as other uncertainties.

Thus it becomes necessary to consider the extent to which the development of the pocket bell and cordless telephone systems will be complementary or competitive with the auto phone and the portable phone. From one perspective, their use could stimulate an increased desire for communications mobility, resulting in demand for more sophisticated mobile communications equipment. Certainly this is the expectation frequently held by product developers. On the other hand, the combination of a pocket bell receiver with readily available public telephones and/or the widespread introduction of cordless phones into office buildings might reduce much of the perceived need for fully portable or automobile phones. This would be even more likely if the paging system included a sophisticated mechanism for identifying the caller, such as that planned for the "Super Bell."

NTT has, of course, carried out extensive market surveys to determine potential demand for its various mobile services, including the cellular automobile telephone. Positing a rather high hypothetical cost of ¥ 50,000 ($167) as a monthly base rate, plus a use charge of ¥ 100 ($.33) per minute, NTT surveyed industrial firms to determine probable purchase patterns.[30] Even at these prices, results showed that approximately 10 out of 100 autos owned by small-sized firms would be equipped with cellular auto telephones. Over 60 percent would involve immediate installation once the service is offered, and the owners of all but 5 percent of the vehicles who want the service would demand it within the first three years of its availability. Firms owning an additional 19.7 out of 100 cars in small industries and 43.0 out of 100 in larger industries indicated that, while unwilling to commit themselves in advance, they would investigate such a system once it became available.

Reactions differed dramatically from industry to industry. Most anxious to adopt the system were large construction firms (67.9 percent), large manufacturing firms (41.0 percent), and larger service industries (31.1 percent). Real estate and financial institutions of all sizes and small service industries were far less enthusiastic.

On the basis of this and other surveys, NTT's central office has projected demands for various mobile services at the anticipated service cost, which are given in Table 13.5. As can be seen, NTT expects the 1987 demand to be about two million for pagers, one million for portable phones, 100,000 for auto telephones, and 500,000 for cordless phones.

The precision of these figures is perhaps less telling than the overwhelming mood of optimism conveyed by NTT, a company with a reputation for a slow and steady pace of growth and for conservatism and accuracy regarding future projections. While the figures are undoubtedly speculative, NTT is in an exceptionally favorable position to determine supply and, to some extent, demand for

Table 13.5: *Present Conditions and Future Projections in Public Mobile Communications Service*

Service Category	Frequency Range Expected	Service to Begin	Service Area	Demand		Remarks
				Present Subscribers	Expected Demand (1987)	
Autophone	830–890 MHz	Fiscal 1978	Nationwide	----	100,000 (all within 50 km of Tokyo)	possibly incl. data transmission
Paging	150 MHz 250 MHz	1968 Oct. 1977	All major cities	640,000	2 million (500,000 in Tokyo)	all systems on 150 MHz to be changed to 250 MHz within 10 yrs.
Railroad	LF band VHF or UHF	1957 1964	Railway lines	100 lines	2,000 lines	LF band not presently in use
Portable phone	VHF or UHF	1985	Major cities	----	1 million	
Cordless phone	VHF or UHF	Fiscal 1976	Buildings	----	500,000– 1 million	radio frequency license necessary

Land Mobile

Table 13.5: *Continued*

Service Category	Frequency Range Expected	Service to Begin	Service Area	Demand		Remarks
				Present Subscribers	Expected Demand (1987)	
Land Mobile *Emergency & Temporary* — *Emergency*	60 MHz 400 MHz UHF	1957 1960 1980	Nationwide	---	---	incl. some isolation prevention units
Temporary	400 MHz (part est.) UHF	1974 (part) 1980	Major cities	About 1,000	---	
Ship	150 MHz 250 MHz	est.	All coastal areas	9,000	28,000	incl. data transmission in future
Air	VHF or UHF	1980	Nationwide	---	1,000	100- to 300-km large zone public service
Satellite	1,600 MHz	1983	Worldwide	---	1,000	incl. data transmission

SOURCE: Private data from NTT, October 1974.

potentially competitive services. The fact that NTT is investing so heavily in automobile telephones indicates the company's own anticipations and preferences for particular systems.

By 1980 this projected level of expansion could easily lead to a Japanese land mobile communications market of $20 to $40 million for pagers, $80 to $120 million for mobile radios, and $40 to $60 million for other land mobile equipment, in addition to a possible $50 million in sales for marine mobile. Cellular auto phones alone should provide a $20 million market. This could result in nearly a $300 million market for land and marine mobile systems in 1980. By 1985-1987, the market may easily be $400 to $500 million, with cellular auto telephones representing more than 10 percent of this total.[31]

The existing mobile communications market in Japan has been virtually the exclusive domain of Japanese firms; in 1974 only $400,000 worth of land mobile radio equipment—or less than two percent of the total—consisted of imports, all from the U.S. On the basis of such figures, and in the absence of any significant lag in Japanese technological development, it would be quite unrealistic to expect the Japanese market to provide major opportunities for Western mobile equipment manufacturers.[32] Rather, the major suppliers of the cellular auto telephone are likely to be NEC, Toshiba, Hitachi, Matsushita, and Kokusai, with lesser shares going to Oki, Mitsubishi, and Fujitsu. This factor, combined with the potential U.S. market for Japanese equipment in land mobile, could pose bilateral political problems.

American manufacturers experienced a severe and ominous disappointment of this type in 1975 when Bell awarded the contract for the first 135 cellular transceivers to Oki Electronics, a U.S.-based but predominantly Japanese firm. As one U.S. marketing executive put it, "In television and comparable areas, the consumer at least could make his own choice between foreign and domestic products. But with the [cellular] market, it is up to AT&T. If AT&T sticks with foreign suppliers, we're dead."[33] Technically, of course, AT&T's position is by no means so exclusive, but there is little doubt that its purchasing patterns could make or break many potential cellular manufacturers, and its initial choice of Oki was seen as a foreboding precedent. It has been reported that Oki offered to supply the items for approximately 40 percent less than the competing American bidders—about $3,700 per unit, as compared with $5,500 to $6,000 per unit from the American bidders. This quick and significant penetration of the U.S. market by low-bidding Japanese firms, plus the expectations of American firms being effectively excluded from the Japanese market, is consistent with growing images of a semiofficial lockout on one side and Japanese government-endorsed dumping on the other. The U.S. electronics trade journals have depicted this situation as another Japanese invasion. Fears have not been radically diminished by AT&T's even more recent decision to award manufacturing contracts to two American firms along with Oki.

The result of these fears has been pressure on the U.S. government to "do something" to prevent Japanese manufacturers from capturing the U.S. market, particularly because the Japanese government also seems to be refusing to "open up" their country's market to American manufacturers.[34] As noted earlier, similar situations have developed in many other areas with a wide array of political and commercial consequences, so it is important to assess the merits and implications of any such problems on mobile communications in the broadest terms.

POLITICAL IMPLICATIONS

American industrialists frequently assert that while the U.S. provides an open and inviting market for Japanese manufacturers, the Japanese severely restrict the efforts of U.S. firms to trade in Japan. The view is widely held that the Japanese government exploits low labor costs, an undervalued yen, low social overhead and military expenditures, and close relationships to big business in order to unfairly assist Japanese exporters in penetrating the U.S. market. The shorthand phrase that has come to symbolize most aspects of this situation is "Japan, Inc."[35] From this perspective, Japan is not a free market dominated by competing firms, but a single monolithic "corporation" combining the resources of government and business, whose goal is rapid expansion at the expense of foreign countries. In brief, Japan is accused of adopting a merchantilist strategy.[36] The gloomy prospects for American manufacturers in the Japanese mobile communications market, combined with the success of Oki in quickly penetrating the U.S. market for mobile telephones, is seen by many as another manifestation of this broader situation.

As is often true of cliches, there is a core of validity in the notion of "Japan, Inc." Japan's conservative, probusiness government has worked closely with big business firms in pursuing high economic growth through the development and protection of domestic industry and the expansion of Japanese exports. There have been government pressures through informal consultation, financial and tax incentives, and supportive legislation to reorganize and strengthen those domestic industries likely to improve Japan's balance of payments through increased exports. In addition, during the 1950s and 1960s the government maintained effective formal and informal restrictions—enforced primarily by the Ministry of International Trade and Industry (MITI)—against foreign imports and direct investment in a wide range of areas where domestic industry was thought to be too weak to stand without government protection or considered in other ways essential to the national interest.

The results of these policies are apparent in the success of Japanese exports. The compound annual growth rate of exports during the decade 1960-1970 was 16.9 percent, far above the 9.3 percent world average and well over double the U.S. rate of 7.6 percent. Moreover, Japan's share of the total world export market grew from 3.2 percent in 1960 to 6.2 percent in 1970 (while that of the European Economic Community went from 23.3 percent to 28.4 percent and the U.S. declined from 16.0 percent to 13.7 percent).[37] The U.S. has been a prime importer of Japanese goods, which leaped from 7.8 percent of U.S.

imports in 1960 to 14.7 percent in 1970; in manufactured goods the share rose from 16.8 percent to 26.6 percent, and the figures are still higher in particular sectors, such as electronics.[38]

Since the end of World War II, the U.S. and Japan have enjoyed a close trading relationship. Initially, however, the U.S. was the clear beneficiary in terms of balance of trade. For example, in 1955 total Japanese imports from the U.S. totaled $773 million, while its exports to the U.S. were $445 million, giving the U.S. a nearly 2:1 trade ratio and a $318 million surplus. In 1960 the gap was somewhat reduced in relative terms; the ratio changed to approximately 3:2, but the actual surplus had grown to nearly $450 million. A rapid reversal took place starting in 1965 when Japanese exports to the U.S. slightly outvalued imports. With the minor exception of 1967, when trade was roughly equal in both directions, between 1965 and 1974 Japan continued to sell much more in the U.S. than it purchased. In 1972 the gap was especially large: Japan exported nearly $9 billion to the U.S. while importing $5.9 billion, resulting in a $3 billion surplus.[39]

The shift was even more dramatic in electronics and telecommunications. By the early 1970s, the U.S. was purchasing about four times as much Japanese telecommunications equipment as it was selling to Japan; in the electronics field, overall, the ratio was about 6:1.

In short, the U.S. has experienced substantial impact from expanding Japanese trade; the situation has been exacerbated by Japanese products having appeared to flood into the U.S., while the reverse has not been occurring. Furthermore, it was extremely difficult for U.S. firms to penetrate the Japanese capital market through the establishment of wholly owned subsidiaries, so that marketing directly was also limited. This was especially true in the most technically sophisticated sectors, including telecommunications. Thus, by the early 1970s, the U.S. had lost the position of international economic preeminence that it occupied a decade earlier, and Japan seemed to be a major contributor to the decline. Yet it is by no means obvious that the reasons for this development were either as clear or as conspiratorial as is often alleged.

It must be recognized that much of the comparative strength of the U.S. economy and the primacy of U.S. products during the 1950s and 1960s was the by-product of World War II, which severely disrupted the economies of virtually all other major powers, but most notably France, Italy, Germany, and Japan. Some of the subsequent redress can also be traced to the war insofar as the devastated nations were able to rebuild by incorporating the most modern technologies; this often resulted in foreign plants that were more efficient than their U.S. counterparts. In addition, the dollar was clearly overvalued during the mid-1960s, accounting for much of the trade imbalance that began to occur.

U.S. exports, moreover, have not been primarily in finished products; approximately one-third of all U.S. exports involve raw materials and agricultural products. In contrast, finished goods account for almost all (93 percent) Japanese exports.[40] Bilaterally, consumer and capital goods represent less than one-third of the total U.S. exports to Japan, whereas more than one-half of Japanese exports to the U.S. involve finished products. For example, over three-quarters of Japanese electronics exports to the U.S. are consumer products.

Part of the reason for this, particularly in the area of electronics and telecommunications, is that most U.S. manufacturers have found it far more profitable to engage in production for the expanding government market—most significantly in the defense and aerospace industries—than to manufacture for the civilian and consumer market. Hence, from 1950 to 1964 the percent of U.S. electronics products made for the consumer market declined from 55 percent to 18 percent, while sales to the government increased from 24 percent to 56 percent.[41] A virtual guarantee of high profits and an expanding market at home resulted in little attention being paid to the overseas market during this period. By the time this problem was recognized, much of the competitive advantage that U.S. firms might have enjoyed had been dissipated. It was not until the mid-1960s that electronics firms, led by Texas Instruments, began to seek substantial penetration of the Japanese domestic market through direct investment. Finally, even when certain domestic U.S. industries did seem to be injured by foreign imports, the 1962 Trade Expansion Act provided few effective corrective mechanisms; proving that specific damage had resulted from import competition was very difficult.

Besides these economic dimensions, there are subtle differences in the business orientations of firms in the two countries that are relevant to the trade imbalance. Various studies have shown that Japanese managers and firms are more frequently concerned with expanding their market share than they are in immediate return on investment; American priorities tend to be the reverse. Organizational growth and stability also carry more relative importance for Japanese firms and managers than the concerns with organizational efficiency, industry leadership, and profit maximization that dominate American business thinking. As a result, Japanese firms have been far more aggressive in developing initially unprofitable overseas markets than their American counterparts, even in the absence of formal inducements.[42]

All of these factors must be recognized in assessing the trade imbalance between Japan and the United States. The problem cannot be understood using the simple notion of a sinister cabal of business and government leaders, convincing as such an explanation might appear to those in search of identifiable scapegoats. If it were so easy to import Western technology, protect domestic manufacturers unfairly, and dump the imitated goods abroad at prices far below those at home, other nonindustrialized countries would have adopted the same policies with results quite similar to Japan's economic success.

At the same time, one cannot dismiss as totally unfounded the charges that Japanese markets have been less open than is expected of an allegedly free market in a capitalist country. Japanese exports have been aided by both government support and unwitting Japanese consumers. MITI guidelines and licensing regulations concerning imports have been extremely complex and frequently arbitrarily applied. Japanese industry has benefited from a broad range of legislative efforts, cartelization, patent regulations, and government-related research and marketing organizations. In addition, a host of other probusiness practices widely used in the West have been adopted, including tax write-offs, failure to enforce antitrust regulations, R&D assistance, and easy credit terms by banks to large firms.[43]

It is quite certain that the Japanese government was anxious during the past twenty years to restrict or prevent the introduction of major foreign firms and products over which it could exercise little of its traditional direction and control. Ever since Admiral Perry demanded that the Japanese government open the nation to Western commerce under terms that proved to be economically unfavorable to Japan and an insult to national sovereignty, all Japanese governments have sought to encourage domestic industrial progress while restricting foreign penetration of key sectors of the economy. As a result the central government has been active in overseeing the nation's commerce, and the Japanese economy has never resembled the laissez-faire market of Adam Smith. Government guidance, indicative planning, cartelization, and a strong national consciousness have long been the Japanese norm.

Justifiably or not, Japanese governments have often perceived latent, and even blatant antagonism underlying Western perceptions of Japan. This antagonism, they contend, has led to actions against Japan that would not have been taken against other Western nations. In the prewar period, such feelings were bolstered by the unwillingness of Americans and Australians to include a clause of racial equality in the Treaty of Versailles, by the restrictions against Japanese immigration, and by the tariff barriers against "cheap Japanese goods." The incarceration of Japanese-Americans during World War II, while German-Americans and Italian-Americans were comparatively undisturbed, did little to alleviate Japanese perceptions of this special hostility. These perceptions were not reduced by the revelation that a Canadian Prime Minister noted in his diary, "It [was] fortunate that the use of the [atomic] bomb should have been upon the Japanese instead of the white races of Europe."[44] Even if such attitudes are dismissed as "unfortunate" aspects of past history and it is clear that irrational animosity toward Japan is declining, the perceptions are kept alive by events like the hostile advertising campaigns of the ILGWU a few years ago—a xenophobic campaign whose motif centered on the question "What is America coming to?", depicting baseball gloves and American flags bearing the label "made in Japan."

Japan is a member of IMF, GATT, OECD, and most other organizations oriented toward the removal of trade and investment barriers and the creation of more open economic transactions throughout the capitalistic world. Consequently, Japan has been under strong pressure to liberalize formal barriers to trade and investment for over a decade, and as a result of the Kennedy round of tariff reductions, Japan has reduced its effective tariffs on manufactured imports by higher percentages than those of the other major participants.[45] In 1967, Japan began a five-year series of liberalizations of restrictions against foreign investment that Dan Fenno Henderson calls "the most significant Japanese effort since Perry to internationalize at home and abroad."[46] Because of the imbalances in trade, in 1972 Japan also unilaterally reduced its tariff levels by an additional 20 percent on all industrial products.[47] By February 1975, Japan had residual import restrictions on only seven industrial products (versus six each for the U.S. and the U.K., eight for Italy, 35 for France, and 20 for West Germany).[48] Furthermore, the U.S. has about 40 agreements with Japan under which the latter "voluntarily" restricts its exports to the U.S., the largest number of such restrictions among the capitalist countries. These voluntary restrictions are rarely included in official calculations of barriers to imports. Their inclusion would lead to the conclusion that Japan is significantly less "closed" and the U.S. certainly far less "open" than seems to have been the case at first glance. In fact, one can make a reasonable case that formally Japan is considerably more open than the U.S.[49]

Since the revaluations of the yen beginning in 1971, American companies have been in a far better position to trade with Japan than during the late 1960s, contributing to the disappearance of the incredibly favorable balance of trade that Japan enjoyed with the U.S. in 1972; bilateral trade was more or less equally balanced before the recent reversal.[50]

Thus, before the U.S. government responds to pressures from American manufacturers to utilize political pressure to "open up" the Japanese market, all of these factors and changes must be considered. It appears that major actions by the U.S. government would seem to be both unnecessary and misplaced. Nevertheless, the question remains: what potentials exist for U.S. manufacturers to participate in the Japanese mobile communications market—particularly in the area of cellular telephone systems—and, conversely, to what extent will Japanese firms be successful in the U.S.? Also, based on such assessments, what is the likelihood that political problems will emerge?

Despite Japan's liberalized import and capital investment policies, it is highly improbable that there will be any substantial increase in the proportion of the Japanese mobile market held by foreign products, at least within the next five years. Several factors support such a conclusion. First, the present penetration by outsiders in this area is extremely limited, which implies that the necessary business and marketing contacts have not been established. Most American companies wishing to take advantage of recent liberalizations to market their products directly in Japan would have to start virtually from scratch. (There are a few exceptions, one of which is Motorola.) It would still be possible, however, to gain a substantial share as long as American firms have some major technological or price advantage, but this does not appear to be the case with cellular telephone equipment or mobile communications generally. Thus, any advances by American firms will be slow. Insofar as there seems to be little promise of sizable and rapid economic gain, it is unlikely that many American firms would find the present market attractive if their prime concern remains the return on investment. There is, therefore, the possibility of a vicious circle: American firms looking for high profits in the short term may not enter the market, but the failure to do so may exclude them from a longer term position with potential for moderate yet significant profits.

Even in areas where American technological superiority seems most probable, such as in computer systems for tracking and switching, the Japanese have made great strides and previous American advantages are declining. There are also the inevitable problems posed by differences in language and business practices, and the informal and close relations among existing Japanese firms, all of which have been handicaps to American success in the past.

A far more substantial obstacle arises because Japanese governmental and quasi-governmental agencies purchase about three-quarters of all communications equipment produced, and about 85 percent of this involves NTT as the end user.[51] Both the Japanese government and NTT maintain a policy of "buying Japanese," which will ensure the Japanese domination of its domestic telecommunications market even if more formal barriers are reduced. In the past, whenever NTT has felt the need for a new product it has done virtually all of the research and testing in its own laboratories and then given detailed specifications, research assistance, and coordinating facilities to several Japanese producers for commercial development. By insisting on three or more producers and dividing most systems into numerous component parts manufactured by different firms, NTT has been guaranteed a regular supply of the items it needs without becoming overly dependent on any single supplier. Since Japanese telecommunications manufacturers are exceedingly talented and sophisticated by any international standard, it has rarely been necessary for NTT to rely on foreign products.

The case of pagers provides a good illustration of this procedure. When NTT finished the research for the "Pocket Bell" system, it encouraged production research among three firms—NEC, Toshiba, and Matsushita—and divided the paging market equally among them. As demand exploded, Toyo and Kokusai Electric were franchised, and each of the five firms now holds about 20 percent of the total market. From this powerful position, NTT has controlled the entire Japanese market and there are no imports in the $14 million paging industry.[52] The cellular auto telephone will probably be treated similarly, and NTT will surely retain predominant control over the entire system, even if small portions of that system are purchased from foreign firms.

There are a few areas where some foreign penetration might be possible, most notably in the more technically sophisticated areas and in the supply of systems for private rather than government use. A substantial market exists for private telephone systems that interface with the NTT network through a Private Branch Exchange (PBX), particularly in office buildings. However, unless American firms are prepared to offer a more attractive combination of service and cost than their Japanese competitors, they are apt to run into the unarticulated but undeniable barrier of nationalistically based purchasing. With an established presence and a generally good reputation, Motorola would appear to be in the best competitive position at present. Motorola has a patent pending for its Dyna-Tac system and may be able to gain some portion of the Japanese market, but unless Dyna-Tac proves greatly superior in cost and technology to Japanese alternatives, significant sales seem unlikely.

Thus there is little reason to expect that U.S. firms will be able to tap any substantial portion of the Japanese mobile market in the near future on purely economic or technical merits. Given the broader array of political, cultural, and economic interests common to the U.S. and Japan, though, there seems to be no reason why the U.S. government should exert any particular pressures on Japan to change this situation. The total Japanese market has been opened considerably to the import of capital, technology, and manufactured goods during the past few years, and certainly there is scant evidence of major differences between Japan and the U.S. in the overall degree of market openness. While the size and significance of the Japanese market in mobile communications is large by Japanese standards, it is a minute portion of the total trade and capital transfers between Japan and the U.S. and is very small relative to the broader field of electronics or telecommunications.

Since American firms have little hope of entering the Japanese markets through trade or the establishment of wholly owned subsidiaries within Japan, a far more promising route for long-term market participation involves joint ventures between U.S. and Japanese firms. Even though U.S. firms have no demonstrable technological lead in cellular mobile systems, they do have technical, capital, and marketing assets that could be useful to Japanese counterparts. Incremental cooperation through joint ventures, rather than futile attempts at instant takeover and one-sided victories, is clearly the best prospect for substantial benefits in the long run.

Increasingly, American firms seem to be adopting this strategy. Nine of the ten largest international electronic firms now have outlets in Japan, and 58 joint ventures in electronics were underway as of the 1974 fiscal year, 52 of them with U.S. firms.[53] More can be expected. If the U.S. government takes any position on the subject, it should be one that favors such joint ventures, particularly when one takes into account the detrimental effects that multinational firms have had on the U.S. economy through foreign-based production and wholly owned subsidiaries.[54]

What of Japanese participation in the American market? What is the significance in particular of the Oki success in selling mobile telephones to Bell? It is questionable whether the Oki bid represents the first thrust of a Japanese sweep of the cellular field. First of all, Oki is not primarily a manufacturer of mobile equipment; it is not expected to be a significant contractor in Japan for NTT's cellular system and its success surprised as many Japanese manufacturers as American. Second, the bid was for a mere 135 units. While some perceive this bid as an overture for expansion into the U.S. mobile market, Oki may simply be willing to withstand a minor financial deficit in order to attract attention in the American market to its other products. Third, the device involved was designed for Illinois Bell's limited Chicago equipment and is largely devoid of the complex logic circuits necessary for an entire cellular system. Fourth, AT&T's acceptance of the Oki bid will allow it to use this comparatively insignificant contract to gain access to any technological advantages that Oki might be able to demonstrate. There seems little reason to believe that the Oki "success" represents a major threat to U.S. manufacturers of mobile equipment. Indeed, although Oki has since won a regular manufacturing contract with AT&T, two U.S. firms have parallel contracts.

Nevertheless, Japanese firms that have already established a solid foothold in the American electronics market are far more likely to make further inroads into the U.S. market for mobile cellular than will U.S. firms in Japan.[55] Furthermore, the Carterfone decision will be of major significance in allowing private firms access to areas that have hitherto been the exclusive domain of AT&T, and Japanese telecommunications firms can be expected to enter this lucrative market in a determined manner. However, the devaluation of the dollar has removed some of the past advantages held by Japanese firms marketing in the U.S., and it seems improbable that there will be an inundation of Japanese products in the area of mobile communications and the autophone on a scale that would call for government intervention.

Even though American fears of Japanese imports are often more extreme than economics would warrant, protectionist cries from American firms are not easily quieted. Certain manufacturers may press for governmental assistance in one form or another, possibly in the form of higher tariffs or outright restrictions. Past experience, however, suggests that outright protectionism in electronics and communications is improbable, since American governmental efforts are generally oriented to after-the-fact adjustment assistance rather than direct protection. In view of the size of the U.S. telecommunications market, it is unlikely that a good case could be made for genuine damage to American manufacturers solely on the basis of developments in the mobile field. But political pressures have been successful in protecting particular interests in the past, especially from regionally based industries such as textiles, and most recently television sets. It is conceivable that a western/southwestern combination could prevail on Congress or the Department of Commerce to seek additional "voluntary restrictions" in mobile communications, particularly if AT&T continues to purchase from Japanese firms and the Carterfone decision allows Japanese firms to make major inroads into the rapidly developing "interconnection" business. Such a response would be both unwise and undesirable in the light of overall efforts to create a more competitive international marketplace and to provide the consumer with the widest choice of products and services at the lowest possible prices.[56] Beyond that, unilaterally imposed restrictions would clearly be counter to the more general political, social, and economic interests that unite the two countries.

Because mobile telecommunications equipment is such a small portion of total U.S.-Japanese trade, and even a small component of communications trade, it is not apt to receive much congressional or executive attention on its own. Problems that emerge are probably soluble at a relatively informal level, unless mobile communications somehow becomes a "bargaining chip" in some broader political maneuver between the two countries. Its economic insignificance would make it an item that is easily "tradable" for purposes totally unrelated to telecommunications. One can only hope that special interests of American mobile manufacturers do not predominate over more important, general interests.

A far wiser course of action for all concerned would be to recognize the mutual interests and benefits that could be pursued through joint ventures between Japanese and American firms. Collaborative efforts could provide an exceptionally advantageous position for firms in both countries through participation not only in Japanese and American markets but throughout the world with its massive potential mobile communications market. Japan and the U.S.

have positions of leadership in the field of mobile cellular systems and their firms will have a distinct marketing advantage over those of most other countries. Manufacturers and government officials in both countries should avoid petty political skirmishing over what could only be a comparatively minor irritant between the two countries, given the more fundamental mutuality of their political and commercial interests.

This chapter concludes a section in which we have explored some economic, health, legal, and trade implications of the expanding use of mobile communications. We shall now extend our discussion to prospects for the future. Chapter 14 discusses the form the mobile communications industry is likely to take and attempts to foresee problems of regulation. The subsequent chapter, after a more general look toward the future, considers the planning that will be required for orderly evolution within the land mobile field.

NOTES

1. Some of the most salient aspects of these problems are discussed in Jerome Cohen, ed., *Pacific Partnership: United States-Japan Trade: Prospects and Recommendations for the Seventies* (Lexington, MA: D.C. Heath and Co., 1972); I. M. Destler et al., *Managing an Alliance: The Politics of U.S.-Japanese Relations* (Washington, D.C.: Brookings Institution, 1976); Allen Taylor, ed., *Perspectives on U.S.-Japan Economic Relations* (Cambridge: Ballinger Publishing, 1973).

2. Ibid., plus Hugh Patrick and Henry Rosovsky, eds., *Asia's New Giant: How the Japanese Economy Works* (Washington, D.C.: Brookings Institution, 1976).

3. Gary Saxonhouse, "The Textile Confrontation," in Cohen, *Pacific Partnership*, pp. 177-97, inter alia.

4. Dodwell Marketing Consultants, *Japanese Companies' Overseas Investments* (Tokyo: Dodwell, 1974), p. 8.

5. *Nihon Kokusei Zue 1976* [Sketches of the State of Japan, 1976] (Tokyo: Dainihon Insatsu, 1976), pp. 463-66.

6. Ibid., p. 462.

7. United Nations, Statistical Office, *Statistical Yearbook,* New York, various years.

8. Ibid.

9. Tsusho Sangyosho Kikai Joho Sangyokyoku Jidoshaka, *Showa Rokujunen no Jidosha Sangyo* [The Automobile Industry in 1985] (Tokyo: Nikkan Kogyo Shimbunsha, 1975), p. 10.

10. Nihon Denshin Denwa Kosha, *Zusetsu Denshin Denwa Saabisu* [Illustrated Guide to Communication and Telephone Service] (Tokyo: NTT, 1975), pp. 19-20.

11. The outline of this plan is presented in *Denshi Tsushin Gakkaishi,* August 1975, p. 1033 ff.

12. An official outline of Japanese governmental categorizations for mobile communications is found in Yuseisho, *Showa Yonjukyunen Tsushin Hakusho* [Communications White Paper, 1974] (Tokyo: Okurasho Insatsukyoku, 1975), pp. 189-94. See also F. Ikegami, "Mobile Radio Communications in Japan," *IEEE Transactions on Communications,* August 1972, pp. 738-46; Shimizu Mitsuo and Suzuki Takayuki, *Tsushin Nettowaaku Gaikyo* [General Outline of the Communications Network] (Tokyo: Omusha, 1974), Chapter 14.

13. Yuseisho, *Tsushin Hakusho,* p. 216.

14. The Electronics Association of Japan, *Electronics in Japan '75* (Tokyo: ELAJ, 1975), p. 16.

15. Ibid., plus Shimizu and Suzuki, *Tsushin Nettowaaku*, pp. 305-7; Ikegami, "Mobile Radio Communications," pp. 738-39, inter alia.

16. The information on the Pocket Bell system is based on material found in the items cited in n. 15, plus *Denpa Jiho* [Radio Frequency Wave Report], January 1974; *Denpa Jiho*, February 1975; Yamaguchi Kaisei and Fujita Fumio, *Denki Tsushin Kogaku II* [Electrical Communications Engineering II] (Tokyo: Omusha, 1974), pp. 140-46; Kuwabara Moriji, "Tasai na Saabisu ni Chosen suru Musen Tokujutsu" [Wireless Technologies Which Demand Variegated Service], *Denki Tsushin*, October 1974, pp. 42-44; Nihon Denshin Denwa Kosha, "Idoseitai Tsushin Saabisu ni tsuite" [On Mobile Communications] (Tokyo: mimeo, n.d., 1971?); Yoshio Kikuchi and Shinichiro Sada, "A Small Sized Pocket Bell Receiver," *Japan Telecommunications Review*, July 1972, reprinted as NTT Technical Report R-No. 62-OJ; plus interviews with NTT and NEC personnel.

17. Shimizu and Suzuki, *Tsushin Nettowaaku*, p. 305; NTT, "Idoseitai Tsushin."

18. Based on interviews with officials of NTT.

19. Based on materials cited in n. 16, plus Mitsuru Komura and Hisanori Shiroie, "Cordless Telephone System," *Japan Telecommunications Review*, April 1973, reprinted as NTT Technical Report R-No. 69-OJ.

20. NTT, *Zusetsu Denshin Denwa*, p. 9.

21. Ibid., plus Shimizu and Suzuki, *Tsushin Nettowaaku*, pp. 300-3.

22. The discussion of the automobile telephone is based on Shimizu and Suzuki, *Tsushin Nettowaaku*, pp. 305-9; Ikegami, "Mobile Radio Communications"; Morinaga Takahiro, "Ido Tsushinmo" [Mobile Communications Networks], *Denshi Tsushin Gakkaishi*, November 1970; NTT, "Idotai Tsushin"; Mitsuru Komura, Tateki Inada, and Ken Osaka, "Mobile Radio Telephone System for Emergency Use," *Japan Telecommunications Review*, January 1974, reprinted as NTT Technical Report R-No. 70-OJ; Ito Sadao, "Jidosha ni okeru Ido Tsushin" [Mobile Communications in the Automobile], *Denshi Tsushin Gakkaishi*, April 1974; plus interviews.

23. The discussion of the cellular system is based on the items cited in n. 22, plus Ito Sadao, "Jidosha Denwa Hoshiki no Kaihatsu" [The Development of an Automobile Telephone System], *Tsushin Kogyo* [Communications Engineering], April 1975, pp. 13-24; NTT, *Development of a High Capacity Land Mobile Radio Telephone System* (Tokyo: mimeo, n.d.); "Happyaku MHz-tai ni okeru Rikujo Ido Tsushin ni kansuru Jitchi Chosa Hokoku" [Report on the Results of Investigations Concerning Land Mobile Communications in the 800-MHz Band], document submitted as research material #2-4 to the Denpa Gijutsu Shingikai; Komura Mitsuru and Ito Sadao, "CCIR ni okeru Ido Tsushin Gyomu no Kenkyu Doko to Jidosha Denwa Hoshiki no Kaihatsu Doko" [Research Developments in the Mobile Communications Business in CCIR and Developmental Progress in the Automobile Telephone]; Nihon ITU Kyokai Jikko, ed., *Kokusai Denki Tsushin Rengo to Nihon* [The International Telecommunications Union and Japan], April 1975, pp. 37-43; plus interviews. *See* also Ito and Yasushi Matsuzaka, "800 MHz Band Land Mobile Telephone System—Overall View," *Review of the Electrical Communications Laboratories*, vol. 25, nos. 11-12, November–December 1977, pp. 1147-56.

24. An interim report by the council was published in 1974; its final decisions are summarized in a series of articles in *Denpa Shimbun* [Radio Frequency News] during Spring 1976.

25. On Dyna-Tac see *Electronics*, April 12, 1973. Motorola has additionally produced a number of promotional documents on the system available through the company.

26. Yamaguchi and Fujita, *Denki Tsushin Kogaku;* Kuwabara, "Taisai na Saabisu"; "Idotai Tsushin"; Shimizu and Suzuki, *Tsushin Nettowaaku*, p. 309; plus interviews with NTT officials.

27. How close final commercial design will be to these figures is unclear, but the low-level power of the battery should indicate that a larger unit would be needed unless major changes are made in the power source's capabilities.

28. Bell Laboratories, "High Capacity Mobile Telephone Service Technical Report," Unpublished document, Bell Laboratories, Holmdel, N.J., December 1971.

29. Based on *The Market for Communications Equipment and Systems in Japan*, a report prepared for the U.S. Embassy, Tokyo, by A. T. Kearney International, Inc. (September 15, 1975).

30. Unpublished survey data made available by NTT officials.

31. Based both on extrapolation from expected usage and several public and private market surveys.

32. This position is presented most forcefully by Clifford A. Bean, "How American Companies Can Capture the Overseas Market," *Communications*, May 1975, pp. 33-38. On the broader question of U.S. company's prospects see also Louis Kraar, "Japan is Opening Up for *Gaijin* Who Know How," *Fortune*, March 1974, pp. 146-57; Ministry of Foreign Affairs, *Keys to Success: Foreign Capitalized Corporations in Japan* (Tokyo: Ministry of Foreign Affairs, 1973), inter alia.

33. Quoted in *Electronics*, August 7, 1975, p. 12.

34. Ibid; see also *Electronic News*, June 9, 1975, inter alia.

35. James Abegglen, "Japan, Incorporated: Government and Business as Partners," *Changing Market Systems*, 1967, pp. 228-32; Eugene J. Kaplan, *Japan: The Government-Business Relationship* (Washington, D.C.: U.S. Government Printing Office, 1972). For a good critique of the notion see Gerald L. Curtis, "Big Business and Political Influence," in Ezra Vogel, ed., *Modern Japanese Organization and Decision-Making* (Berkeley and Los Angeles: University of California Press, 1975).

36. On the virtues of free trade see particularly Gardner Patterson, *Discrimination in International Trade: The Policy Issues, 1945-1965* (Princeton: Princeton University Press, 1966); Bela Balassa, *Trade Liberalization Among Industrial Countries* (New York: McGraw-Hill, 1967); Edward L. Morse, "The Politics of Interdependence," *International Organization*, Spring 1969, pp. 311-26; Robert Gilpin, *U.S. Power and the Multinational Corporation* (New York: Basic Books, 1975), especially Chapter 8; Shimano Takuji, ed., *Sekai no naka no Nihon Keizai* [The Japanese Economy in the World Context] (Tokyo: Gakuyo Shobo, 1975); Isoda Keiichiro, *Nihon Boeki no Mirai* [The Future of Japanese Trade] (Tokuo: Toyo Keizai, 1972), inter alia.

37. Organization for Economic Cooperation and Development, *Policy Perspectives for International Trade* (Paris: OECD, 1967), p. 166.

38. James C. Abegglen and William V. Rapp, "The Competitive Impact of Japanese Growth," in Cohen, *Pacific Partnership*, p. 20.

39. Based on U.S.-Japan Trade Council, *United States Exports to Japan* and *United States Imports from Japan* (various dates).

40. Abegglen and Rapp, "Competitive Impact," p. 20.

41. U.S. Arms Control and Disarmament Agency, *The Implications of Reduced Defense Demand for the Electronics Industry*, September 1965, p. 7.

42. See George W. England and Raymond Lee, "Organizational Goals and Expected Behavior Among American, Japanese and Korean Managers—A Comparative Study," *Academy of Management Journal*, December 1971, pp. 432-36; Lewis Austen, *Saints and Samurai* (New Haven: Yale University Press, 1975); and William A. Fischer, *Postwar Japanese Technological Growth and Innovation: A Comparative Review of the Literature* (Washington, D.C.: Program of Policy Studies in Science and Technology, George Washington University, 1974), pp. 27-30.

43. See Fischer, *Postwar Japanese Technological Growth*, pp. 52-75; Haruko Fukuda, *Japan and World Trade: The Years Ahead* (Westmead, England: Saxon House, 1973), pp. 75-94; Kaplan, *Japan;* Yoichi Okita, "Japan's Fiscal Incentives for Exports," in Isiah Frank, ed., *The Japanese Economy in International Perspective* (Baltimore: Johns Hopkins Press, 1975); and Patrick and Rosovsky, *Asia's New Giant.*

44. Quoted in *The Japan Times*, January 5, 1976.

45. United Nations Conference on Trade and Development, *The Kennedy Round: Estimated Effects on Tariff Barriers* (UNCTAD, 1968); Fukuda, *Japan and World Trade*, p. 79.

46. Dan Fenno Henderson, *Foreign Enterprise in Japan* (Chapel Hill, N.C.: University of North Carolina Press, 1973), p. 237.

47. Fukuda, *Japan and World Trade,* p. 77.

48. Economic and Foreign Affairs Research Association of the Ministry of Foreign Affairs, *Statistical Survey of Japan's Economy: 1975* (Tokyo: Ministry of Foreign Affairs, 1975), p. 53 (as calculated under four digits of BTN).

49. Data supplied by Ministry of International Trade and Industry. See on this general problem Patterson, *Discrimination in International Trade,* pp. 293-300; and John Lynch, *Toward an Orderly Market: An Intensive Study of Japan's Voluntary Quotas in Cotton Textile Exports* (Tokyo: Sophia University Press, 1968).

50. *Statistical Survey of the Japanese Economy,* p. 44.

51. For a study of this problem in comparative perspective see Organization for Economic Cooperation and Development, *Government Purchasing in Europe, North America and Japan* (Paris: OECD, 1966); also Balassa, *Trade Liberalization,* p. 65.

52. Based largely on interviews with industrial manufacturers of products for NTT.

53. Henderson, *Foreign Enterprise,* p. 366.

54. On this problem see most notably Gilpin, *U.S. Power and the Multinational Corporation,* though of course the criticisms have been widely made.

55. Some estimates are that the U.S. market will be in the neighborhood of $10 billion within 15 years. *Business Week,* June 30, 1975, p. 32.

56. See especially Patterson, *Discrimination in International Trade,* but also the other works cited in n. 36.

Part IV
PERCEPTIONS
OF
THE FUTURE

Chapter 14
THE EMERGING MOBILE COMMUNICATIONS INDUSTRY
Structure and Regulation

Erwin A. Blackstone
Harold Ware

W e shall now attempt to anticipate the future structure of the mobile communications industry and problems associated with its regulation. In order to provide an appropriate context for our anticipatory analysis, it is necessary to discuss briefly spectrum use and allocation, trends in regulatory policy, and projections of demand for services.

SPECTRUM, REGULATION, AND DEMAND

Spectrum: Processes of Allocation and Efficiency of Use

The market is not used to allocate spectrum; consequently, the spectrum, unlike most other resources, is not priced, and forces that act via the price mechanism to promote efficient allocation are absent. Instead allocations of spectrum are made by regulatory decisions that do not permit exchange among users that would ensure spectrum is employed for its most valuable uses. Accordingly, neither the spectrum's worth nor the appropriate degree of conservation in its use has been established. If a market for spectrum existed, the (more capital intensive) systems that conserve spectrum would be more apt to be utilized in areas where spectrum congestion is now a problem; users would tend to employ spectrum efficient systems, involving more expensive equipment, where the price of spectrum was high.

In Docket 18262 the Federal Communications Commission (FCC) has made some improvements in its mode of allocation. Spectrum in the 900-MHz band has been allotted to specific technological systems rather than to a type of user.

[347]

This method reduces the chances that some users will have idle spectrum while others must engage in excessive conservation or tolerate poorer quality service. However, despite such improvements, a market still does not exist, and the value of spectrum in its many uses and potential uses remains unclear. As a result society (as represented by the FCC) must employ political and administrative means to allocate spectrum. Thus, an assessment of mobile communication's regulatory policies requires a comparison between the spectrum efficiencies of the systems under consideration.

Quantifying the spectral efficiency of systems is difficult because no single measure is adequate for all purposes; nevertheless, the number of mobiles per unit of spectrum can be used for some analysis if one is aware of its limitations. Since such a measure conveys nothing about the quality of the service provided, using this measure to compare the "efficiencies" of systems used for different services can be misleading. One can, however, use mobiles per MHz to compare different systems providing the same services with equivalent service parameters. The material below is based on hypothetical analyses that take this approach.

Comparing the spectral efficiency of the various systems providing major mobile communication services will be facilitated by computing an index of Relative Spectral Efficiency (RSE):

$$RSE = \frac{\text{mobiles per MHz for a given system}}{\text{mobiles per MHz for a trunked system}}$$

The data shown in Table 14.1 indicate the large differences between various systems with respect to this single efficiency measure. When based on Richard Lane's calculations, the differences are even larger. For example, according to his analysis, cellular systems providing dispatch service can accommodate almost 9 times as many mobiles/MHz as 16-channel trunked systems and almost 22 times as many as conventional systems.[1] Surprisingly, our data also indicate that, in comparison to trunked systems, cellular systems provide greater gains in spectral efficiency for dispatch service than for mobile telephone service (MTS).[2] (It should be noted that these relationships are based on the assumption that cellular dispatch service will continue to be characterized by the same service parameters as those typical of other means of dispatch service. Since the new system may change the nature of dispatch, this assumption could lead to spurious results.)

In general there are two competing factors at work: the stringency of the service quality parameters and the complexity of the systems. Other things being equal, the higher quality *services* use more spectrum per subscriber, but this can be offset by using more complex *systems*. One should also keep in mind, however, that the more spectrum that is available, the greater the efficiencies to be realized via trunking, sharing, and spectrum reuse. Moreover, a large total allocation to land mobile radio (LMR) can promote efficient results by raising the number of firms that can compete in the industry. Thus, the reallocation of spectrum to mobile communications (as well as aspects of the suballocations within the mobile area) is consistent with recent procompetitive trends in regulating the communications industry.

Table 14.1: *Relative Spectral Efficiency of 900-MHz Systems for Proposed Two-Way Services*

	Dispatch Service*	Mobile Telephone Service
Cellular	2.80	2.75
Trunked	1.00	1.00
Conventional	0.64	--**

SOURCE: Derived from Tables 2.3 and 2.4, Chapter 2.

NOTE: The underlying estimates assume that a total of 10 MHz is available per system; thus, the advantages of cellular are reduced.

*Values computed for 2-frequency simplex technology.

Not applicable because achieving the desired level of service is too costly. Lane's estimates imply that the desired parameter values could only be achieved with a conventional system in which each user had his own dedicated channel. [Richard N. Lane, "Spectral and Economic Efficiencies of Land Mobile Radio Systems," **IEEE Transactions on Vehicular Technology VT-22 (November 1973):98.]

Transfer of Spectrum from Television Broadcasting to Land Mobile Communications

Docket 18262 involved a direct transfer of spectrum from television to land mobile. It is difficult to make an economic analysis of the costs and benefits of this transfer because the uses are very different; the latter is currently an input into the production process, while the former is primarily a final consumer product. Only rough estimates can be made of the potential economic benefits of LMR, and the benefits of UHF-TV—resulting from entertainment and trade stimulation through commercials—are even more difficult to assess. Furthermore, the allocation of spectrum to UHF-TV for expanded programming involves a number of issues that are not primarily economic. Nevertheless, a number of comments can be made concerning the economics of the spectrum allocation issue.

When appraising the merits of the spectrum transfer, it should be remembered that as late as 1973 a total of only 42 MHz were allocated for land mobile use.[3] At the same time, 492 MHz were allocated to television channels, including VHF (lower frequency) and UHF (higher frequency). Each television channel requires 6 MHz, which is twice the total allocation for all public mobile telephone service prior to Docket 18262. One television channel occupies the same spectrum space as 2,400 MTS users or 25,000 dispatch users on a 16-channel trunked system. This very large allocation to television resulted from policies adopted by the FCC in the early days of television to encourage diversity and local programming. In particular, the FCC decided that many UHF stations were necessary to achieve these objectives. The policy was not successful. UHF frequencies often went unused, and UHF stations were frequently unprofitable;[4] there is still substantial excess capacity in the UHF allocation. A very conservative estimate of the capacity of the 70 UHF channels is 3,850 stations,[5] about 90 percent of which were not utilized in the early sixties.[6] Additionally, there are only two stations operating in the 900-MHz band, and they could be reallocated to lower frequencies for a total cost of roughly $200,000, a cost that is minute compared to the net gains estimated in Chapter 10. Because the UHF television band has been an underutilized resource, and the net benefits, both quantifiable and unquantifiable, of increased mobile communications are likely to be very large, the spectrum reallocation decisions resulting from Docket 18262 seem justified, at least on economic grounds.

Even if one accepts the desirability of having several local stations, UHF broadcasting is not necessarily the best method of achieving this. Cable offers a good substitute for UHF television; already an economically viable alternative, technological improvements will make it more attractive. The economics of cable involves decreasing average cost since the cost of adding another channel is very small.[7] In contrast, the extra cost of UHF television, because it is less efficient, is greater than VHF television. These cost factors imply that cable television makes local programming more likely. On the other hand, cable television's capacity to increase the number of alternatives for viewers could reduce the viability of some local stations by diluting their audience and lowering advertising revenue.[8] Our calculations suggest that the cost of transferring to cable in areas where there is a shortage of spectrum would be significantly smaller than the expected long-term benefits from use of land mobile communications.

Comparisons between the allocation of spectrum to television and to land mobile communication must also examine the relative social benefits. While externalities exist for both UHF and mobile communications, the potential economic benefit of mobile services for emergency use alone seems a very persuasive factor in favor of redistributing spectrum—especially considering television can utilize cable and land mobile has no similar good substitute.

Our analysis of the gains from reallocation is only suggestive. For example, some may argue that consumers are receiving large amounts of consumer surplus from over-the-air television and might be willing to pay much more for such entertainment. Also, television and radio broadcasting generate consumer benefits that exceed the cost of their production and use.[9] Our analysis is based solely on economic grounds and therefore does not fully account for a number of cultural, regional, and consumer issues that are relevant to the reallocation. Given the spectrum efficiencies of the new mobile communication technologies planned for the 900-MHz band, it seems unlikely that further reallocation will be necessary for a considerable period. Yet our analysis suggests that even if more spectrum was devoted to land mobile, and cable displaced over-the-air television, the economic gains would be significant.

Trends in Regulation

To understand the significance of the decisions resulting from Docket 18262 for the future structure of the industry, it is necessary to consider the regulatory context in which those decisions were made. Over the past two decades, the FCC has gradually narrowed the scope of the monopolies of the wired telephone companies; the resulting increase in competitive pressure on these companies has had important consequences for the pricing patterns in the industry.

The Carterfone decision, which concerned a device that permits a mobile radio unit to be connected to the wired telephone system, led to competition in the interconnection of the terminal equipment.[10] A large market for various attachments such as private branch exchanges (PBX) and answering devices that were previously the exclusive preserve of the telephone companies[11] has been opened to rival suppliers.

The "above 890" decision, made in 1959 by the FCC,[12] authorized the operation of private microwave systems for long-distance communications, and is another area where competitive pressure has increased. The existence of this alternative resulted in much lower rates being offered by the Bell companies,[13] but the benefits were available only to large companies that could afford and utilize a private microwave system.[14]

Competition in long-distance communications services was also increased by the 1969 Micro-wave Communications, Inc., decision, involving a limited common carrier microwave radio service designed to meet the interoffice and interplant communication needs of small firms.[15] In 1971, the FCC allowed additional companies providing specialized common carrier microwave systems entry into the market to compete with the available private line services.[16] AT&T responded by lowering rates on some high density routes and raising them on some low density routes. Many issues associated with specialized common carriers are still being contested.[17]

In the wake of these developments, the telephone companies have complained of "cream skimming"; they have responded by reducing prices and accelerating their rate of innovation in the areas under competitive challenge.[18] The FCC now has the difficult task of deciding whether the new "competitive" rates are compensatory.

The increased competition has caused erosion of the traditional telephone industry price structure, which has involved internal subsidization, with some users receiving services below cost while others pay prices above cost. AT&T claims that eliminating such practices will result in large increases (as much as 75 percent) in the cost of residential service,[19] but this claim is disputed by others.[20]

Prices in a monopolistic telephone industry can depart considerably from costs by being based, for instance, on the value of service, which involves price discrimination. Such pricing has some undesirable consequences in terms of efficiency and equity; users receive false signals, leading them to consume too much of the subsidized services.[21] However, the new conditions of entry and competition should bring prices closer to marginal costs and may lead to a more efficient resource allocation through the influence of price on use.

Demand at 900 MHz

The question of potential market size is crucial to ascertaining whether the above trends can be extended to the mobile communications industry. Unfortunately, estimating demand for 900-MHz systems and services that have yet to be deployed is fraught with serious problems. First, since the services that the new systems can perform are of a higher quality than those provided by the current technology, simple projections based on past demand are of questionable value; this is especially true for cellular services. Second, cellular systems have such a large capacity that it is possible for MTS to become a private consumption item. Consequently, advertising may be significant in determining consumer preferences and influencing demand in a manner that is impossible to predict with confidence. Third, systems and services of differing quality are only imperfectly substitutable. This is important because system quality affects demand and the quantity demanded through its influence on costs and, hence, prices.

There have been many estimates made of the number of potential buyers for various systems, estimates that are based on a disparate set of assumptions concerning demand and supply parameters. Some of the results are reproduced in Table 14.2. For reasons given above, one should be skeptical about the reliability and validity of many of the figures.

It is also relevant to consider demand projections for one-way paging, in part because some of those users might be induced to employ MTS or dispatch service if the prices become low enough. In 1975 there were about 400,000 to 500,000 pagers in use.[22] That number has almost doubled in only two years; there are now 800,000 units being employed.[23] Systems Applications, Inc. projects that by 1982 a total of almost 1.5 million pagers will be used in a common carrier mode, and Arthur D. Little predicts the number of paging units in operation in 1985 will be 2.9 million.[24] The Arthur D. Little study assumes that the paging unit's average price will decline from the current price of about $190 to $125 by 1985 (in 1974 prices) as a result of economies of scale, heightened competition, and use of more advanced semiconductor technology.[25] Obviously, such a reduction will increase the attractiveness of paging vis-à-vis two-way services if the latter's prices do not also decline. But the same factors that would tend to reduce paging prices should similarly affect two-way prices (since the two types of services are substitutable).

In view of the disparities in the figures presented above, it would be useful to provide our own projections. Unfortunately, a survey of adequate reliability would have required resources beyond our means and, given the problems mentioned at the outset of this discussion, a casual study would have clarified little. Nevertheless, we can shed some light on the subject of demand by considering the delay that must be avoided in order to compensate for the cost of a month's mobile radio service. Table 14.3 shows the number of hours per month that must be saved (or delay avoided) through the use of mobile communications to cover the service's cost; the table is broken down by hourly wages of the users and by the cost per month of one- and two-way systems. Any

Table 14.2: *Some Estimates of Demand for Two-Way Service*

Service and Market Area	Source	Number of Subscribers or Users (× 1,000)	Year	Assumed Price Per Month (Where Available)
MTS, national	Systems Applications, Inc.[a]	1,000	1980	$40
		2,340	1990	$40
		5,500	2000	$40
	Martin Marietta[b]	112	1975	$60
		5,600	1985	$25
MTS, Chicago	Electronic Industries Association[c]	----	1970	$80–90
		25	1980	$40
		45	1990	$30
MTS, New York City	Electronic Industries Association[c]	76	1990	$30
Dispatch, national	Martin Marietta[b]	3,800	1975	----
		17,000	1985	----
Private dispatch, national	Systems Applications, Inc.[d]	7,600	1980	----
		21,200	1990	----
		36,900	2000	----
Common carrier dispatch, national	Systems Applications, Inc.[d]	44.5	1980	----
		150	1990	----
		500	2000	----
Private dispatch, Chicago	Electronics Industries Association[c]	152	1980	----
		192	1985	----
		234	1990	----

a. U.S. Department of Commerce, "Land Mobile Communications and Public Policy," prepared by Systems Applications, Inc., vol. 1 (Springfield, Va.: National Technical Information Service, 1972), pp. 131-32.
b. Cited in Paul F. Kagan, "Go-Ahead Signal," **Barrons**, June 10, 1974, p. 11.
c. Electronic Industries Association, Communications and Industrial Electronics Division, Land Mobile Section, "Comments," Formal submission to FCC Docket 18262, July 7, 1972, pp. 7-11.
d. U.S. Department of Commerce, "Land Mobile Communications," p. v.

additional hours saved make use of mobile communications profitable. Our computation ignores the costs of other items such as fuel, vehicle depreciation, and public (or other) telephones and, therefore, overstates the delay necessary to make adoption of mobile communications profitable.

This table indicates the small number of hours that have to be saved to make mobile communications profitable; hence, it suggests that the potential market for such communication services might be considerable, especially because wages are likely to increase and the cost of mobile communications is apt to decline.

Having considered both the costs (Chapter 4) and benefits (Chapter 10), we

Table 14.3: *Break-Even Times for Use of Mobile Communications*

Attributed Hourly Wage of the User or Potential User	One-Way Paging $20/Month (Hours)	Two-Way Service $90/Month (Hours)	Two-Way Service $60/Month (Hours)	Two-Way Service $40/Month (Hours)
$ 5	4	18	12	8
10	2	9	6	4
20	1	4-1/2	3	2
40	1/2	2-1/4	1-1/2	1
80	1/4	1-1/8	3/4	1/2

NOTE: These figures do not include consideration of additional consequences of delay such as other individuals not being able to perform their work.

conclude that the net benefits may be large enough to support some of the more optimistic "demand" estimates cited in Table 14.2 as well as projections referred to in Docket 18262. Since different people will evaluate the potential future benefits of various systems in different ways, no particular system seems, at this point, likely to predominate. However, the discussion of system suitability presented in Chapter 10 indicated that the benefits of more complex systems may be sufficient to compensate for their higher cost. Evidence and arguments concerning modes of use and attitudes of potential purchasers already presented in Chapters 5 and 9, and others to be presented in Chapter 15, suggest the possibility of significant expansion in the currently underdeveloped MTS market.

Thus, it appears that the initial market is apt to be substantial enough to allow equipment prices to fall. There are several reasons for this conclusion: such a market makes it feasible to introduce specialized, automated production techniques; experience in producing many similar units reduces costs along "learning curves";[26] input costs may decline when purchases are in bulk; and the knowledge that there is a large market may promote greater expenditures of research and development (R&D) funds to increase efficiency.

If the above forces are not sufficient to lower the price of equipment, it is possible that over a longer period more firms will enter the industry in search of profits. The actual entry or threat of entry by more producers will raise the probability that competitive forces may lower prices. If the market continues to grow, more firms may enter the field and/or firms already in the industry may expand their operations in order to achieve greater economies of scale. Existing firms might also expand to try and block competitive entry, but such maneuvers are difficult to accomplish in a growing market.

Over the long run, there are other factors that may affect the MTS market size. As the number of people possessing mobile units increases, these units become more desirable to others, and the influences of business competition, fashion, and status become significant. (See Chapters 5 and 9.) The demand curve for MTS could also shift outward because of the substitution effect: as prices for wireline phones are pushed up by higher resource costs, consumers may replace them with mobile phones. The costs of required resources, such as copper, are rising, as are the costs of land and labor, which are more significant for landline than radio systems. In addition, cost reductions through "learning" are more difficult to realize in the older industry, while in the emerging, capital intensive LMR industry, costs are likely to decline for some time, as we noted earlier. Finally, the demand for mobile communications will rise along with the demand for communications services in general, because the costs associated with a primary substitute—transportation—are climbing rapidly.

Although the preceding analysis is certainly highly speculative, it strongly suggests that the market for mobile communications is likely to be large and will expand as the technology becomes more visible to society. With this perspective, we can now turn to a detailed analysis of the emerging land mobile communications industry.

ALLOCATION WITHIN THE 900-MHz BAND
AND ITS IMPACT ON THE INDUSTRY

A recapitulation of the principal decisions in Docket 18262 relevant to the future structure of the industry will provide some helpful background for our discussion. Of the 115 MHz that were allocated to mobile communications, 40 MHz were designated for high-capacity common user cellular systems, and 30 MHz for dispatch users employing either conventional or trunked systems. The remaining 45 MHz were held in reserve. There were also a number of other important elements to the decision, some of which set new precedents, including: (1) allocation of spectrum by technological system rather than by type of use—a departure from previous practice; (2) granting of an effective monopoly over mobile telephone service to cellular operators; (3) creation of a special class of common user systems in the 30 MHz allocated to dispatch, and reliance on competition rather than regulation to protect consumers within this class; (4) encouragement of competition by (a) limiting vertical integration in both trunked and cellular operations, (b) requiring complete separation of a cellular subsidiary from any parent wireline telephone company, (c) granting cellular operators the right to offer dispatch service as well as mobile telephone service, but prohibiting cellular firms from providing fleet call (by which all vehicles in a given group are called at once), and (d) prescribing an open entry policy for the dispatch market.

In order to evaluate the likely impact of these decisions on competitive potential of the various sectors of the land mobile radio industry, it is necessary to identify those sectors and specify their relationships. For our purposes, it is especially important to distinguish between providing service and manufacturing equipment.

Within the categories of manufacturing and service, it is possible to construct a more detailed classification involving subsectors based on a system's type or the functions it performs. In the case of equipment manufacture, we have not found it useful to employ subcategories based on type of system, since many of the components and techniques necessary to construct the various systems have common features. We have found it desirable to distinguish between the manufacture of radio and computer components, because producing the more sophisticated computers required for cellular and trunked systems may be beyond the capacity of many smaller and less integrated firms in the communications industry.[27]

With regard to the service networks, a classification based on whether the system is trunked, cellular, or conventional has some validity but can be misleading because each of these systems offers services that are interchangeable to some extent. The degree to which the different systems are alternatives is an important factor in policy analysis, and a classification scheme based on the type of system can obscure issues involving substitution.

Consideration of these factors has led us to use the scheme depicted in Figure 14.1. The prime distinction made is between manufacturing and service; classi-

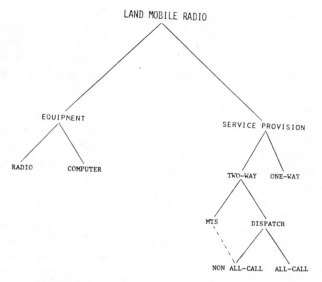

Figure 14.1: *Relationships within the land mobile communications industry*

fication within the service component is based on the character of the service provided.

Several additional points need clarification. First, we must explain our use of the term "competition," since the word has a variety of meanings in common usage. The traditional economic definition of competition is quite restrictive and involves a definite set of conditions (economists view competition as an ideal, at least in a static sense). Industries that are truly competitive have no barriers to entry and involve numerous small firms (relative to the market) producing the same product, with no control over price. These structural factors imply an industry with efficient production, optimum allocation of resources, and absence of supernormal profits. In reality perfect competition does not exist; however, there are industries that are workably competitive, that is, entry barriers are sufficiently low, the number of firms is sufficiently large, and product differentiation is sufficiently limited so that the benefits associated with perfect competition are essentially realized. Henceforth, when we use the terms "competition" and "competitive," we are implicitly referring to "workable competition" and "workably competitive," respectively. Unfortunately, an industry is often labeled "competitive" when it does not possess the proper attributes; frequently the absence of monopoly is equated with the presence of competition.

In many cases, a multifirm industry containing some firms that are large in comparison to the market should have its practices labeled as rivalry rather than competition. The larger rivals can exert considerable monopoly power if restraining factors, such as the threat of entry, the threat that industrial buyers might produce their own equipment, etc., are absent. We do not mean to imply, though, that large firms are always disadvantageous from a societal viewpoint.

Large firms may be necessary for taking advantage of economies of scale, and they can spend large amounts on research and development that may bring about gains in efficiency through technological change. Many innovations have originated in concentrated industries, yet within these industries it is often the small- or medium-sized firms that produce the change. Thus, even in the dynamic sense, it is necessary to carefully scrutinize an industry to ensure that size greater than required for an efficient scale of production does not result in monopolistic abuse and lethargy.

It is important to note that some of our analysis will be based on data derived from past and contemporary forms of the industry, although allocation of new spectrum in the 900-MHz band will surely lead to an expanded industry with a more complex structure. One might be misled by current or historical data, but it is a risk that cannot be avoided.

Many of the decisions in Docket 18262 are based on the FCC's contention that competition is likely in some sectors of the land mobile communications industry. We will attempt to ascertain whether this contention is justified. In addition, we will examine the specific regulations proposed for those sectors where the FCC feels competition is unlikely, as well as for those sectors that may require assistance to ensure that competition emerges. Many argue that even if competition is improbable, regulation is undesirable for a variety of reasons, some of which were noted in Chapter 3.[28] The avoidance of regulation would be consistent with the policy trends discussed in the previous section.

The FCC has sought to establish an environment wherein those desiring dispatch service in the 900-MHz band can obtain it from a group of competing sources very much like those in intercity microwave service. The commission has allowed cellular operators (who have an MTS monopoly at 900 MHz) and independent (noncommon carrier) systems employing conventional or trunked technology to provide dispatch service to eligible subscribers. Dispatch service can also be provided on private (user owned) systems. A key question is whether introducing new 900-MHz systems will make this sector of the land mobile radio industry sufficiently competitive. With regard to MTS, the crucial issue is a different one—whether the deployment of the new technologies would result in this component of the industry becoming sufficiently monopolistic to justify the imposition of government franchised and regulated monopolies.

To answer these questions an examination of the potential costs and benefits, as described in Chapters 4 and 10, is useful. Indeed, doing so is virtually mandatory given the tentative nature of existing demand estimates. By supplementing those estimates with the above material, we are able to use the essential concepts of the standard method for determining an industry's competitive potential: examining and comparing cost structure and market size. If the industry demand curve intersects the Long Run Average Cost (LAC) curve at a point where the latter is declining, monopoly is likely to develop, while competition by numerous small firms would yield unnecessarily high costs. On the other hand, if an efficient scale of production can be achieved at low output levels (relative to industry demand), then competition is possible.

Although insufficient data exist for a rigorous analysis, it is possible to obtain useful results by employing these notions. They will be used to deal first with the mobile equipment manufacturing industry and then with the dispatch and mobile telephone sector of the land mobile radio service industry.

THE MOBILE EQUIPMENT MANUFACTURING INDUSTRY

Participants in the equipment manufacturing sector were enumerated in a June 10, 1974, article in *Barron's*. Four firms—Motorola, General Electric, RCA, and E. F. Johnson—accounted for 90 percent of its output, and Motorola dominated the industry, having produced 64.2 percent of its output in 1973.[29] (The recent surge in production of CB units has altered the picture. However, CBs are less expensive and are not substitutes for most of the other mobile devices under consideration for commercial and emergency use. Nevertheless, firms involved in producing CBs are potential entrants into other mobile equipment areas and could thus exert a competitive influence.) It is clear from these figures that the mobile equipment industry has not been competitive in the sense defined earlier.

Based on the statistics alone, one cannot conclude that there are economies of scale inherent in the industry that account for its high degree of concentration. In fact, numerous "small" firms with market shares of less than 3 percent have been able to survive; six firms had market shares ranging from 0.4 percent to 2.8 percent, and there are at least seven more firms with not more than 0.4 percent of the market. The existence of this competitive fringe will help to limit the market power of the larger, dominant firms. No single factor or group of factors is apparent, however, that will guarantee the mobile communications equipment industry will be competitive. Thus, the FCC's "hands-off" policy is only a necessary condition for the emergence of a competitive industry; it is surely not a sufficient one.

What, then, are the influences working for and against the future development of a competitive mobile communications equipment industry? We will look first at the procompetitive factors.

The expanding demand indicated earlier is an important factor favoring competition. This has been recognized by the FCC, and was discussed extensively by the commission in Docket 18262.[30] Figures given in Table 14.2 indicate that the total growth in the market for mobile units, excluding CBs, could amount to 18.6 million units over the next decade. In order to achieve this growth, and accepting an FCC estimated growth rate of 10 percent per year,[31] we would need an annual production rate that increases from the present figure of 1.2 million units to greater than 2.3 million by 1980 and 4.3 million by 1985.

Expanding demand is favorable to competition for several reasons. In the first place, new firms can enter the industry with less fear of retaliation from those already established. Also, the larger the market, the greater the number of firms that can be absorbed. Furthermore, a large and expanding market enables entry of new firms that are big enough to take advantage of economies of scale; these entrants will not be at a cost disadvantage and will be better able to survive.

This last point leads to the important question of how many plants of the minimum size necessary for efficient production can be sustained by the projected market. Let us make the conservative assumption that there will be a demand for at least 2 million mobile units per year during the 1980s. We can deduce from the figures in the *Barron's* article that a viable firm must produce from 2,000 to 350,000 units per year. (These figures were derived by dividing the total revenues of the manufacturing firms by $1,000, which is roughly the cost per mobile.) Clearly this estimate is very crude, but the results are consistent with other pertinent sources of information. For example, AT&T has stated that a firm would have to produce 20,000 units in order to realize scale economies involved in manufacturing the cellular mobile phones. Since the largest figure (350,000) refers to Motorola, which operates several plants, and the figure 2,000 refers to a firm producing units that are simpler than future units and may therefore entail fewer economies of scale, we can justifiably narrow the potential range of efficient production to between 10,000 and 100,000 units per year. This implies that the industry should be capable of sustaining 20 to 200 firms. The data also suggest that firms operating several plants would be able to produce larger quantities efficiently, thereby decreasing competition. Nevertheless, one can safely conclude that the combined effect of an expanding market and limited economies of scale will be procompetitive.

Firms require substantial capital and expertise to enter technologically sophisticated industries, and numerous obstacles exist to prevent successful participation by new firms. However, there are many organizations that could easily be suited to the mobile market. Numerous large firms exist that are involved in manufacturing other kinds of communications equipment; as the market expands, some will diversify and enter the commercial and emergency mobile equipment industry. Small firms now part of the "competitive fringe" could become another source of genuine competition as the market expands. Finally, the fact that technology in the LMR industry is changing so rapidly means that established firms may have less of an advantage over emerging ones.

Unfortunately, the highly technological nature of the LMR industry could also inhibit competition, so we will now examine this and other anticompetitive influences. High technology industries are often more concentrated because large firms can more easily afford the required expenditures on research and development. Motorola, for instance, has already invested substantial amounts of money in its advanced 900-MHz systems, and this will surely help to maintain its dominant position in the industry. Another related factor is the existence of patents, which can combine with R&D investment to form a formidable barrier to entry.

The FCC has stated that it will set standards for mobile users. If these standards require a particular technology be employed, rather than requiring that specific service parameters be met, then the entry barriers could be almost insurmountable. Thus, together the preceding considerations make it clear that the competitive *potential* cited by the FCC is just that; it is not a certainty.

A number of policies should be pursued to increase the likelihood of competition. First, the Justice Department should monitor the industry for practices that inhibit competition. Second, FCC standards should be concerned with performance rather than technical means for implementation. Where this is not possible, the FCC should adopt policies to prevent the abuse of patents on key techniques.[32] Third, foreign competition should be given a fair chance; restrictive tariffs or quotas should be avoided if possible. Fourth, the FCC should maintain its restrictive position on vertical integration.

In summary, workable competition is not a foregone conclusion in the mobile equipment industry and the benefits commonly associated with competition will not necessarily be realized. If "rivalry" develops, specific policies will be needed to ensure that it is channeled into productive and innovative activities.

THE MOBILE COMMUNICATIONS SERVICE INDUSTRY

The current structure of the mobile communications service industry is compatible with competition—there are a large number of small firms. For example, in Los Angeles in 1976 there were, besides Pacific Telephone, four companies offering radio telephone service and ten providing paging service. There are also about 700 independent radio common carriers (RCCs) that supply mobile telephone service through interconnection with the wired system. Given the currently limited number of mobile telephone subscribers, many of these companies are small and probably have little capital;[33] they have typically provided service for only 90 subscribers per system.[34] In addition, the wireline telephone companies operated 1,300 mobile telephone systems, which have also been small. On the average, they provided service for 60 units per system, and about half of the 1,300 systems were operated by AT&T companies. The firms providing two-way service had insufficient spectrum to be able to exploit potential economies of scale. Paging service of the one-way variety was also provided by radio common carriers.[35] Paging, which has had sufficient spectrum to develop advanced systems, has proven to be a viable competitor for many mobile users.

The mobile communications service could be competitive, but the franchised common carriers have been able to use the regulatory apparatus as an effective barrier to competitive entry and perhaps as a means of avoiding competitive pricing via tariff regulation. In this respect the situation is like the common carrier transportation industry wherein regulation has led to inefficiency by stifling competition. The situation is similar in other respects to the grocery industry before the advent of supermarkets—the number of firms is conducive to competition but inefficiency prevails.

Dispatch service has only shown limited growth in the common carrier mode, and most firms desiring this service have had to operate their own systems. Such dedicated systems have resulted in the inefficient use of spectrum for reasons described earlier in this volume. In order to encourage spectral efficiency, the FCC decided to promote capital intensive shared systems. This will tend to cause the industry to become more concentrated.

For example, MTS at 900 MHz will be provided only by cellular firms, although limited competition for telephone service will be provided by RCCs at lower frequencies. Consequently, nonbusiness users will face little choice when purchasing mobile telephone service. Cellular companies, which have been allocated approximately the same amount of spectrum originally designated for the entire mobile communications industry prior to Dockets 18261 and 18262, can probably service in excess of a third of the total market (depending, of course, on demand). Similarly, trunked companies may control a substantial share of the dispatch market. Small firms that previously could not afford mobile communications, because they had too few mobile units to make a system profitable, will now be able to employ common user or common carrier systems; this should increase the competitiveness of small firms vis-à-vis their larger rivals.

On the other hand, in some markets the vast increase in spectrum made available for commercial use can offset the impact of this FCC policy. Excessive concentration is not likely to be a problem in large markets with sufficient demand to support several trunked dispatch systems, many conventional systems, and a cellular system. In smaller markets, though, concentration may be high in the dispatch field. Finally, in mobile telephone service provided by cellular systems, concentration will be high in all markets. However, the possibility of turning to the less versatile but cheaper substitutes available from other mobile services will limit the market power of cellular MTS.

The latter possibility complicates the analysis of regulation and competition within the service sector of the industry. Analyzing these topics would have been much easier if we could clearly subdivide them into categories under headings such as dispatch service, mobile telephone service, and operation of the cellular system. Unfortunately, the problems in these different areas are interconnected to a high degree and clear separation is impossible. Nevertheless, for reasons of exposition, in the more detailed discussion below we shall try to achieve a partial separation.

Dispatch Service

It has been noted that the decisions in Docket 18262 reflect the FCC's contention that the dispatch industry will develop in a sufficiently competitive form to ensure efficient results. Moreover, the rates for dispatch (except for cellular) will be unregulated by the FCC; the commission has argued that easy entry and a competitive structure will assure reasonable prices.[36] This point of view led the FCC to create the new, unregulated class of dispatch firms designated Specialized Mobile Radio (SMR) systems.

Several groups and organizations in the mobile industry, especially the RCCs, allege that the FCC's failure to regulate the SMRs will lead to abuses and put them at a disadvantage. During the late sixties and early seventies, routine applications for uncontested spectrum took many months to process, while contested applications took several years to resolve. These administrative problems may have provided motivation for devising the new SMR category, but they cannot completely justify the freedom from regulation. The effectiveness of competition as a regulatory force depends on its extent within a given industry, and we must examine whether there will be enough competition in the dispatch industry to eliminate the need for regulation.

Our discussion of this topic must be even more speculative than that of the equipment industry. Because dispatch service has been private, we have little data to rely on, and the RCCs have confined most of their operations to mobile phones and paging. The problem is further complicated by the fact that regulated operators of the cellular MTS will also be allowed to operate dispatch services; no comparable situation exists today. Despite these difficulties, we can utilize the data and methods presented earlier in an effort to gain some insights about the dispatch market.

The figures presented in Table 14.2 indicate that by 1990 the national market could contain 20 million dispatch units. If, as is presently the case, demand is concentrated in the larger SMSAs, each of the 20 most populous urban areas may contain from 150,000 to 500,000 dispatch users (see Chapter 15 for a fuller discussion).

Our analysis of system costs in Chapter 4 indicated that the average total cost (ATC) for trunked systems declines until approximately 2,000 to 3,000 users are served within a system and then remains fairly constant. Furthermore, since spectral economies are exhausted if more than 20 channels are assigned to a trunked system, the FCC has limited these systems to 20 channels, which effectively limits trunked dispatch to user capacities of about the same number—2,000 to 4,000. Operating at capacity, trunked systems have ATCs of about $1,000 to $1,100. Conventional systems have a capacity of 50 to 80 mobiles (the actual number depends on the source of the estimate and the service parameters assumed), and even at low levels of utilization, costs are reasonable; a conventional system could serve 5 mobiles with an ATC of $1,600. As capacity is reached, ATC drops to roughly $1,000. For cellular systems, average cost curves level off at about 20,000 users. At this point, ATC is about $1,300 per dispatch subscriber. However, these systems have a very large capacity; potentially, they could serve all of the users in even the biggest markets.

Based on the cost estimates and market potential, it is likely that large urban areas are capable of sustaining competition. None of the systems exhibit a cost structure that would lead to a natural monopoly in a market of the size projected. Moreover, all of the systems offer roughly comparable services. Cost differences tend to compensate for quality disadvantages; that is, systems with lesser capabilities are also less expensive.

While all of the systems should be able to compete on the basis of their net benefits, both the probable association of a cellular system with wireline telephone firms and its large capacity imply that aggressive promotion could result in its obtaining a substantial market share. The larger cellular firms would be able to spend more on advertising than their rivals, and those associated with wireline companies could exploit the public's familiarity with and acceptance of telephone companies such as those owned by AT&T; those advantages could generate substantial market power.

Nevertheless, it does seem likely that the relatively low entry barriers facing potential trunked operators and the alternative of setting up or sharing a private conventional system will limit the exploitation of this potential market power. In addition, the lower prices anticipated for service on noncellular systems will help to compensate for both the real and created product differences associated with cellular. Since the dispatch market will, in many areas, be reasonably competitive, prices should approach marginal and average cost; firms offering dispatch service will have an incentive to cut costs to increase their profits, and this will encourage the cellular firm to try and do likewise. The regulatory agency should not prevent such a beneficial occurrence.

It is possible that only one or a few trunked dispatch systems will be able to operate in a given area, so that problems of monopoly could arise even without the cellular systems. Such problems might be especially serious because typical subscribers to a dispatch service lack any countervailing monopsony power. In the smaller markets, the economic and spectral economies of scale involved in operating trunked systems imply that these operations could possess significant monopolistic power. However, here again the possibility of setting up low cost conventional service will provide a useful safeguard. The small number of subscribers and less congested spectrum in such markets will allow the use of lower frequencies, which will tend to reduce costs and extend range, thereby increasing the potency of this option.

It has been suggested by the White House Office of Telecommunications Policy (OTP) that the granting of licenses be tied to spectral efficiency. OTP proposed "a schedule of license fees reflecting in part the scarcity value of the spectrum being used." They argue that this would be particularly useful "in those areas where spectrum or channel congestion is a major problem."[37] Such a policy would favor the trunked operator, because it would lessen the potential for users to set up their own small-scale systems.

The factors discussed above suggest some desirable policies. First, the decision to regulate should depend on the size of the market being considered. The FCC should not try to prohibit all state and local agencies from regulating, but should instead ascertain which markets can or cannot support competition and base its regulatory stance on these findings. Given the embryonic nature of the industry, its competitive potential, the importance of technological change, and the costs of regulation (direct and indirect), regulation should be avoided if at all possible. Second, if license fees are based on spectral efficiency, they should not be imposed on a uniform, nationwide basis. Rather, as implied in the OTP statement, they should reflect regional differences in spectrum scarcity.

Issues Concerning Both Dispatch and MTS

SUBSIDIZATION. The dispatch sector presently comprises the bulk of the mobile communications market, and this situation is likely to continue for some time. The FCC decision allowing cellular operators to offer a limited variety of dispatch service was intended to spur development and deployment of cellular systems, so that cellular firms could attain economies of scale sufficient to make MTS attractive. Unfortunately, this policy also raises the possibility that cross-subsidization may be used by cellular operators in hopes of dominating the dispatch market.[38] Cross-subsidization is a problem when a firm has a regulated monopoly in some markets; such a firm can exert its monopoly power in "protected" markets to gain the funds necessary for predatory pricing in ostensibly competitive markets.

One possibility is subsidization by the wired telephone companies of their cellular affiliates, since much of the cellular mobile message is transmitted over landlines. However, this type of subsidization will be hampered by two factors. First of all, trunked and conventional systems are simpler and less costly to operate. Thus, they are likely to be deployed on a major scale much earlier than cellular, and their services can be provided at lower rates. These considerations imply that it may be extremely expensive to drive them out by subsidization. Furthermore, the FCC's decision to require separate cellular companies and records will make subsidization more difficult to conceal.

The above subsidization problem raises the question: Should wireline companies be permitted to operate cellular systems? Initially, this is probably desirable because wireline companies have the expertise, the resources, and, evidently, the motivation to speed deployment. But, once cellular is established, issues like subsidization and heightened monopoly power may become important—the mere size of the wired telephone companies could conceivably inhibit rivalry by smaller companies. Also, the difficulty of determining which firm is actually the low cost producer may be exacerbated because of joint wire and radio operation (see, for example, the current problems of cost separation in FCC Docket 18128). Motorola has opposed operation by wireline companies,[39] pointing out that telephone companies are normally barred from offering cable television service because it is a potential substitute for some wired telephone services. Similarly, if the penetration of MTS is low, such service is apt to be complementary to the wired telephone service, but if penetration is high, MTS may well be a substitute for landline telephone service.

Subsidization can also occur within the mobile field. The cellular firm will have a monopoly over MTS but face competition from trunked and conventional systems in the dispatch market; it could, therefore, subsidize dispatch service at the expense of MTS. Approximately 16 dispatch users could be accommodated in place of each MTS subscriber, though this factor would be less significant when spectrum is not fully utilized. Yet, according to AT&T data, the addition of dispatch service will reduce MTS costs for a system with less than 20,000 subscribers.[40] At the initial stage, then, when the marginal cost for dispatch service is very low, the cellular firm has both the incentive and perhaps the means to subsidize dispatch service. It is much less clear whether the problem of subsidization will be serious once the MTS market has grown. (In some markets the growth potential of MTS is limited, so that the incentive to reduce ATC by subsidizing dispatch users may persist for some time.) As noted earlier, by the time the MTS market is large, it may be too costly to drive out noncellular dispatch operators. Moreover, since the noncellular dispatch field is relatively easy to enter, driving out such firms would not preclude reentry or new entry once prices were raised by the cellular firm. If many noncellular firms are driven out, however, potential entrants may be deterred; therefore, antitrust agencies should ensure that unfair practices do not go unprosecuted.

Whether or not subsidization actually occurs, the coexistence of a regulated firm with other unregulated competitors will surely bring allegations of cross-subsidization and cream skimming. When considering these issues it is important to note that some pricing practices could develop that may appear to involve subsidization but, in fact, result from profit maximizing behavior without an intent to monopolize. The marginal benefit of a communication system rises with the number of users in the system; thus, cellular firms will be able to charge higher rates as the system expands. This could lead to a variation of rates with time which would be similar to that involved in subsidization.

One type of subsidization remains to be discussed. Cellular systems are likely to be more profitable in large metropolitan markets. The question of whether some of the profits should be expended so that rural areas can have service at prices less than the additional costs of serving them is important in many areas of telecommunications. It will probably be a significant issue in the mobile field, especially with regard to providing widespread mobile telephone service, because cross-subsidies involving separate and distinct mobile systems will be more difficult to rationalize than those in the more integrated wireline network. Indeed, regulators will face a difficult task when deciding how large of an area can be served by each cellular operation and the extent of geographic markets in order to establish uniform rates.

There are a number of other points that are relevant here. First, it is not clear that the cellular industry will be sufficiently monopolistic to practice internal subsidization to a substantial degree. Second, it is claimed that such a policy erects additional entry barriers, since eliminating cellular's monopoly power would cause higher rates for rural users. Third is the equity issue—why should rural users not pay the full costs of their service? If they have lower income, it can be argued that one should subsidize their income directly and let them choose how to spend it.

However, arguments can also be made for internal subsidization: expanding service into the rural areas increases the service's value for urban users—who may want to use MTS when they are in rural areas and may wish to call rural users from their urban base. In addition, the consideration of externalities favors subsidizing rural service. The fact that cellular systems should reduce notification delay in emergencies is especially significant in rural areas where landline telephones are less accessible.[41]

VERTICAL INTEGRATION.   The FCC tried, by means of its 1975 decision, to foster competition for mobile equipment and to avoid the problems associated with vertical integration that it faces in the wireline industry. It prohibited the cellular firm (which is quite likely to be AT&T in many markets) from manufacturing mobile equipment used in the cellular system. The cellular firm can, however, produce its base station equipment and can possibly use prototype equipment that it constructs for the developmental system. On the other hand, if the mobile communications industry is reasonably competitive, as it may well be, vertical integration will be unimportant. Nevertheless, it may be desirable to separate production of cellular mobile equipment from the company providing the communications services in order to stimulate innovative rivalry. Given the existence of firms like Motorola, General Electric, and RCA, innovative rivalry is apt to be quite strong in the cellular (and noncellular) mobile unit market. Consideration might be given to simply allowing consumers to "hook up" any satisfactory mobile units (as in *Carterfone*); this would reduce the dangers associated with vertical integration but still allow manufacture by the cellular operator.

The FCC has also prohibited manufacturers of trunked equipment from establishing more than one trunked system. This eliminates the somewhat improbable threat that manufacturers could, through vertical integration, apply a profit squeeze to nonintegrated dispatch services. If the radio equipment industry continues to be concentrated, such a profit squeeze could occur if the manufacturers charge high prices for their dispatch equipment and low prices for their dispatch service.

It can be argued that limiting entry of manufacturers may delay the introduction of spectrally efficient trunked systems. However, the RCCs could easily diversify into this area and provide a substantial pool of experience that should be utilized; the FCC should encourage and facilitate their expansion into this market. Perhaps consolidation and joint ventures in this field should be allowed if firms concerned are too small to bear the capital costs involved in trunked systems—costs that could easily amount to several million dollars.[42] Low interest government loans and technical assistance may also be of benefit.

In its submission to the FCC, the Justice Department recommended what they called mandatory "interoperability" of the equipment (which we presume means that the mobile units of any one system can be used in any other system), especially if manufacturers are permitted to provide service. This recommendation is important even if the manufacturers are excluded from the dispatch service industry. If interoperability were coupled with a policy that prevented service companies from requiring their customers to rent or buy the mobile unit from the base station operator (through so-called "tying agreements"), this would help to encourage entry into the industry, because of the reduced capital costs. Also, it would be easier for a new firm to capture clients from the existing firms. With interoperability, if a subscriber owned his own unit or could rent one directly from a manufacturer or an independent distributor, he could easily switch from one base station operator to another. Thus, the prohibition of "tying agreements" combined with universally compatible mobile units could greatly enhance the competitiveness of this industry.

Finally, we need to be concerned with the possibility that the cellular operator, sheltered by government franchise, could eventually dominate a given market area because of the fragmentation of trunked and conventional dispatch operations. This danger has been reduced by prohibiting the cellular system from providing the capability for "all-call," although this capability does not figure greatly in the anticipated expansion of the dispatch market. We believe that the FCC will have to carefully monitor this situation.

Mobile Telephone Service

Unlike the landline telephone system, where fixed costs are extremely high in relation to variable costs, the major cost component of the cellular MTS system is the mobile unit itself. The mobile unit determines the marginal cost component of the cellular system. Therefore, one should not expect MTS to be regulated in the same way as standard landline services.

The wireline phone systems can be viewed as a natural monopoly, at least in a given local area, because of the tendency for declining costs over its entire market. Clearly, granting a nationwide monopoly for MTS cannot be justified on these grounds. However, local monopolies are to be granted to cellular mobile telephone systems. Is this action called for? To answer this question we must first consider whether the cellular system should have been designated as the sole provider of MTS at 900 MHz, and also whether only one cellular firm should be allowed to offer MTS in each market.

In the absence of the FCC prohibition of noncellular MTS, one might presume that each of the three two-way systems would be competing for MTS subscribers. While such competition may be likely in the dispatch sector, the basic market conditions are quite different for MTS. The smaller projected demand and especially the cost structures apt to exist in the MTS sector imply that cellular firms could, without the specific policies of Docket 18262, gain substantial market power. (This is particularly true because AT&T could use its substantial R&D portfolio and facilities, power over interconnection, access to capital, and the regulatory procedures to quickly preempt the mobile field at 900 MHz.)

In terms of equipment costs per subscriber, conventional is the most expensive means of providing MTS. The analysis presented in Chapter 4 suggests that, at all but the lowest levels of demand, cellular MTS equipment will cost the operator less per subscriber than trunked equipment used for the same purpose.[43] Therefore, in markets where MTS demand is concentrated, cellular firms could dominate those using other means of providing MTS service.

It is bad policy to ratify a given outcome simply because it is likely to occur. The relatively low private cost of cellular may indicate, however, that an efficient resource allocation will result if cellular is utilized for MTS, provided that the private costs accurately reflect social opportunity costs. In this case, since a major input (spectrum) is not priced, care must be taken to ensure that resources are allocated efficiently. If the cellular system is to be chosen, then the cost of cellular equipment plus the cost of spectrum utilized by the system must be less than equivalent sums computed for the other systems. (This will result in a socially desirable outcome assuming, of course, that all of the systems share the same externalities.) Under existing methods of spectrum allotment these sums cannot be calculated, but we can use our information on spectrum efficiency to determine which system would have the lowest real cost. Table 14.1 shows that cellular is more spectrally efficient than the other systems considered, so that any advantage in terms of private costs will be augmented, while such disadvantages will be lessened, if spectrum costs could be added to equipment costs in determining real, social costs. Thus, a good case can be made for choosing cellular as the sole means of providing MTS at 900 MHz.

The advisability of allowing only one cellular operation per market must now be examined. The presence of economies associated with larger cellular systems is of central importance to this and other major issues. Hence, we will examine the economies of scale question in some detail.

According to AT&T:

> The available spectrum has a strong effect on cost per user and therefore on system viability. Additionally, the maximum system capacity is controlled by the size of the spectrum allocation. Note that the trunking efficiency is lower for smaller channel allocations, so that the maximum capacity decreases somewhat more rapidly than in direct proportion to the available spectrum. Any significant reduction in available spectrum, therefore, would result in higher per-user costs and lower spectrum efficiency. Moreover, the reduction in system capacity would severely limit the growth potential of these systems in major urban areas.[44]

The primary reason for costs being higher with a smaller spectrum allocation is that fixed costs must be shared by fewer mobiles due to lower spectrum efficiency. In recognizing that some economies of scale would be lost if the spectrum allocation were reduced from 64 MHz to 40 MHz, the FCC stated that:

> AT&T has argued that an allocation of less than 60 MHz for a cellular system would result in an increase in the cost of the service. The Commission has studied this point carefully, but we do not feel that the cost penalty of reduction from 64 to 40 MHz would be significant. Using AT&T's own data, the proposed reduction in available spectrum for the cellular system will raise the shared facility cost per mobile unit by about 20%. In as much as the shared facilities' cost represents only about one-half to one-third of the total cost to the user, the 20% cost penalty is reduced to less than 10% in his overall bill.[45]

One must be somewhat cautious when examining the AT&T estimates because of their own interest in obtaining as much spectrum as possible for MTS. For example, AT&T later claimed that if certain design changes were incorporated, reducing the spectrum allotment would increase the base station costs not by 20 percent, as originally estimated, but by 50 percent;[46] if these figures were correct, the increase in the total cost per user could be 25 percent, a substantial amount. But the FCC was not persuaded by what they considered to be a weak argument.[47]

Nevertheless, there seem to be substantial potential economies in having one system with a greater spectrum allocation rather than multiple competing systems. Even Motorola, potentially a large supplier of trunked and conventional systems, reported the results of a study that indicated there would be a significant cost penalty associated with smaller spectrum allocations.[48]

The economies of scale being discussed apparently diminish beyond 64 MHz. Since a very large number of subscribers can be accommodated using 64 MHz, government sanctioned monopolies may be desirable in many markets. In any case, factors discussed above could result in monopolies emerging without a regulatory mandate, so that the question of the desirability of regulation must be dealt with.

In considering this issue it is important to determine, as precisely as possible, how much monopoly power would be possessed in the absence of regulation. Before examining the factors that would limit cellular's market power, or otherwise render it undesirable to establish a regulated monopoly, we will discuss additional considerations supporting the FCC's decision to regulate cellular MTS.

Besides economies of scale, there are a number of factors that augment the cellular firm's monopolistic power. The capital required to establish an efficient-sized firm in markets large enough to accommodate such firms would be $10 million—20,000 users X $500 per user in common equipment. (Obviously, the capital entry barrier would be raised if users did not buy or rent mobile units independently.) The economies of scale in this area are such that entry by a new cellular firm would have to be on a big scale, perhaps involving a large percentage of the total subscribers. This could cause prices to fall substantially. Moreover, there is apt to be considerable excess capacity in the initial stages, making entry even more difficult.

If the industry were totally unregulated, two or more firms might start cellular systems in the same area, which could lead to larger investments in base equipment and possibly some additional inconvenience when calling subscribers of the other system. The possibility of ruinous competition is substantial, because initial demand is likely to be low and the marginal cost per user ($950) is much smaller than the average cost ($1,500 even at levels of use near capacity). If prices were driven down to marginal costs, both cellular firms could suffer serious losses. Such a possibility would make investment in cellular systems far more risky. The FCC's decision to establish regulated monopolies in this area was no doubt partially motivated by a desire to avoid this situation.

Consideration of interconnection also provides some support for the desirability of regulated cellular monopolies. The usefulness of a mobile communication device is greatly increased if it is connected to the wired telephone system. In the case of conventional systems, automatic interconnection would add significantly to their capital cost; the FCC has allowed only manual interconnection for conventional and trunked systems at 900 MHz, in contrast to requiring unlimited interconnection for cellular systems used for mobile telephone service. Thus, the conventional system is disadvantaged as a result of regulation. It is probable that this regulatory restriction was designed to stimulate the development of cellular systems, but there are technical reasons that favor the decision. If conventional and trunked systems were allowed automatic interconnection, the lengths of calls might increase markedly, thereby reducing the number of users that could be served. Such systems might still be acceptable for dispatch use but not for uses requiring longer message lengths.

These regulatory rules act as a surrogate for real pricing of the spectrum. A system like cellular, which conserves a scarce resource that is not priced, is disadvantaged unless such rules are established. Moreover, the cellular system generates external benefits that are not realizable with other systems, and without intervention, the socially desirable resource allocation may not be attained. The greater potential for nationwide service of the cellular system is an argument in favor of its stimulation through regulation.

A related question is whether as much as 40 MHz should have been committed to use by cellular firms; some have claimed that the allocation was too large. The RCCs contended that the cellular system will not have any significant impact on meeting the demand for public MTS for 10 to 15 years and argued that the FCC should have made sufficient spectrum available for RCCs to offer more noncellular MTS in the near future. It is clear, however, that permitting trunked systems to offer MTS would reduce the incentive for cellular investments. Motorola wanted the initial spectrum allocation for the cellular system limited to 12.5 MHz, with more to be granted only as needed. Since that firm was interested in protecting its substantial share of the dispatch equipment market, Motorola's proposal reflected its contention that, due to the limited demand for MTS, cellular firms might try to dominate the dispatch market via cross-subsidization. (While Motorola is also interested in selling to cellular operators, its autonomy, market power, and profit rates may be reduced if large cellular operators capture the bulk of the dispatch market.)

The FCC rejected proposals to allocate less than 40 MHz to cellular on the grounds that costs would have been substantially increased, which would, in turn, raise the risks and lower the probability that the system would be expeditiously deployed.

Despite these arguments for granting cellular MTS monopolies and regulating them, contrary positions were advocated in the proceedings associated with Docket 18262. The White House Office of Telecommunications Policy, for instance, argued against the granting of a monopoly, emphasizing the need to maintain flexibility in a rapidly changing market.[49] The Antitrust Division of the Justice Department argued that "the Commission should insure that other providers of land mobile services are allowed to interconnect with the fixed telephone plant on a reasonable and nondiscriminatory basis."[50]

Substantial reductions in investment per user are possible by relaxing some of the service parameters such as wait time,[51] and this is relevant to the issue of competition. As an example, trunked systems operating under less stringent service parameters could provide MTS at levels of spectrum efficiency approaching those of the cellular system. By not allowing such competition at 900 MHz, the choices available to consumers are likely to be less diverse. Cellular could have been granted considerable spectrum without forbidding others to offer MTS at 900 MHz.

Indeed, in earlier hearings the FCC allowed trunked systems to offer MTS at 900 MHz, at least on an interim basis.[52] Such a policy would have meant more prompt relief of the claimed shortage of mobile communications. The FCC evidently decided that it was more important to prevent future shortages by supporting the deployment of the high-capacity cellular system. Thus, cellular will now face direct competition for MTS only from radio common carriers at lower frequencies. Of course, the possible substitution of one-way paging or dispatch service also exists, alternatives that may sufficiently limit the monopolistic power of cellular firms to eliminate the need for regulation.

Another argument against regulation is that supervision of the expanding mobile communications industry will place new strains on the already overburdened regulatory agencies. In particular, the FCC, with a current annual budget of roughly sixty million dollars, may not be able to operate effectively unless its resources are increased. It should be kept in mind that several communications firms spend far more than the total FCC budget to achieve public acceptance of their objectives.[53] Regulation of the industry is increasing in magnitude and complexity, and mounting problems have led some to propose drastic policies, such as dismantling the Bell system, and even public ownership. It is clearly possible that adding MTS to the regulatory load could result in reducing the overall effectiveness of communications regulation, while conferring little if any benefit on the land mobile field.

A final argument against regulating mobile communications concerns the influence on innovation. The belief is often expressed that regulation causes unacceptably low rates of technological change (see Chapter 3).

One can conclude from the arguments presented above that it is desirable for the mobile communications industry to develop at 900 MHz without outside interference, beyond that which is necessary to ensure (technical) operational efficiency. All systems could be allowed to provide any service—trunked could offer MTS and cellular could provide fleet call. Under such a regime, those systems providing services satisfactory to consumers would prosper; regulation would only be instituted if monopolistic abuses actually developed.

The detailed analysis of the potential structure and regulation of MTS has shown that the FCC's decision is defensible but that it is not the only justifiable course of action. In light of the rapid technological changes, possible loss of competitive pressure, and costs associated with regulation, the issue of a cellular monopoly deserves continuing analysis. If the FCC's goals in this area are not realized within a reasonable period of time, further thought should be given to removing the restrictions on services offered by various systems and allowing competition. Meanwhile, considering the importance of progress in this industry, it is important to find policies that will promote innovation. To this end we will briefly recapitulate the major negative influences that must be overcome.

When a monopoly is granted or only a few producers are franchised, the incentive to innovate is reduced because the firms possess a "captive market." Moreover, since the amount of profits allowed by regulatory agencies is often a fixed percentage of investments, even cost-reducing innovations will be less likely to occur. Even worse, regulated monopolists have an incentive to seek more costly ways of doing business as long as their rate base can be expanded in the process. And, where profits are computed as a percentage of a "rate base," producers will usually opt to amortize their capital slowly; this further reduces the rate at which technological improvements are introduced.

In the past, vertical integration in the ordinary telephone industry eliminated the incentive for outsiders to produce innovative telephone equipment because there was no market for their products. The FCC has sought to prevent this from happening in parts of the mobile communications field by forbidding those companies that operate the cellular system from producing all of their own equipment. Several other FCC policies—for example, guarding against a nation-wide monopoly for MTS and fostering competition for dispatch services—will help to promote innovation.

Additional policies can be employed to reduce the negative influence of regulation on innovation:

1. The mobile unit could be excluded from the rate base of the cellular operators. This policy would allow a substantial part of the equipment to be depreciated by the independent users or rental agents; it would help avoid slow depreciation rates and thereby speed innovations. By using the cost of base stations to establish rates, there would be an added incentive to expand the geographic coverage of the system. However, vigilance must be exercised to prevent expansion that is not justified by demand but is an artifact of the rate base method of determining profits.

2. Prohibit operators from requiring their subscribers to use a specific mobile unit. This would complement the preceding policy and stimulate the introduction and development of widely diversified equipment rather than of a few standard units.

3. Mandate shorter depreciation periods than those that have been allowed for wireline carriers. This is feasible because cellular MTS operations must be independent of the wireline firms.

4. Base the right to expand and renew local franchises on performance relative to MTS operations in other areas. To be meaningful, this must be coupled with policies ensuring that independent entities are permitted to operate cellular systems. If AT&T, for example, operates a large majority of the cellular systems, useful comparisons may be impossible. (Utilizing the dispatch industry, which is likely to be more competitive, as a basis for comparison may help to solve this problem.)

5. FCC regulations should not require the detailed standardization of equipment. Such a policy would offset the monopoly power that derives from patents. This could be important since Bell and other wireline companies, who cannot continue to produce all of the equipment, will be allowed to

produce the developmental equipment and their patents might provide them with some advantages in equipment manufacturing and distribution.[54] An alternative would be to require mandatory sharing of critical patented techniques and equipment designs with appropriate compensation, although legislation would be required to provide the FCC with the necessary authority. The compensation would be less costly than the possible welfare loss from the market power that patents confer. Unfortunately, this policy could diminish the incentives of innovation.

6. Allow price discrimination based on intensity of use. If rates increased as a function of the duration and frequency of calls, improved spectral efficiency could be realized, the number of subscribers could be increased, and unit costs might fall. By stimulating demand, this could promote innovation. (Moreover, usage sensitive pricing will help avoid internal subsidization.)

Conclusion

W. G. Shepherd's classification scheme for this life cycle of a regulated utility implicitly indicates the importance of periodically assessing the structure and regulation of such industries. He suggests the following sequence: (1) the starting-up period when the utility is invented; (2) the period of growth where the utility establishes its markets and develops its pricing policy (it is usually in this period that regulation is introduced); (3) a third stage involving consolidation and perhaps some retrenchment as new technological competitors are introduced; and (4) a final stage in which the industry is subject to strong competition, when regulation is not needed and may in fact inhibit the utility's ability to respond properly to the new competitive situation. As Shepherd states, "Utility conditions tend to be temporary and to change rather than to be permanently fixed. The main task for public policy is to anticipate these changes and to supply sufficiently basic treatments rather than to be carried along by events."[55] If one is too slow in instituting governmental controls, excessive monopolistic power may develop, while maintaining government regulation for too long could artificially and needlessly perpetuate monopoly.

The mobile communications industry is probably in the first stage; there is now only limited common carrier dispatch service, and existing MTS is primitive compared to that which the new systems can offer. The decision to regulate MTS at this point may not conform with Shepherd's cycle, which places the introduction of regulation in the second stage. One would hope the FCC's policy has accurately anticipated the industry's probable structure, rather than being "carried along" by those seeking the creation of franchised monopolies where competition may have been possible. On the other hand, it will be necessary to monitor the other sectors of the industry to ensure that effective competition develops.

NOTES

1. Richard N. Lane, "Spectral and Economic Efficiencies of Land Mobile Radio Systems," *IEEE Transactions on Vehicular Technology* VT-22 (November 1973):98.

2. This factor is more pronounced in Lane's analysis.

3. Joel S. Engle, "Guest Editorial," *IEEE Transactions on Communications* COM-2 (November 1973):1171.

4. See detailed discussion in Harvey J. Levin, *The Invisible Resource: Use and Regulation of the Radio Spectrum* (Baltimore: Johns Hopkins Press, 1971), p. 342.

5. Ibid, p. 347.

6. There were 168 commercial and 97 noncommercial stations in operation. Ibid., p. 345.

7. Clair Wilcox and William G. Shepherd, *Public Policies Toward Business,* 5th ed. (Homewood, Ill.: Richard D. Irwin, Inc., 1975), p. 457.

8. For a more complete discussion, see Wilcox and Shepherd, *Public Policies,* pp. 457-59.

9. Ibid., pp. 453-54.

10. "In the Matter of the Use of the Carterfone Device," Docket 16942, *Federal Communications Commission Reports,* 2nd series, vol. 13, pp. 420-27 (hereinafter *FCC Reports*).

11. Jonathon C. Rose, "Baby Dinosaur and Antitrust Policy," *Communications,* April 1976.

12. "In the Matter of the Allocation of Frequencies in the Bands Above 890 MHz," Docket 11866, *FCC Reports,* vol. 27, p. 359.

13. John D. Dingell, "The Role of Spectrum Allocation in Monopoly or Competition in Communications," *Antitrust Bulletin,* Fall 1968, p. 939. See also Wilcox and Shepherd, *Public Policies,* p. 444.

14. "Thus, the smallest users, who can least afford the high cost of communications, are forced to pay the highest prices of all." Dingell, "The Role of Spectrum Allocation," p. 939.

15. "In the Matter of Applications of Micro-wave Communications, Inc.," Dockets 16509-16519, *FCC Reports,* 2nd series, vol. 18, p. 953, cited in Leonard W. Weiss and Allyn D. Strickland, *Regulation: A Case Approach* (New York: McGraw-Hill, 1976), p. 185.

16. U.S. Congress, Senate, Committee on the Judiciary, Subcommittee on Antitrust and Monopoly, *The Competition Improvements Act of 1975: Hearings Before the Subcommittee on Antitrust and Monopoly,* 94th Cong., 1st sess., 1976, p. 1041.

17. See, for example, "FCC Affirms Rule MCI Can't Offer Executive Service," *Wall Street Journal,* May 26, 1976, p. 40.

18. Senate, Committee on the Judiciary, Subcommittee on Antitrust and Monopoly, *The Competition Improvements,* p. 1042.

19. "AT&T's Bold Bid to Stifle Competitors," *Business Week,* March 15, 1976, p. 87.

20. See "Ma Bell's Consumer Reform Bill," *Consumer Reports,* January 1977, p. 42.

21. See also a study by the staff of the New York Public Service Commission which found that terminal equipment was priced below cost. Alfred E. Kahn and Charles A. Zielinski, "Proper Objectives in Telephone Rate Structuring," *Public Utilities Fortnightly,* April 8, 1976, p. 21.

22. U.S. Department of Commerce, "A Study of the U.S. Radio Paging Market," prepared for the Office of Telecommunications by Arthur D. Little, Inc., Contract no. OT-117, April 30, 1975, pp. 1-2.

23. *Time Magazine,* May 2, 1977, p. 67.

24. U.S. Department of Commerce, "A Study of the U.S. Radio Paging Market," pp. 1-2.

25. Ibid., pp. 2-5.

26. According to industry spokesmen, each doubling of output over time reduces cost by about 12 percent.

27. A firm that managed to capture a large share of the radio equipment market might find it desirable to produce its own computers. It is also possible that computer firms may enter the mobile communications industry.

28. Richard A. Posner, "Natural Monopoly and Its Regulation," *Stanford Law Review* 21 (February 1969):548-643.

29. Paul F. Kagan, "Go-Ahead Signal," *Barron's,* June 10, 1974, p. 22.

30. "Memorandum, Opinion and Order," Docket 18262, *FCC Reports,* 2nd series, vol. 51, March 1975, p. 969.

31. Ibid.

32. There is some question about the FCC's power to force manufacturers to share patents or engage in cross-licensing. Perhaps legislation will be needed in this area. However, the issues are complex; policies that involve patent sharing may lessen competition and incentives for innovation.

33. Kagan, "Go-Ahead," p. 22.

34. American Telephone and Telegraph, "Comments," Formal submission to FCC Docket 18262, American Telephone and Telegraph, Basking Ridge, N.J., July 20, 1972, p. 6.

35. Ibid., p. 7. AT&T states that the figures for the radio common carriers include pagers as well as two-way devices.

36. "Memorandum, Opinion and Order," *FCC Reports,* vol. 51, pp. 959-60, 974.

37. Executive Office of the President, Office of Telecommunications Policy, "Conclusions and Recommendations of the Office of Telecommunications Policy Regarding Land Mobile Radio Service in the 900 MHz Band," Formal submission to FCC Docket 18262, Executive Office of the President, Washington, D.C., 1973, pp. 14-15.

38. Motorola contended that the entry of cellular companies into the dispatch market " . . . would thus set the stage for the displacement of private dispatch systems and ultimate monopolization of the mobile radio service business. Inevitably, competition in the mobile radio equipment market would also be substantially impaired." Motorola, Inc., "Reply Comments," Formal submission to FCC Docket 18262, Motorola, Inc., Schaumburg, Ill., July 20, 1972, p. 100.

39. Ibid., p. 59.

40. Bell Laboratories, "High Capacity Mobile Telephone Service Technical Report," Unpublished document, Bell Laboratories, Holmdel, N.J., December 1971, pp. 4-5.

41. Insurance rates for motor vehicles, accident, and health, and even life, may be lower for those who have MTS. Such a pricing policy may provide strong incentive for more widespread deployment of cellular.

42. In its submission for Docket 18262, Motorola projects that the shared equipment will cost $600,000 for its 20-channel system. In the developmental stage, it is likely that the operators will have to furnish the mobile units whose cost could range from approximately $700 to $1,000 each. For a system with 2,000 to 5,000 users, the total initial cost might be between $2,000,000 and $5,600,000.

43. Lane's analysis indicates that the reverse is true, perhaps because he assumes a more intensively used advanced cellular system with many small cells. Lane, "Spectral and Economic."

44. Bell Laboratories, "High Capacity Mobile," pp. 4-7.

45. "Second Report and Order," Docket 18262, *FCC Reports,* 2nd series, vol. 46, p. 756, par. 13.

46. "Memorandum, Opinion and Order," *FCC Reports,* vol. 51, p. 948, par. 11.

47. Ibid.

48. Motorola, Inc., "Reply Comments," p. A-19. For a 24-MHz allocation, the shared investment per user would be 1.42 times that for 64 MHz, or $785 instead of $560. The total investment per mobile would rise to $1,735 instead of $1,510, an increase of about 15 percent. Ignoring a small decrease in the cost of the mobile unit that would have to operate on only 300 channels instead of 800, the monthly charge would rise, according to the calculation, from $59 to $67. The study was done by Snavely, King, and Tucker, Inc., an economic consulting firm.

49. Executive Office of the President, Office of Telecommunications Policy, "Conclusions and Recommendations," p. 2.

50. U.S. Department of Justice, Antitrust Division, "Comments," Formal submission to FCC Docket 18262, Washington, D.C., August 17, 1973, p. 7.

51. Lane, "Spectral and Economic," p. 101.

52. "Memorandum, Opinion and Order," *FCC Reports,* vol. 51, p. 946, par. 5.

53. See, for example, Ray Connolly and Stephen E. Schrupaki, "Can the FCC Cope with Changing Technology?," *Electronics,* July 25, 1974.

54. However, AT&T has pointed out that: "Under the provisions of the 1956 Final Judgment, the Bell system is required on application to grant non-exclusive licenses under Bell system United States patents. The Bell system may require as a condition of such licenses that the applicant grant to it at reasonable royalties non-exclusive licenses under the applicant's U.S. patents for equipment useful in furnishing common carrier communications services. Reasonable, non-discriminatory royalties may be charged for licenses granted by the Bell system." Henry G. Fischer and John W. Willis, eds., *Pike and Fischer Radio Regulations,* vol. 13, p. 2143.

55. See W. G. Shepherd, "Alternatives for Regulating Cablecasting," *Public Utilities Fortnightly,* August 28, 1975, pp. 20-24.

Chapter 15
PLANNING FOR THE FUTURE OF LAND MOBILE COMMUNICATIONS

Raymond Bowers
Alfred M. Lee
Cary Hershey

I n the preceding chapters, the future of mobile communication technology and its probable impacts on society have been discussed within specific contexts. Now we would like to discuss that future in more general terms. First a brief review of the relevant social, political, and economic factors will give us the basis for assessing the outlook for mobile communications. We shall then discuss some uncertainties concerning the future and present scenarios that describe various future states. Finally, an overview of the societal costs and benefits to be expected and an enumeration of substantive issues that require attention will be followed by a discussion of the scope of the planning efforts necessary to deal with them.

Our views on the evolution of land mobile communications are presented cautiously, because the theories and limited data at our disposal are an inadequate basis for confident prediction. Therefore, our appraisal will be speculative to some degree, though we will make it as informed and disciplined as our knowledge and understanding will allow.

While many factors will influence the evolution of the technology and the nature of its social consequences, the large and sustained growth rates of mobile communications utilization indicate that we are not dealing with a transient phenomenon. Indeed, it is our conclusion that conditions favor continued and even accelerated growth in the deployment and adoption of the new devices, and to meet demand, very high capacity systems—as exemplified by cellular technology—will be needed. So we shall begin our speculations about the future by briefly discussing some of the more important factors that will influence the diffusion and impact of the technology, in order to expose the assumptions and conceptions that underlie our belief. These include the changing geography of where we live and work, the role of electronic communications in productivity and the influence of altering leisure and consumption styles.

FACTORS AFFECTING THE FUTURE OF LAND MOBILE COMMUNICATIONS

Regional Geography and Mobility

The future of the field we are considering is inextricably bound to the scale, character, and necessity of physical travel in our society. From this standpoint, the most significant aspect of urban development since World War II has been the decentralization and suburbanization of economic and residential activity. Between 1950 and 1973, for example, the suburban population almost doubled from 41 to 79 million people.[1] Yet, while the businesses and population concentrated in the central cities have proportionately declined, the central cities still remain very important commercial and cultural loci. The relationship of the cities to the ever-expanding circle of development around them has provided the major impetus for our contemporary "automobile society."

The extraordinary scale of automobile travel within metropolitan areas is evident in statistics on automobile usage. Each day, in urban areas, approximately 53 million one-way automobile trips averaging 10 miles in length are undertaken by persons traveling to and from work and for other business purposes; 46 million one-way trips whose mean length is 5.5 miles are taken for family business, such as visits to doctors and shopping centers; and 32 million one-way trips of about 13 miles are the result of social and recreational travel.[2] This only partially reflects a national pattern of increasing motor vehicle utilization; the number of registered motor vehicles in the U.S. rose from 32 million in 1940 to 135 million in 1974.[3] The congestion of central cities and the spread of suburban areas continues to increase the time spent in cars.

The assumption seems inescapable that the more people travel for such purposes, the greater will be the potential usefulness of mobile communications. The present spatial configuration of urban areas makes physical mobility a necessity, and growth at the urban fringe is continuing. It is conceivable that this trend may be slowed or reversed through, for example, the development of highly planned communities that locate business, commercial, and residential activities in close proximity. However, such projects have encountered severe difficulties, including high initial costs and lagging acceptance, conditions that have led to many new community projects failing over the last decade. One cannot be too optimistic that such planned communities will become common during the next two or three decades.

Similarly, high capital costs and uncertain consumer acceptance of rail-based mass transportation within metropolitan areas makes it unlikely that many additional cities will pursue these alternatives to the private auto. While the automobile is causing many problems in our society, its convenience for personal transportation is undeniable. Nevertheless, if large-scale efforts are made to develop economical and convenient public transportation systems, such alternatives might gain acceptance. Also, providing mobile communication devices for passenger use, where feasible, could add to their attractiveness, since riders could use their travel time more effectively.

Another development that may affect the amount of driving in the future is the expected increase in the cost of energy. People may be forced, because of high prices or enforced rationing, to curtail their automobile travel, especially for social purposes. Car pooling for work commutation could become commonplace, and more efficient scheduling of shopping and other trips might become a necessity. Introducing advanced forms of telecommunications may permit certain employees to work directly out of their homes, since wired contact with the parent organization could be maintained.

It is important to recognize that such reductions in automobile related mobility, if they occur, would not necessarily lessen the demand for mobile communications. One can foresee possible incentives, in the form of tax advantages and government or private subsidies, to promote mobile communications as an energy-saving technology. In short, use of mobile communications in the future may be less related to the overall amount of travel in society than to the need to make travel more efficient. The prospect that demand will be curtailed because new patterns of metropolitan development have reduced the need for mobility seems much less likely; the causes of geographical spread have very deep roots. Nor is it plausible that pressures to make travel more efficient will decline.

Productivity

Increasing productivity is an important goal of American society. "Productivity" expresses the relationship between the value of inputs (labor, capital, energy, etc.) and the value of goods and services produced. Thus it is a major element in promoting economic growth, controlling inflation, and maintaining U.S. competitiveness in international trade.[4] Increasing costs and the profit motive will ensure continued efforts to improve productivity.

Our economy's service sector, which has the lowest rate of productivity, will be an especially important target for development. Such improvement is crucial because by 1980 over two-thirds of the labor force is expected to be engaged in the production of services such as health, safety, education, transportation, banking, insurance, retail sales, and home and office repairs.[5] Assuming that major efforts will be made to organize services along industrial lines, telecommunication technologies are apt to figure substantially in attempts to make them more efficient. Service organizations are increasingly resorting to more sophisticated fixed communications and information-processing systems in order to transfer information and data concerning orders, inventories, and clients; this change in itself will stimulate the use of advanced forms of mobile communications, since many services have an important mobile component. Furthermore, the mobile component will need to be integrated, preferably without the use of human intermediaries, into the "instant information networks" that are becoming common in many manufacturing and service organizations. As a result, we can expect greater reliance to be placed on those mobile communication technologies that help reduce response time, utilize available resources more productively, conserve energy, and improve the interaction between the mobile worker and those involved in centralized activities.

Leisure and Consumption Styles

The centrality of work in American lives is unlikely to diminish. Even if the overall amount of time people spend working does not change significantly, the distribution of work will probably continue to evolve in important respects—and thus have implications for leisure and consumption styles. For example, we are likely to see greater consolidation of free time with more frequent and/or longer vacations, more sabbatic and educational leaves, increased use of early retirement, more flexible work schedules, more working at home, and greater frequency of part-time and irregular work. And for some workers, the workweek is likely to be shortened. Most of these developments imply less rigid work/leisure schedules and less regularity in people's whereabouts at specified times. Consequently, the ability to reach others, and to be reached, will be limited if we continue to rely solely upon the fixed phone.

During leisure time, a person's choice of actions is greater than when one is working, latitude that William Burch refers to as "choosing time."[6] If work time does decrease in the future, it does not necessarily follow that unencumbered free time will therefore increase. Some observers, such as Daniel Bell, foresee a greater scarcity of actual "free" time for pure recreation, because it will take longer to purchase goods and services and to care for our technological possessions; hence, the scheduling of leisure time will no doubt assume added importance.[7] This new pattern could create more personal uses for mobile communication technologies, with the result that the distinction between technologies needed for work and those needed for leisure would become steadily less clear.

As for the purely recreational aspects of leisure, a few possibilities relating to the future of mobile communications can be noted. The deteriorating energy situation may reduce mobility and modify leisure patterns, but increasing emphasis on leisure activities still seems likely, as fewer people will be required to produce goods and services for the rest of society. It is worth observing that energy costs, particularly those for gasoline, have been much higher in Western Europe and Japan for some time and yet investment in leisure and recreation continues to grow. In the United States, many leisure pastimes require substantial mobility.

In general, Americans appear reluctant to limit their use of automobiles— whether viewed as a necessity, a status symbol, or both, the private automobile is a major means of access to places of recreation. Moreover, the range of choices and spontaneity that a car permits are highly valued in American culture. An automobile outing is often an end in itself; the popularity of Citizens Band radio suggests that for many combining communications capacity with mobility can add to that pleasurable experience. The very existence of new mobile communication technology may well increase the intrinsic desire to be mobile in the future.

One should also consider the increasing popularity of vans and recreational vehicles for outings and vacations. Such vehicles have facilitated part-time or full-time mobile living, which seems especially attractive to nomadic youth, retirees, and workers who are employed for only part of the year. These "on-the-roaders" are major users of Citizens Band radios. In part, CBs are used to form temporary communities while transient. More sophisticated mobile communication technologies, like the cellular system, may increase the incentive for some to adopt mobile lifestyles, because it will be easier to maintain contact with the fixed elements in their lives.

Attitudes of Purchasers

It is difficult to find reliable indicators for assessing the potential diffusion of the more advanced mobile communication technologies. Many believe that the high cost of the more sophisticated systems precludes their adoption as consumer items. Although cost is obviously an important, and maybe even the most important, factor, we must be wary of making it the sole determinant. The commercial history of color television, high fidelity equipment, and the "fully equipped" automobile suggest the American consumer is motivated by more than a desire for the lowest costing device that will perform a given function. Furthermore, one should keep in mind that initial rates for telephone service were also high. In the 1880s, a call from a pay telephone cost 15¢ and charges for subscriber services were approximately $150 for 1,000 messages—equivalent today to 50¢ for a pay telephone call and $500 for the subscriber service—costs that did not prevent the rapid growth of telephone service. Introduction of cellular technology could reduce mobile telephone service costs by about half from their current level. Because the technology of mobile service is less mature than that of the fixed telephone, and is deployed on a much smaller scale, one can expect the cost disparity between the two services to keep decreasing. This will surely contribute to increased adoption.

The potential for expanded use of mobile communications as a highly desirable convenience, rather than a necessity, seems very great indeed. While many of the functions facilitated by mobile technology can be carried out using fixed telephones, especially coin telephones, the mobile devices provide extra convenience and speed—and Americans have frequently demonstrated a willingness to pay substantial sums for such qualities. In addition, fashion, purchasing habits of peers, and status implications can stimulate desire for devices that are not strictly necessary to perform a given task. Americans often embellish their automobiles with articles of convenience and entertainment; the merchandising of "optional extras" has certainly set a precedent for the purchase of two-way communication devices.

The systems we are analyzing combine two technologies that have had a profound influence on our society—the automobile and electronic communications. Since the American appears to be addicted to, and even in love with, both the automobile and the telephone, he may find it difficult to resist a combination of the two. Moreover, the attribute of complete portability, made possible by the more advanced systems, is likely to be very attractive to many.

We have discussed a number of trends in American society that point to increasing use of mobile communication devices. Another perspective on the prospects for future use can be suggested by specifying conditions that might severely restrict implementation rates of the more sophisticated devices. It seems to us that some combination of the following conditions would have to prevail in order for growth rates to be low:

1. The devices prove to be ineffective in increasing productivity, or the net benefits are less than those achievable by other methods.
2. An unprecedented tendency emerges among organizations to deemphasize the use of electronic communications.
3. The recreational activities associated with Citizens Band radio become unfashionable.
4. American standards concerning privacy and intrusion change substantially, becoming more rigid and less permissive.
5. Adverse economic conditions seriously constrain the availability of capital and reduce consumer purchasing power.
6. The industry does not vigorously promote the technology and limits its use of advertising to stimulate new desires.
7. Increasing energy costs greatly reduce mobility in our society.
8. The forces causing urban sprawl and the geographic dispersion of business activity are substantially reduced.
9. Governmental regulations discourage diffusion by explicit or implicit, conscious or unconscious policies.
10. Concern over the health hazards of nonionizing radiation increases greatly. Very low-level athermal physiological effects are verified.
11. Significant improvement in public (e.g., coin operated) telephone service is made, especially when used in conjunction with paging devices.
12. The development and acceptance of more sophisticated fixed technologies, such as video phones and teleconferencing devices, reduce the need for physical mobility.

Clearly, the conditions specified above are not wholly independent of each other. On balance, however, it does not seem probable that they will prevail, although we are least confident about the importance conditions numbered 5, 7, and 12 may have on the future of the technology.

These, then, are the fundamental social, political, and economic factors that lead us to conclude that the growth rate of mobile communications will continue to be high. But our arguments are far from conclusive, whether taken individually or collectively, and others may reach a different conclusion based on their evaluations of the factors involved.

UNCERTAINTIES: SCENARIOS FOR ALTERNATIVE FUTURES

While we are persuaded that the evidence favors continued and even accelerating growth of mobile communications' use, the future cannot be predicted with certainty. As private and public agencies attempt to influence the technology's evolution, they will face a number of interrelated uncertainties that must be included in their analysis and planning. Such uncertainties can be subdivided into four categories: demand, performance, institutions, and impacts.

Projections of future demand are based on many unproven assumptions; one cannot be certain whether demand for land mobile services will stabilize or experience rapid growth. Nor can one be certain about the demand for particular services—that is, whether users will want paging, telephone, or dispatch-like services, or even some new kind of service. Composition and size of demand will be affected by the scope and nature of the technological and service options available in the market. Obviously, cost will also be influential, yet the cost of systems is another area of uncertainty.

One cannot know in advance whether new technologies deployed on a large scale will function as expected; only through the accumulated experience gained from operating several systems can performance be determined with any reliability. Implementing several of the technologies requires major capital resources supplied over long periods, so that at some stage in the regulatory and industrial decision-making process, a commitment must be made to a particular system. Nevertheless, it can be expected that once the decision is made and implementation underway, new technological alternatives will be proposed, requiring reconsideration of the original plans. A delicate balance is needed: too frequent evaluation can prevent the deployment of any large-scale system, while too infrequent evaluation may result in the wrong system being deployed. The rapidity of technical innovation will cause major uncertainties in the planning process.

The institutions with an interest in the technology's evolution will change. New users of the technology may enhance the political influence of existing lobbying organizations. On the other hand, specialized interests may emerge around particular technologies and applications. It is unknown how much pressure will be exerted on the regulatory agencies and the Congress by those groups, or whether they will be operating in harmony or at corss-purposes.

It is also unclear how the relationships between government institutions will change. New agencies, reflecting new national priorities, will appear, while existing agencies may develop new interests in the land mobile area. The mobile communications policy arena will undoubtedly be broadened, but the agencies involved and the form and substance of their linkages remain uncertain.

Our assessment has outlined many possible impacts of new mobile communication technologies. Some consequences can be anticipated, but many will not be foreseen. Some impacts resulting from the manufacture, deployment, and regulation of this technology will not be particularly influenced by utilization patterns; others will be highly dependent on the particular uses and users. Major uncertainties characterize all evaluations of potential effects.

Because of these uncertainties, analysis and planning must embrace a wide range of possibilities. The technique of scenario analysis is one means of achieving this.[8] By this method questions can be formulated and examined with respect to a set of "states" designed to span a variety of future conditions that need to be considered.

In the course of our work, we have explored many scenarios that could include alternative futures. For most purposes, however, we have found it adequate to consider three states. They portray very different diffusion patterns, each of which will influence the nature and magnitude of the demand, the impacts, and the institutions involved. These scenarios were most frequently used in an implicit fashion, but in some cases—for example, in our analyses of capital requirements (Chapter 4) and of impacts on the trucking industry (Chapter 6)—we used them explicitly as a basis for our calculations. We are reproducing the scenarios here partly because we expect them to be useful to those who may extend and refine our analysis.

The three scenarios are as follows:

STATE 1: LOW DEPLOYMENT. Cellular systems are deployed at a small number of sites (perhaps two to five markets) principally for vehicular communications. Cellular technology is used to provide dispatch and mobile telephone services, though much of the demand for dispatch is accommodated by cheaper, automated trunked systems.

The comparatively large capital expenditure necessary for system implementation, when combined with the small number of users per market, creates a high shared investment cost per user and, hence, a high service price. This costly system requires no major technological breakthroughs for its implementation.

The system is used primarily in situations that are relatively insensitive to cost considerations; for example, principal use is by managerial personnel, whose time is accorded a high opportunity cost, and by public law enforcement agencies and emergency service organizations for command and control activities. Only a very small number of households per market utilize mobile telephones because of the high cost. It is also assumed that cellular technology will not expand into new markets because demand for mobile communications can be accommodated by existing systems that utilize new techniques for increasing spectral efficiency.

This low deployment level requires no major allocations of capital (for dividing cells) beyond those envisaged in initial planning projections. Visibility and general public awareness of the technology is low, perhaps analogous to present perceptions of pocket pagers.

STATE II: MEDIUM DEPLOYMENT.   Alternatively, over the next 15 to 20 years, dynamic economies of scale resulting from production advances could substantially reduce costs. One can then envisage a scenario in which systems are deployed in five to ten regional markets and services are offered at a moderate cost. While trunked systems are still competitive in the provision of dispatch services, new users prefer cellular dispatch because of its equivalent price and unsaturated channel-loading characteristics.

The moderate costs encourage utilization of services for both business and personal purposes. Numerous organizations adopt the technology because of its demonstrated potential for increasing productivity within the public and private sectors of the economy. In addition, an average of one in ten households have mobile telephones available for personal use.

This level of demand would necessitate either an increase in allocated spectrum or the addition of cell division and frequency reuse equipment to existing systems, requiring a supplemental capital expenditure. Such demand could also prompt development of portable units and add-on equipment, as well as stimulate systems to cover extended areas, including transportation routes between several pairs of close markets. The scale and physical presence of cellular technology would be significant, and its visibility to the general public would compare with that of CB equipment at the present time.

STATE III: HIGH DEPLOYMENT.   This scenario assumes achievement of both the production and design advances necessary to make low cost cellular services possible. As a result, there is a very large increase in the number of cellular subscribers per market and also in the number of markets where systems are deployed. Rapidly falling costs over a 20-year period have made trunked systems less competitive because of the comparable price and superior service characteristics of cellular telephony. Mobile units become widely deployed in business operations as useful and economic organizational tools and emerge as consumer items to meet household needs. Several models, both portable and vehicular units, that incorporate a variety of add-on options to suit various purposes are available for rental or purchase.

The burgeoning demand for mobile services requires supplemental spectrum allocations and multiple cell divisions to increase channel reuse capabilities. Such popularity might also stimulate deployment of cellular equipment along the major highways connecting metropolitan centers. This scenario suggests a wide public awareness of cellular technology's utility, perhaps comparable to present perceptions of FM radios or tape recorders.

If the present trends continue, the medium and high scenarios could be realized in two or three decades. Given a longer period, a possibility exists that cellular devices could become as pervasive and important as today's fixed telephone. Should this occur, virtually everyone would have a mobile unit. Mobile communications would be available in all major population centers and along major highways at a cost only slightly higher than that of wireline telephone service. Usage would be commonplace in most aspects of daily life. The range and magnitude of potential impacts make this distant future scenario worth attention; we shall refer to it as the "universal" state. In Table 15.1, the principal characteristics of the three scenarios used in our work, as well as the characteristics of the universal state, are summarized.

These scenarios have been formulated to take a variety of factors into account—factors that will both influence the deployment of cellular technology and result from that deployment. The level of deployment reached in two or three decades will depend on the industry's ability to improve the production and design of equipment, the continuation of stimulative policies by the regulatory agencies, and increasing the acceptance of the devices by consumers. Other important factors include the competition from alternative technologies, popularity and convenience of fully portable units, geographic patterns of deployment, availability of adequate capital, and influence of business use on personal use.

The range and magnitude of impacts will vary significantly between different scenarios. The importance of specific public policy questions, for instance, depends greatly upon the particular future state that is envisaged. One example is that issues concerning international trade or the public nuisance aspect of the devices will be relatively inconsequential if the number of users stabilizes at the low scenario state. Our discussion of economic issues earlier in this volume contains many other examples of policy issues that change as the scenario is altered. Costs and benefits generally increase with the level of deployment (though the relation is not necessarily linear); many of the public safety benefits associated with individual use of the devices require a medium or high deployment level for their realization.

It should be noted that the medium scenario we have described is not an arbitrarily chosen intermediate state between the low and high deployment levels. Rather, we intend to describe an important "transitional" state that provides the conditions for subsequent rapid growth. Once the medium scenario has been achieved, the cellular system will have significant public visibility and it is likely that the technology's potential for use in areas such as public safety and business will be fully appreciated. The recent experience with Citizens Band radio illustrates the possibilities. The extremely rapid growth of CB in recent years has led officials to consider it as a viable alternative for motorist aid systems. Similarly, in the medium scenario, we expect insurance companies and other organizations to become aware of, and seriously interested in, the potential of other mobile communication devices for reducing the costs associated with accidents and hazards. Thus the medium scenario represents an important stage in the evolution and diffusion of the technology.

Having given our assessment of the course development of mobile communication technology will probably take, and described some scenarios that embrace the many uncertainties associated with the technology's future, we now turn to a recapitulation of the benefits and costs that can be expected.

Table 15.1: *Characteristics of Scenarios*

Scenario Variables	I Low	II Medium	III High	Universal
Avg. no. dispatch users/market (thousands)	5-25	100-250	150-350	Very High
Avg. no. MTS users/market (thousands)	15-25	100	500	Very high
Avg. no. households/market (percentage)	1	10	50	Most
Total equipment costs (1976 dollars)*				
Dispatch	$2,000	$1,700	$1,100	Slightly higher than fixed phone
MTS	$2,400	$2,000	$1,200	Slightly higher than fixed phone
Service costs/month				
Dispatch	$45-75	$30-55	$25-40	
MTS	$75-100	$45-75	$30-45	
Geographic availability of cells				
No. of markets	2-5 high density urban markets	5-10 regional centers	Up to 20 centers	Majority of 200 SMSAs; extension to rural
Transport routes	No highway deployment	Between several pairs of close markets	Major corridors	Most major highways

Mobile unit design				
For auto	Yes	Yes	Yes	Yes
Portable	No	Partially (e.g., autos as repeater)	Yes	Yes
MCTS competition	Yes; provides cheaper dispatch	Yes, but saturation without new allocation	Less competitive because of superior MTS service	Less
Major technological advance needed	None	Production	Production and design	Production and design
New spectrum allocation above initial	Not necessary	Necessary if no subdivision	Yes	Yes
Type of use				
Personal	Sparse	Small	Most upper income households	Pervasive use in both categories
Business	Essential and managerial operations	High mobility business operations	Large scale	
Social presence analogous to current level of	Pagers	CB units	FM radios or tape recorders	Fixed phones

*Does not include all costs associated with joint use of fixed telephone facilities.

THE BENEFITS AND THE COSTS

In earlier chapters, we have argued that substantial benefits are likely to result from increased deployment and use of mobile communications. Extrapolation from our present experience makes it clear that these devices can facilitate: increased productivity in commercial activities; improvements in public transportation through better vehicle control and passenger-responsive systems; and improved public safety services by reducing client response times and increasing the level of coordination. The growth of Citizens Band demonstrates that there is a significant market for systems permitting more private and recreational uses and, although such uses are less amenable to quantitative analysis, they should be included in the list of public benefits.

Moreover, the benefits can be both private and social. The private benefits will be accrued by the possessor or purchaser of the device, who will be able to communicate with fixed and mobile sites while on the move. The social benefits will be enjoyed by a broader public through more prompt delivery of services and quicker responses to emergencies. As mentioned earlier, mobile communications should have special significance within the large and expanding service sector of the economy, where productivity gains have lagged behind those in the manufacturing sector; even small improvements in this area could result in substantial net economic gains. We have emphasized many times in this volume that the magnitude of benefits cannot be evaluated with precision. Yet it is certain that these communication technologies will influence a wide variety of commercial and public activities, though the size of the "leverage" must be a matter for speculation.

In extending the deployment of mobile communication services, the society will incur a number of costs. Of these, the most easily quantifiable is the direct economic cost borne by those who adopt the technology, a cost that will vary depending on the type of system, and that will change as new technology is incorporated and new manufacturing processes are adopted. The basis for these costs has been described in Chapter 4. We can summarize costs to the consumer in the following way. Citizens Band requires purchasing a unit that costs $100 or more; there is no monthly charge for use of the service. One-way paging involves a relatively inexpensive technology with monthly charges of about $15 to $25. Conventional dispatch service is two or three times higher. Current mobile telephone service is much more expensive, in the range of $90 to $120 per month, because the number of users is limited by spectrum constraints, thus preventing the realization of economies of scale. It is thought, however, that mobile telephone service provided through a cellular system will be considerably less expensive, perhaps half the cost of current systems. While such costs are high when considered as consumer items, they are well within the reach of commercial and public service organizations.

Industries deploying new systems will have to raise substantial capital, the amount of which will depend on the deployment level. To review some figures from Chapter 4, ten cellular markets, each serving 250,000 dispatch users and

100,000 mobile telephone users, would involve a total capital investment of somewhat more than $4 billion, based on 1972 cost estimates, or about $7 billion in 1976 dollars. This example assumes a medium level of deployment—3,500,000 units. Such capital requirements, while substantial, are within the range of new capital investments commonly made in the communications industry.

In principle, the direct costs we have been considering can be quantified, although the precision and reliability of the resulting data may be questionable. But there is a broader range of potential costs or "disbenefits" that are much harder to analyze and virtually impossible to quantify. For example, placing communication devices in transportation vehicles and even in our pockets will surely affect the social environment; the constraints affecting electronic communications will be reduced and the boundaries of social interaction will be extended. Traveling could become a less isolated and more secure experience. Yet privacy and solitude will be decreased because people will be "on call" a greater portion of the time. People will be expected to respond to the communication needs of others while they are traveling as well as when they are at fixed sites, multiplying the demands made upon them.[9] If this happens, we will have taken a further step toward the evolution of the communicating and chattering society. The concept of the wired man, a favorite topic of science fiction, becomes a little less fantastic; only the purist will notice that the means will involve wireless devices. Furthermore, the devices are potential nuisances in public places, like theaters and restaurants; etiquette issues will be raised regarding where and when it is acceptable to communicate via the mobile phone while in the company of others.

Despite many citizens' need for more information and more accessible communication, some feel unable to cope with the levels to which they are already subject. In the case of a busy executive, for instance, adopting a pocket or vehicular telephone might improve decision making within his organization, but possessing the device may put him under more stress. The phenomenon of "information overload" has been dealt with by others; its effects, however, are not irrelevant to the private citizen. In addition, unlike the virtually universal fixed telephone, the mobile telephone will be a technology that many citizens will not be able to afford, assuming any but the "universal" scenario. If strong marketing efforts are undertaken to promote the technology, people who cannot afford it may feel deprived in a sphere—telephonic personal communication—in which relatively low cost formerly provided equality of access. Thus, we note the possibility of a two-tiered mobile communications stratification: cellular mobile telephone for high income people and Citizens Band for others.

As discussed in Chapter 5, new technologies can be the agents of organizational change in those industries and public agencies that adopt them. In many respects they can give the worker who uses them greater freedom. At the same time, however, they also provide the means for greater organizational centralization and control. Some of the gains in efficiency and coordination might, therefore, be offset by the field-worker's reduced autonomy. Surely the level of job satisfaction experienced by these people will be affected, and not always for the better.

As with other technological innovations, the effect of advanced mobile communication systems on worker alienation will depend primarily on how the technology is used. If an organization develops a pattern of rigid and hierarchical modes of communication, the technology's effect may well be to increase alienation of subordinates and to consolidate valuable information in the hands of a few select individuals. If, on the other hand, an organization's mobile communication system is designed to assist in disseminating information and involving workers in the decision-making process, negative reactions to the technology may be mitigated. The central point, then, is that cellular technology should not be viewed as an intrinsically alienating device; it is only a medium through which people communicate.

We can also anticipate, from our analysis in Chapter 8, the possibility of problems arising from the utilization of sophisticated mobile communications in the public service sector. Public safety agencies could easily resort to advanced technology in circumstances where more manpower and increased human interaction are called for. The more sophisticated and complex dispatch systems, especially those requiring advanced computer technology, could become the political pawns of the public safety bureaucracy without potential clients really being benefited. Moreover, if these new technologies are expensive, and beyond the financial resources of small communities, investing in them will almost certainly necessitate greater dependence on the federal government.

Some of the new technologies, particularly those involving vehicle monitoring, can potentially provide the state with new means for surveillance of its citizens. For instance, the cellular system will require a centralized computer that knows in which cell a mobile unit is located. To be sure, even the most advanced system currently contemplated will only be able to come within one or two miles of the exact location, so it might be easy to conclude that this has little bearing on surveillance. Nevertheless, past experience should make us extremely cautious in this area; the possibility of abuse and misuse clearly exists.

Increased deployment of mobile communication devices will increase the ambient level of electromagnetic radiation. A central point of the discussion in Chapter 11 is that, unlike the situation with ionizing radiation emitted by the nuclear power industry, we have already increased the nonionizing radiation level above that of the natural environment by many factors of ten. Further deployment of personal communication devices will increase it more. An analysis of the radiation emissions expected from the mobile units themselves suggests that such emissions will be inconsequential if judged by the American safety standard, and small in terms of the Soviet standard. The one exception is hazard to the user of the portable device held close to the head during operation; this case requires additional study. Other safety and environmental concerns, such as driving hazards when these devices are used in vehicles, will also require careful attention.

Finally, no contemporary assessment of costs and benefits would be complete without reference to energy consumption. It is far from clear whether more widespread use of mobile communications will ultimately increase or decrease energy consumption. Some uses will certainly result in fuel savings: more efficient scheduling and deployment of trucks, public service vehicles, and buses, and the partial substitution of automobile travel by communication are examples. On the other hand, mobile communication technologies, particularly the more sophisticated and versatile ones, have the potential of further enhancing the automobile as a means of transportation. For the businessman, an automobile equipped with a high quality communication system could easily become an extension of his office.

Energy and environmental considerations have spurred many efforts to make public transportation more attractive. But the effectiveness of such efforts could be undermined unless public transportation is equipped, wherever feasible, with advanced means of communication for the passengers' use. Even if the direct costs for providing this service seem to be high, they may be justified by secondary social benefits. Indeed, a case can be made for partial subsidization of such services, given that increasing the automobile's advantage as a mode of transportation does not seem to be in the long-range public interest.

It is tempting to arrange the costs and benefits of the new technology in two columns for easier summation, since some may seek what present jargon calls "the bottom line." This procedure would be unproductive, however, because individuals and groups attribute different weights to the specific costs and benefits based on their values and preferences. Nor can consideration of concepts such as the "general welfare" or the "public interest" endow the process of aggregation with meaning: "Almost without exception, technological developments will affect some people and interests adversely and others beneficially, and there is no agreed-upon algebra by which one can really subtract the pains from the pleasures in order to arrive at an index of social desirability."[10]

Still the process of identifying costs and benefits is extremely important. Policies can be adopted that will minimize the costs. Also, the details of the balance sheet will be greatly influenced by governmental and business decisions that affect the diffusion of the technology.

Regulatory decisions will play an especially important role in the balancing of costs and benefits. When decisions like the allocation of spectrum are being made, the choices reflect, either explicitly or implicitly, our values and preferences. While the alternatives will rarely be clear-cut, we can give priority to communications for emergency and public services, or to uses that will increase the efficiency of our systems of private production, or we can emphasize recreational and private uses.

If the development of mobile telephone systems is emphasized, we will, at least initially, be giving priority to the managers of our production and service organizations and to the affluent citizen. Favoring the development of two-way dispatch systems would reflect the value we place on productivity. In the past, such priorities have been prominent in the area of mobile communications. But the growth of Citizens Band, the "people's" mobile communication device, presents us with a different kind of choice.

Utilization of the airwaves for private and recreational purposes is, of course, not new. Yet, while some similarity to the standard broadcasting situation exists, important differences are also obvious. For instance, in standard broadcasting the role of the citizen is passive, and the medium is often controlled by large commercial enterprises. The dominant characteristic of Citizens Band, however, is "participation," and its rapid growth clearly reflects an unfilled need in our society. In terms of public policy, the critical issue is how to respond to this need. Although these recreational uses are not directly related to the economic issues of productivity, they are related to other needs of the populace, such as the allocation of resources for public parks and other leisure-time facilities. In fact, the disconnection from productivity, in the narrow sense, may mean that these private needs do not receive adequate attention and resources from the political sector. It is extremely easy, for example, to rationalize assigning the lowest priority to private and recreational uses when allocating spectrum. We must guard against such a tendency.

This overview of costs and benefits will lead some to ask whether the technological developments that have been the focus of our work are being stimulated by a perceived and important social need or by what has become known as the "technological imperative." Certainly, the technologies that we are examining can serve many useful purposes in our society. As is true of most new technologies, they were not developed as a direct response to some of our most basic social problems, such as unemployment, the high cost of medical care, or even those ills associated with the state of the physical environment; yet perhaps by indirect means they will be able to contribute to the amelioration of these problems. Instead, the technologies we are analyzing are products of a highly industrialized society that emphasizes productivity and the saving of time, and—with the exception of Citizens Band devices—they are largely oriented to serving business and governmental ends, rather than those of a cultural, private, or social character. It would be erroneous to suggest, though, that these technologies are being thrust upon an unwilling public and reflect an artificial stimulation of wants. To be sure, many of these devices are manifestations of our highest technological skills, but the public has demonstrated many times its large appetite for them. The quality of service provided in many of the areas, particularly current mobile telephone and dispatch services, has been far from satisfactory, essentially because of spectrum constraints. Still waiting lists grow, and the public has made it plain that it prefers poor service to none at all.

The "push" of the technology and the industry is unmistakably present, but there is also substantial public "pull." The benefits of the new technology, however, will also exact social and economic costs; foresight in the deployment and control of the new systems is necessary. Nevertheless, it is difficult to sustain the argument that a national policy should be developed to discourage the continued proliferation of these devices.

SOME POLICY ISSUES

As public and private agencies attempt to foster land mobile communications in a manner that will be most beneficial to our society, they will confront numerous problems. Unfortunately, these problems cannot be translated into a set of questions that, if answered, will provide solutions for the indefinite future. Rather, the problems involve a set of issues that will require continuing attention and reappraisal as part of a planning process.

Many of these issues have been discussed in detail in preceding chapters, and it is impractical to repeat them here. Consequently, we shall restrict ourselves to a review of the principal themes by listing some examples of policy questions along with the principal agencies that should be engaged in their consideration.

The policy questions are of diverse types—some involve basic strategic considerations, while others are less extensive in their scope and importance. The relative significance of these questions and their potential answers will depend on the breadth of diffusion that is envisaged for the new technology. As a result, detailed consideration of them will require use of our scenarios, or some other means of describing alternative futures. In presenting this summary, we remind the reader of the hazards involved in attempting to reduce complex issues to a set of short questions. Not only is there a danger of oversimplification, but at any given time, it is inevitable that certain important specific questions will be overlooked.

With these reservations in mind, below are listed some important areas and examples of specific policy issues:

ECONOMIC EFFICIENCY IN LAND MOBILE COMMUNICATIONS
[Federal Communications Commission (FCC), state regulatory agencies, Department of Justice (DOJ), the courts, the Congress]

—Is government intervention required to achieve an efficient allocation of resources?
> Are there significant externalities?
> Will a competitive industrial structure develop without intervention?

—What form should intervention take?
> Are taxes or subsidies desirable?
> Is regulation warranted?
> Will antitrust action be needed?

—For how long should governmental intervention be employed if it is required?

—What is the most desirable industrial structure?
> Should all systems be allowed to offer all services?
> How much competition between the systems is desirable and how should such competition be encouraged?

—What policies should regulators employ to promote static and dynamic efficiency?
> Should an exclusive license be granted to one cellular company in each market?
> What criteria should be employed in the granting and renewal of local franchises?
> What criteria should be used for establishing the rate base of cellular

operations? Should the cost of the mobile unit be excluded from the rate base?

Should operators be prohibited from requiring use of specific equipment?

Should detailed standardization of equipment be required, or should performance standards be used to encourage diversity of equipment?

Should usage-sensitive pricing be introduced to encourage efficient use of the spectrum?

Should differential pricing be instituted that discriminates between private (personal) and business uses of mobile telephone service?

Should the subsidization of rural service, using revenues from urban sources, be allowed?

Should state regulatory agencies set performance standards for various services to ensure the quality of those services?

USE OF SPECTRUM
(FCC)

—Should the spectrum reserved at 900 MHz be used to experiment with technological alternatives to the cellular system before widespread deployment is allowed?

—What policies can be employed to ensure that the general public has adequate access to mobile services for private use?

  Should Citizens Band radio receive a larger proportion of spectrum resources and should it be considered an integral part of the entire land mobile area?

—How cautious should the FCC be about deploying new FM systems in the face of rapidly developing alternatives?

—What criteria should be used to decide between dividing cells or adding spectrum when new subscribers must be accommodated by cellular systems once they are deployed?

—Should trunking be required on smaller radio common carrier services?

FCC PROCEDURAL AND ORGANIZATIONAL MATTERS
(FCC, the Congress)

—What procedures are necessary to ensure effective enforcement of regulations as devices proliferate (e.g., should there be automatic transmitter identification)?

—Should public participation in FCC proceedings be increased?

  How can it be achieved and how should it be funded?

—What should be the range of considerations (e.g., social, economic, political, technological) examined by the Office of Plans and Policies in its planning and analytical efforts?

  Should the composition of professional staff in that office be altered?

INTERNATIONAL ISSUES
[FCC, Department of State (DOS), Department of Commerce (DOC), Congress]

—What policies, if any, should be introduced to minimize the expected trade imbalance, with respect to mobile communications equipment, between the U.S. and Japan?

—Should any measures be taken to forestall the possibility of a foreign-based (probably Japanese) company securing control of the land mobile system in one or more cities/regions of the U.S.?

—Since anticipated land mobile use of the 900-MHz band is constrained in 35 percent of the continental U.S. because of proximity to international borders, what agreements should the U.S. be willing to make at the World Administrative Radio Conference to satisfy domestic needs and international frequency allocations?

—What steps, if any, should be taken to encourage cooperative ventures with Japan in land mobile development, both to aid U.S. manufacturers within Japan and to take advantage of the technological lead that the U.S. and Japan now appear to have in the land mobile field?

—What provisions should be made to allow for satellite development and usage in conjunction with land mobile communications?

—If radio methods, as exemplified by land mobile technology, offer less developed countries the opportunity of bypassing the deployment of wired telephone systems, should the U.S. undertake special efforts to develop land mobile components for such purposes and seek an export market in this area?

LEGAL ISSUES
[FCC, DOJ, Law Enforcement and Assistance Agency (LEAA), Federal Reserve Board (FRB), state banking and insurance agencies]

—Will increased use of mobile communications by law enforcement authorities require more stringent controls on electronic surveillance (such as the tracking potential of the cellular system) than the law furnishes at present? Should there be legislation enacted that would limit access to locational information?

—Should some form of the criminal offense misprision of felony be revitalized, considering the increased ease with which observed crimes can be reported via mobile communications?

—Will the mobile telephone be involved in the development of cashless financial transactions? Will changes in the law of commercial credit and other banking transactions be needed?

—Should a relatively low negligence standard be established for physicians using mobile telephones in order to encourage its use for medical purposes?

TRANSPORTATION SERVICES
[Interstate Commerce Commission (ICC), DOJ, FCC, Department of Transportation (DOT), LEAA, Department of Energy (DOE), Department of Health, Education, and Welfare (HEW), American Trucking Association, insurance companies, Amtrack]

—Should the ICC consider the potential benefits of present and future communications when making rules concerning allowed routes and service areas?

Should the ICC establish rates that reflect the savings made possible by mobile communication and ensure that such savings be passed on to the shipper and the general public?

—Is deregulation more likely to promote the adoption of new communication techniques by the trucking industry?

—Should ICC regulations, and perhaps antitrust policies, be amended to allow consolidation within the trucking industry in cases where economies of scale related to the use of mobile communications can be achieved, and where reduction in costs to the shipper and general public are likely?

—Should the federal government sponsor experiments to evaluate the effectiveness of the cellular system as an antihijacking tool?

—Should DOT sponsor additional demonstrations to evaluate the potential efficiencies afforded by mobile communications in demand-responsive bus systems, as well as other public transportation systems?

—Should further incentives be used to promote Citizens Band radio as part of a motorist aid system?

—Should federal subsidies be provided for installing cellular systems along a selective, strategic set of interstate routes?

—Should the government develop incentives and/or regulations to encourage the deployment of mobile communications on buses and trains for passenger use?

DELIVERY OF PUBLIC SERVICES
[FCC, Department of Housing and Urban Development (HUD), HEW, LEAA, Congress, DOJ, the courts, state regulatory agencies, local governments, associated public safety communications officers]

—Should the federal government increase its promotion of integrated mobile communication systems for emergency public service delivery agencies by expanding its financial support?

—Should the federal government extend its promotion of mobile communications use by citizens seeking assistance in emergencies?

—Should the government promote specific types of and certain add-on features to the cellular system that enhance the usefulness of mobile services in emergency situations?

—Should preferential pricing of mobile services be instituted to encourage voluntary and publicly oriented uses by organizations such as citizen patrols?

—Should new methods of coordinating responses to service requests be developed, since the new technologies will facilitate the making of requests and may therefore greatly increase the demand for service?

SAFETY AND HEALTH HAZARDS
(FCC, Consumer Product Safety Commission, National Highway Traffic Safety Administration, state motor vehicle agencies, Environmental Protection Agency).

—Should state and/or federal agencies issue standards regarding location and mounting of mobile communications equipment?

—What design and safety standards are needed for mobile communications equipment?

Should abbreviated dialing be required for automobile telephones?

Should radiation emission standards be established for portable units, like they are for microwave ovens?

—What safety standards and rules are necessary for governing motorists' use of mobile communications equipment?

—Is the level of research concerning the athermal effects of nonionizing radiation adequate?

TOWARD BROADENED PLANNING PERSPECTIVES AND OBJECTIVES

The diversity of the policy issues described above reflects the very wide range of private, public, and commercial activities that the new technologies will permeate and influence. Continuous planning and analysis will be required to deal with these issues; not only are we unable to give permanent answers to the questions posed, but the questions themselves will have to be reformulated as conditions change.

The decisions associated with Docket 18262 suggest a heightened awareness within the FCC of the need for planning and a longer-range view of regulatory problems in this area. Yet despite these encouraging signs, one cannot overlook the well-established tradition of short-term and reactive decision making associated with federal regulation of communications. This tradition will continue to influence regulation unless conscious efforts are made to prevent it from doing so.

The recent shift toward a more anticipatory approach is timely. Earlier in this chapter, we provided several arguments that suggest continued, and perhaps accelerating, rates of growth in the use of mobile devices. Moreover, the scale of national resources that will be devoted to the land mobile area—including spectral, economic, and human resources—will be too large to permit a laissez-faire attitude or purely reactive decision making. The encouragement of stable and orderly growth, the attempt to maximize social benefits and reduce undesirable side effects, the achievement of rational and equitable allocation of spectrum to meet the needs of diverse uses and users will require a high level of anticipation by regulatory decision makers. Clearly, the many uncertainties associated with the future of land mobile communications discussed previously will complicate the process of planning, but they in no way reduce the need for it.

One critical area requiring foresight consists of those regulatory decisions that give a relative advantage to a particular technology or system. There is a need to ensure not only that a wide variety of services are provided by means of a large-scale system such as cellular, but also that alternative and perhaps less costly small-scale systems retain an appropriate share of the market. The goal of public policy should be to give users the broadest choice of devices that is compatible with an orderly evolution of the land mobile communications complex. Concerning the cellular system, the FCC can reduce the possibility of recurring congestion and spectral "feast or famine" by its decisions on frequency reuse, through subdivision of cells and other techniques, and the allocation of new spectrum. Only by continued attention to such matters can the full potential for social benefit, implicit in the decisions associated with Docket 18262, be realized.

We do not mean to imply, by repeated use of the word *planning,* the possibility of a "blueprint" that will prescribe the evolution of land mobile communications. Indeed, our earlier discussion of uncertainties makes it clear that such a concept is utterly unrealistic. It is, in any case, incompatible with American political and economic processes for allocating resources. Rather, we are seeking increased institutional capacity for continuing attention and reappraisal of policy issues and regulatory decisions.

Several steps can be taken to improve this capability. First, the range of expertise that agencies employ to carry out such planning can be broadened. Much current analysis in the telecommunications area, and this is especially true of the FCC's work, reflects the technical knowledge and perspectives of engineers, lawyers, and to a much smaller degree, economists.[11] Yet, if significant social impact is to be considered, it is essential to utilize the skills and insights of other professionals, such as sociologists, political scientists, planners, etc. We hope that the breadth of our work—to a large degree made possible by such inclusions—will provide persuasive evidence of the benefits that can result.

Second, while one should strive for a substantial quantification of costs and benefits, it is also important to include those areas not amenable to quantitative analysis. For example, impacts on privacy and social patterns should be considered even if it is difficult to do so. The importance of quantitative analyses should not be minimized, but it is clear that many important impacts of a new technology cannot be expressed in quantitative form; at the same time, one must resist the temptation to give undue attention to some factor just because it can be quantified.

Third, the initiation and continuance of broadly based long-range studies must become part of the normal business of the executive and regulatory agencies. Otherwise the agencies will continue to have their agenda set by the demands of immediate problems and be too dependent on industrial interests for information and the specification of alternatives.

The particular perspective of our work in mobile communications has brought us to these conclusions. We are aware, however, that such views have been proposed by many others and are gaining wider acceptance, especially within the government. The formation of the Office of Plans and Policies within the FCC in 1973 is an indication.[12] Similarly, the Office of Telecommunications in the Department of Commerce (1970) and the Office of Telecommunications Policy in the Executive Office of the President (1970) were established because of the need for new planning and coordination capabilities. Offices concerned with telecommunications can also now be found in a number of executive agencies such as DOT and HEW.

In late 1977, as part of President Carter's executive reorganization, the Office of Telecommunications Policy was eliminated and its functions transferred to a new Office of Communications and Information within the Department of Commerce, headed by an assistant secretary. As expected, the President has argued that this reorganization will facilitate government decision making on communications matters. One can only hope that this is correct and that transferring OTP out of the President's Executive Office does not reflect a change in the perceived importance of telecommunications planning. There can be no doubt that many issues in the communications field will require attention at the White House level, and some will require the attention of the President.[13]

While the FCC will continue to be the main source of communications policy, the characteristics of land mobile systems are such that effective coordination with other regulatory agencies and executive departments will be of crucial importance.[14] For example, the benefits that could result from more extensive mobile communications use in the trucking industry are unlikely to be realized if ICC rules, which place restrictions on allowed routes and service areas, are not modified to account for this potential. Nor are the possibilities for increased public safety apt to be realized without some coherence in the decisions and activities of the FCC, HEW, and Department of Justice. The Department of Transportation is now vitally involved in renovating urban public transportation; new mobile communication technology has a significant role to play in this process as well.

The consequences of poor coordination are likely to be especially serious in the energy field. As we have already mentioned, new communication technologies may save energy in some situations and increase energy use in others. Therefore it seems essential that electronic communications, especially of a mobile character, be considered in the development of an energy plan.

Agencies that should be involved in this planning and coordination fall into three categories, though the distinctions are not always clear-cut. First there are those agencies already charged with significant responsibility for telecommunications research and planning (e.g., the Departments of Commerce and Defense). A second group consists of those agencies that are major utilizers of the technology (e.g., LEAA, HEW). Yet a third is composed of agencies with broad areas of purview, such as DOE and EPA, whose own goals might be achieved by increased utilization of mobile communications.

The federal government's influence also extends well beyond making regulations and directly stimulating technological development for civil purposes. For example, the federal government alone operates more than half a million vehicles; because of this massive market, government procurement decisions can influence future development of the technology, the level of deployment, and safety standards. If the General Services Administration decided to equip its automobiles with Citizens Band radios for safety reasons, the impact could result in further public adoption of such devices.

Moreover, military and space programs are continually extending the limits of mobile communication technology. They are the source of many advanced techniques applicable to the civil sector, but one can ask if the private sector derives the maximum possible benefits from the research. This is the old problem of "spin-off."

In analyzing this problem it is important to recognize that many of the firms performing communications research and development for military and space programs are also involved in the civilian market. So without organized efforts to transfer information from one sector to another, considerable "leakage" of technological methods already occurs. During this project, we have not examined whether further explicit procedures to increase the interchange are desirable. It should be kept in mind, however, that the kind of economic and performance specifications involved in choosing military and space technologies differ markedly from those in the civilian sector. As a result, extensive transference between the sectors should not be expected.

Thus, virtually every agency of the government is involved with telecommunications in one way or another—some as users and others as promoters or regulators. Such a pervasive involvement might lead some to wonder about the feasibility of a central agency to deal exclusively with telecommunications, much like the new energy agency. Yet, although our work has convinced us of the need for more coordination and planning, it has not led us to believe that a separate agency would be desirable or effective.

This chapter's central theme has been planning for the future of land mobile communications. We have discussed the multifaceted character of the planning and proposed a partial agenda. Clearly some capacity for the analysis that is needed already exists within the legislative branch, the regulatory agencies, the executive branch, the universities, and private sector research organizations. But it is hard to persuade oneself that these are adequate, given the large scale of the resources that industry has to support its positions.

Of course, such planning and analysis are only means to an end: to better inform the actions and decisions of government. Whether our governmental agencies can adequately cope with the new demands that rapid growth of land mobile communication will place upon them is an even larger question. Better analysis is a necessary, but not a sufficient, condition for improving governmental actions. In this broader context, land mobile communication deserves no unique consideration; it takes its place alongside many other segments of the communications field—a field that will continue to affect the lives of all of us.

NOTES

1. U.S. Department of Commerce, Bureau of the Census, "Population by Residence and Race: 1950–1974," *Statistical Abstract of the United States* (Washington, D.C.: U.S. Government Printing Office, 1975), p. 17.

2. U.S. Department of Commerce, "Telecommunications Substitutability for Travel: An Energy Conservation Potential," prepared for the Office of Telecommunications by Charles E. Lathey, OT Report 75-58, Washington, D.C., January 1975.

3. U.S. Department of Transportation, *Summary of National Transportation Statistics,* report prepared for the Office of the Secretary by William F. Gay (Washington, D.C.: U.S. Government Printing Office, 1976).

4. See U.S. Department of Labor, Bureau of Labor Statistics, *Productivity and the Economy* (Washington, D.C.: U.S. Government Printing Office, 1971), p. 1.

5. Daniel Bell, "Five Dimensions of the Post-Industrial Society," *Social Policy* 4 (July–August 1973):104.

6. William R. Burch, Jr., "Images of Future Leisure: Continuities in Changing Expectations," in *The Sociology of the Future,* eds. Wendell Bell and James A. Mau (New York: Russell Sage, 1971), p. 161.

7. Daniel Bell, *The Coming of Post-Industrial Society* (New York: Basic Books, 1973), pp. 472-75.

8. H. Kahn and A. Wiener, *Toward the Year 2000–A Framework for Speculation* (New York: Macmillan, 1967).

9. For a discussion of the relation between the telephone and solitude, see Alan H. Wurtzel and Colin Turner, "Latent Functions of the Telephone: What Missing the Extension Means," in *The Social Impact of the Telephone,* ed. Ithiel de Sola Pool (Cambridge, Mass.: MIT Press, 1971), p. 246.

10. National Academy of Sciences, "Technology: Processes of Assessment and Choice," report prepared for the U.S. House of Representatives, Committee on Science and Astronautics (Washington, D.C.: U.S. Government Printing Office, 1969).

11. Of the FCC's professional staff, excluding those engaged in support activities, approximately half are engineers and half are lawyers. Less than one percent of the total staff are economists, and representation of other social science disciplines is virtually zero.

12. The Office of Plans and Policies has a budget of somewhat less than a million dollars out of the FCC's total budget of $60 million.

13. See David Burnham, "Nation Facing Crucial Decisions Over Policies on Communications," *New York Times,* July 8, 1977.

14. For a more detailed discussion of governmental organization questions, see Glen O. Robinson, ed., *Communications for Tomorrow: Policy Perspectives for the 1980's* (New York: Praeger Publishers, 1978).

Part V
AN ASSESSMENT
OF
THE ASSESSMENT

Chapter 16
REFLECTIONS
ON THE STUDY

Raymond Bowers

A s mentioned in the introduction to this volume, our work has had two interrelated goals: first, to conduct a broad assessment specifically on land mobile communications and, second, to carry out the analysis in a manner that would provide some guidance to those conducting future studies in related areas. Consequently, we shall conclude with some brief comments on the procedures used in executing our task, but restricting ourselves to observations that seem relevant to persons who might undertake comparable studies.

Preliminary planning for our work began in the fall of 1972 with a series of weekly group meetings designed to appraise the current state of technology assessment and to consider whether another "case study" in the area of telecommunications would be useful in enhancing our ability to conduct broad assessments. Choosing telecommunications as a focus for our work enabled us to take advantage of our previous experience in microwave technology and videotelephony.[1] Moreover, there was an unquestioned assumption that the actual and potential societal impacts of telecommunications are of great importance. Most of the faculty members who coauthor this volume participated in these early planning meetings, though a few joined the project at a later stage. Four other faculty members also acted as consultants during the planning phase but did not participate once the project was underway.

One of the group's early decisions was to continue working in the mode of a "technology initiated" assessment, in contrast to undertaking one that grows out of an examination of some national problem.[2] Our rationale for this choice, apart from our previous experience with the approach, was that new regulatory and other policy issues frequently result from the proposal to introduce a new kind of technology. While a strong case can be made for paying more attention to the wide range of technologies that might be brought to bear on a particular national problem—that is, undertaking "problem initiated" assessments—many of the national policy questions concerning telecommunications are, in fact, "technology initiated."

In our first sessions we also surveyed the current state of telecommunications in order to choose the area most suitable for our purposes. Land mobiles were chosen for reasons previously described (see the Introduction). Once this was accomplished, considerable time was spent defining the scope of the study—whether, for example, to embrace all mobile communications or to concentrate on just one aspect. Here our decision was to focus on the cellular method, desiring to keep the study's scope within reasonable bounds; this was also consistent with our objective to explore the potential consequences of a particular technology. Of course, we were aware that the cellular system should not be studied without devoting significant attention to related technologies, especially those that are in some sense competitors. Throughout this volume, the reader will note the attention given to related communication devices such as paging systems and Citizens Band radio.

There are substantial risks associated with basing a study on a particular technology, since the one chosen may ultimately be of little significance or rendered obsolete by new technical advances. This can be an especially serious problem if the study is based on the unique characteristics of a particular technology. Therefore, while the cellular system was the central focus of and starting point for our work, our analysis emphasizes the general characteristics common to many new technologies being proposed in the land mobile field. As a result, our conclusions are usually relevant to various means of increasing the level of mobile communications' use.

In designing, organizing, and executing the study, careful attention was devoted to the conflicting demands of integration and reduction. To achieve a coherent and holistic view of the implications of the technology being considered, integration was important; at the same time reduction was an essential step in making the research manageable. Thus, there was a tension between our need to hold the totality of the problem under investigation in mind while dividing the subject into component areas that could be matched to the expertise of the group's individual members.

THE PARTICIPANTS IN THE STUDY

Few of the participants worked on the project full time. Most were professors with other research and teaching responsibilities, graduate students pursuing advanced degrees in related areas, and undergraduate research assistants. An important exception was one person who was appointed to a full-time position halfway through the study. Formerly a graduate student working on the project, he rapidly acquired major responsibilities and senior status within the group upon assuming the full-time position. Two other research assistants worked full time for extended periods at the midpoint of our work. Yet from many members of the research team, this project received major attention for the bulk of a three-year period. Most members of the team, hereafter referred to as "component investigators," had the following responsibilities:

—to participate in defining the subject of the study and the scope of the research.

—to specify important and feasible research areas from a disciplinary perspective, leading to a component study (partially identifiable as the present chapters).

—to redefine topics for research after conducting preliminary investigations concerning feasibility and availability of data. Redefinition of research plans often resulted from suggestions made by other group members.

—to collect empirical data when appropriate.

—to draft reports on research activities in preparation for writing "working papers," which were regarded as first drafts of material intended for the final volume. These documents were circulated to all members of the group and to several external experts for criticism.

—to work closely with the principal investigator in revising material for use in the final report.

All members of the group were responsible for coordinating their activities with those of other members and participating in the process of synthesis. The principal investigator did not concentrate on any single component area. Rather, his efforts were devoted to achieving coherence between the components and activities aimed at synthesis and overview; he read all of the material being produced to identify gaps and inconsistencies in the emerging documents.

As expected, the interests and expertise of the various members influenced the choice of research topics. During the planning phase, considerable thought was given to how we should deal with important topics that were outside our members' competence. In two areas (transportation and the political dimensions of regulation), we were successful in finding colleagues with substantial expertise who agreed to join the project. Other topics were covered by encouraging members to extend themselves beyond their conventional disciplinary boundaries. Still other important areas, such as social psychology, could not be adequately treated because of our inability to recruit researchers with appropriate backgrounds.

The conduct of the research involved frequent interaction between the project members. Group meetings—held once a week during most of the academic term and often twice a week during the summer when research activities increased their tempo—played a critical role in maintaining coherence and communication. Sometimes the hours spent in group meetings seemed excessive; yet, in retrospect, most members agree that the regular assemblage was an essential element in our organization. The topics covered included: strategy for the study; the work of individuals; administrative matters; and, occasionally, reviews of our work with visitors. The format ranged from highly structured business sessions with a definite agenda to broad ranging seminars, often of a tutorial character.

Twice the entire group met with an external review committee to evaluate the state of the research.

ORGANIZATION AND MANAGEMENT

As might be anticipated in an undertaking of this magnitude, we faced a number of problems in organizing and managing the project. Such problems can be expected in any organization, but some are likely to be more serious in a university setting. For example, the principal investigator in a university context, unlike an industrial manager, has virtually no direct authority over the team members; he must rely on their interest in the research and sense of responsibility to the group to ensure adequate and timely performance of tasks. In addition, everyone had other obligations and pressures to which they were subject. For the junior members of the faculty, having their work evaluated by their departmental colleagues was an ever present reality, one whose importance increased when consideration for tenure was imminent. And work on technology assessment, a relatively new interdisciplinary activity, will seldom be considered equal to conventional disciplinary work by a regular academic department.

Another complicating factor, and one not restricted to universities, was that most component areas were covered by a single senior investigator. Put another way, we had no "spare parts." Consequently, if a component investigator was unable to produce the material in his area of responsibility, for whatever reason, it was extremely difficult to fill the gap. Unfortunately, our experience provides no simple prescriptions for dealing with such situations. It would be too costly, for example—both financially and organizationally—to attempt to avoid this problem by multiple staffing in each area. Problems of this kind are compounded as the study proceeds; beyond the project's midpoint, it becomes impractical to involve new investigators.

We do not mean to imply that the dilemmas mentioned above are so immense that attempts to carry out such research projects within a university are unwise, because they occur everywhere to some degree. On the contrary, the advantages universities have over other research institutions, in the opinion of this author, outweigh the serious difficulties we have discussed. Few other research institutions can match the range of expertise and breadth of intellectual resources that are characteristic of a major university. Furthermore, the university's relative independence from external pressure and its suitability for contemplative and long-range analysis are in its favor when public policy is involved. Finally, conducting research within universities results not only in new knowledge and understanding, but also in the production of trained and educated people able to continue related work at other institutions.

Debating the universities' appropriateness to work in technology assessment will ultimately be useless if the question is posed too simply. Universities, private research organizations, professional societies, and national academies all have the ability to conduct broad assessments; but some will be most suitable for one kind of study while others will be suitable for a different kind. All have a role to play in the evolution of the art, craft, and science of technology assessment.

COMMUNICATION IN A MULTIDISCIPLINARY GROUP

It has often been asserted that serious problems can arise in a multidisciplinary group because its members cannot "communicate." Each field represented in the group has a specialized language and jargon, the use of which can lead to misunderstandings between members. The various modes of expression reflect both a body of knowledge and, perhaps more important, a set of conceptions and assumptions that may be unfamiliar to those outside of the discipline.

In the course of our work, we found that such communication problems—at least in their simpler manifestations—were much less serious than anticipated. After several months of working together, each member acquired enough familiarity with the language and operating assumptions of the others to greatly reduce the problems of exchanging information. Indeed, through the tutorial sessions held within the group's regular meeting schedule, most members gained a sufficient understanding of the substantive content of research being pursued by others to enable them to make useful criticisms and suggestions.

Problems concerning the use of technical terms were slightly more troublesome but were overcome with modest effort. In the area of engineering, for example, all members of the group gained some understanding of basic concepts such as bandwidth and spectrum efficiency; by limiting the reliance on excessively technical language, the underlying ideas often proved simple enough to be accessible to those not trained in the physical sciences. Tutorial sessions and the circulation of articles written for a nonexpert audience were effective means of dealing with this problem. The vocabulary used by the social scientists, on the other hand, relies largely on commonly used words with specialized meanings. Those members with little social science background had to learn that when words such as "competition" were used in an economic context, the meaning intended was far more limited than that prevailing in common usage. Tutorial sessions are less useful for handling this type of problem, although understanding did steadily improve through the interaction in our regular working sessions.

One problem, which might appear to be the result of poor communication, in fact had its origin in successful communication. The standards for judging the adequacy of data and the level and kind of analysis required to substantiate conclusions varies markedly from field to field. These differences reflect the state of each field's development and the nature of the subjects being investigated. As expected, in the early phases of our work, practitioners of one field would inevitably apply their own standards to the work of others. The issue of quantitative versus qualitative analysis arose frequently, reflecting a debate that has been especially intense in the social sciences for decades. The assumptions underpinning the analysis in several areas were vigorously challenged by those working in other areas. The tension thereby generated within the group proved to be a desirable consequence of multidisciplinary work. It gave a broader perspective and a heightened sensitivity to our work, qualities that would have been lost if the several disciplinary contributions had been executed independently; criticism by those outside of a specific discipline proved to be valuable. Moreover, the interchanges certainly led all members of the team to a deeper understanding of the strengths and weaknesses of any single investigative approach.

Good communication also led to open discussions of the political and social values that were influencing some members in their choice of research topics and in their execution of the analysis; this is especially important when dealing with public policy. We had no illusions that technology assessment can or should be "value free" or "fully objective," though we were committed to making our analyses as "even-handed" and devoid of doctrinaire positions as our professional skills would allow. Yet, while the explicit recognition of values and assumptions is desirable, it is difficult to discern whether an implicit set of values and assumptions remains in our work that deserves questioning and challenge. This question must be left to those evaluating our efforts.

METHODOLOGICAL QUESTIONS

Once we had subdivided the research into a number of component investigations, each was carried out by using the appropriate disciplinary methodology. Synthesis was extensive and systematic, but it was achieved by a variety of methods, which are fairly characterized as heuristic or ad hoc. Formal methods—impact trees, cross-impact matrices, digraph techniques, etc.—influenced how we thought about multiple interactions and provided aids to the research, but they were not used extensively or mechanically. We surveyed the literature dealing with general methodology and examined prior assessments in order to find techniques that were applicable to our work.[3] Also, in an earlier phase of the project, significant effort was devoted to developing a computer program that would facilitate information storage and the study of interactions. Technical difficulties and the loss of critical personnel, however, prevented us from fully utilizing this promising technique.

Most of the methodological difficulties that we encountered were foreseen in a report on technology assessment issued by the National Academy of Sciences in 1969.[4] In that report, conceptual constraints and limitations were categorized as follows: shortcomings in modes of analysis, failures of imagination, inadequacies of fundamental understanding, and deficiencies in the data base.

Shortcomings in analytic modes have already been alluded to. We found that formal techniques, such as impact analysis, relevance trees, and trend extrapolation, were definitely useful but only in the broadest conceptual sense; incomplete data and inadequate understanding of linkages prevented their full exploitation. Anticipating effects that could ensue from the interaction of a new technology with other technologies and social processes requires imagination and the ability to recognize a variety of subtle interactions. We know of no formulas for making such projections rigorous or complete, though the analysis can be systematic. We frequently had to resort to informed and disciplined speculation based on the information we had collected, the data we had generated, and the views of experts that we consulted, techniques that constitute the most widely used "methodology" in any intellectual activity.

In addition to being speculative, we have assumed an "evolutionary" model of American society in most of our work and, consequently, our forecasting does not include the possibility of cataclysmic events. Thus possible occurrences, such as World War III, revolutionary social change, or even a major depression, were not considered in our analysis. This was not meant to reflect optimism or conservatism on our part, but rather represents our desire to hold constant as many factors as possible so that we could focus on what seemed to be the most plausible consequences of the technology itself. By excluding such events from our purview, we have followed the practices of most public policy analysis dealing with the civil sector.

There is, at present, an undesirable schism between those developing formal methods for use in technology assessment and those engaged in carrying out assessments in particular areas. Put in oversimplified terms, those engaged in methodology research construct elaborate schemes that are not easily adaptable to practical situations. On the other hand, those executing specific assessments (and certainly this is true of the present project) adopt ad hoc methods that other investigators find equally difficult to utilize. Clearly, there is much to be gained by attempting to relate the work on general methodology to that involving particular assessments. We hope that people studying questions of general methodology can exploit the abundant information contained in this volume to develop their techniques further and perhaps demonstrate how a more extensive application of general methods could have resulted in more rigorous analysis.

Our work has also led us to more deeply appreciate the limits of our present fundamental understanding of the relationship between technology and society in general, and the role of electronic communications in particular. We have attempted to draw on some basic sociological concepts, but this has proved to be a difficult task; enormous gaps exist in our understanding of the sociology of electronic communications.[5] Furthermore, it was often difficult to apply the results of existing research to the particular set of problems we were examining. Nevertheless, this more basic research provided us with important insights and means of organizing apparently disconnected data. Perhaps this is all that should be expected, since even in the physical sciences it is frequently difficult to apply basic theory directly to practical problems of engineering in an unambiguous and fully deterministic manner.

We are especially conscious of the fact that our analysis of the new technologies' effect on individuals leaves much to be desired. While we have made more progress in considering those impacts that result from organizational changes in the business and public sectors, the more profound and direct impacts on individuals remain quite uncertain. As Alex Reid pointed out several years ago, "Although the study of human interactions occupies a central place in a number of disciplines, the human aspects of person-person telecommunications have been curiously ignored. . . . Despite their obvious relevance to each other, the fields of social psychology and telecommunications engineering have made little contact."[6] At various points in this volume, we have touched on the issues of privacy and alienation, but our treatment of these fundamental and subtle aspects barely reflects their intrinsic importance. If our work provides further evidence of the need for sustained research concerning the interaction between man and his communication tools, it will have an important consequence.

THE SCOPE OF THE STUDY: QUESTIONS OF BREADTH AND DEPTH

A fundamental problem, common to all technology assessments, is defining the scope of the study. Questions such as where to set the boundaries, and what to include or exclude within those boundaries, arise not only in the planning stage, but also during the study's execution.

We used specific criteria so that decisions could be made systematically, but we also had to rely on our subjective judgments—taking into account the importance of the area, the particular expertise of our members, and whether the topic was amenable to the kind of methodical research we were qualified to undertake.

There are certain essential topics in most assessments that must be investigated. With these, the problem is not whether to include the topic but rather to determine the depth and scope that is appropriate. In our assessment, analyzing the technological and economic factors was of central importance, as were the implications for transportation. For areas beyond this core, however, choosing topics became more problematic.

While our study is quite broad, significant gaps exist. For example, our discussion of international questions is restricted to problems of trade with Japan; we have not discussed Western Europe, or considered if the new radio technologies could permit less developed countries to "leap-frog" the deployment of wired telephone systems. In the domestic area, we were not able to treat effects on family life, despite some early investigative efforts. And, as mentioned earlier, our discussion of the direct impact on individuals is fragmentary. Some of the omissions resulted from explicit decisions to exclude the topic; others reflect an inability to carry out significant research because we lacked the expertise or because there was no readily applicable body of knowledge.

Financial resources have also placed a constraint on our activities, although we do not mean to imply that greater resources would necessarily have resulted in a superior or more useful assessment. One goal of the experimental approach we have used in conducting specific assessments was to gain insight into the question of appropriate scale. Our previous study on the videotelephone was carried out with substantially smaller resources than the present study. It was largely an attempt to synthesize existing literature and expert views on the subject and, therefore, collecting the primary data did not require as much effort. Also, it dealt with most matters in less detail than the present work, and less effort was involved in relating the information to concepts in the social sciences.

Opinions will obviously differ on how extensive an assessment should be at any given stage in the development of a technology or of the policy-making process. While too much information can obfuscate the issues, it is hard to argue that the resulting problems are more serious than those produced by too little information. A case can also be made, however, for the desirability of carrying out a study in the detail of the present one, but that it is then important to write derivative articles for those needing much less detail. Given that the work is not directed to any single audience, it is extremely difficult to decide how much is too little and how much is too much. Reaching a reliable conclusion will take some time and must be based on how useful these various audiences find our work.

SINGLE PROJECT OR MULTIPLE INVESTIGATIONS

The work described in this volume was conceived and executed as a single project with a definite time of initiation and termination. There is another procedure that, we believe, merits careful consideration, especially by the funding agencies. If it is decided that an area such as the future of mobile communications is likely to be important for policy analysis within the next decade, the funding agencies could stimulate or encourage a series of small research efforts to be carried out in different institutions, each dealing with particular aspects of the problem. With this approach, it would be necessary to synthesize these separate efforts after several years, but those undertaking the synthesis would start with a large amount of relevant information. Of course, the probability of significant gaps developing between the work of individual and, perhaps, isolated investigators is a major defect of this procedure; however, the group responsible for the final synthesis could initiate investigations to fill the more serious ones.

This method of organization would be similar to procedures common in the natural sciences, in which basic research is conducted by placing a much greater reliance on multiple investigators taking different approaches, with periodic attempts at synthesis. One advantage of using the multiple project mode is that the results are less likely to be perceived as a definitive and comprehensive study requiring no further elaboration. Obviously, the basis for deciding upon either the single project or multiple investigation mode cannot be reduced to a general formula; the advantages and disadvantages will have to be weighed for each individual case.

THE DURATION OF THE PROJECT

Preliminary planning for our work began in the fall of 1972. The major research activities did not begin, however, until the spring of 1974. Then, more than three years were spent completing the research, with much of the final year being devoted to substantially revising the written material for the final volume.

In retrospect, several of us feel that the project should have been designed to reach completion in a shorter time. Apart from considerations of efficiency in conducting the research, there are hazards associated with long duration: some members will leave the institution where the work is being performed, and others may lose interest in the topic being investigated.

While our study could have been completed sooner if most of the participants had worked full time on the project and had had more expertise specifically in the field of mobile communications, we believe that reducing the time might have sacrificed quality. Considerable "start-up" time is required, during which the members of the research team learn to work together effectively. Also, we often faced obstacles in our research that could not be overcome by merely applying more immediate effort; the problem often had to be put aside until new approaches emerged. A process of "maturing" occurs in most intellectual activities; our understanding of the subject under investigation was cumulative, increasing steadily during the project, and the processes involved are not readily amenable to acceleration.

Some comparable projects have been designed for completion within 18 months, a period that seems to be favored by some federal agencies. Our experience leads us to be skeptical about whether, in most cases, this is a sufficient period to accomplish a high quality full-scale assessment. For those organizations whose research is supported entirely by federal grants, the problem can be exacerbated, since the team cannot be assembled until the funding is assured and, in the final stages of the endeavor, the project director and some staff are distracted by having to seek funds for the next project. Fortunately, such problems did not arise in our work because we had private foundation funds available that reduced our dependence on federal grants.

It is worth noting that one of the early influential reports in the field of technology assessment, that of the National Academy of Engineering, recommends that "full-scale technology assessments should be performed by carefully chosen, single-purpose, and specially qualified *"ad hoc"* task forces that will be disbanded upon completion of their assignments."[7] While the report describes the merits of such an approach, it is clear from our experience that the "start-up" time required for a group project is a serious drawback. Moreover, this strategy is disadvantageous for developing trained people, new methods, and the institutional capacity to carry out assessments. We believe that this field has as much need for procedures that allow for continuing work and institution building as other areas of public policy. The intellectual difficulties in the field are great, and they are unlikely to be overcome by a series of ad hoc assessments.

AN EPILOGUE. Our project was conceived as an "experiment" in technology assessment. While the limitations of our work are many, the experiment, in our opinion, has been worthwhile and productive. We have demonstrated that it is possible to produce a broad assessment of a newly emerging technology. Aware that a few aspects of land mobile communications had been analyzed in greater depth, we have tried to incorporate the results into our overview. Throughout, our goal has been to conduct a study that emphasizes the holistic, so that more detailed analyses can be placed in a wider perspective. Thus, the range of factors included in our work is, we feel confident, more extensive than that normally considered in governmental and industrial decision making.

More work is required in this area; our efforts really reflect only one step toward gaining a deeper understanding of the implications of these new technologies. As we have said before elsewhere:

> No single technology assessment can be considered final because of the absence of complete data, uncertainty about the methodology, limitations of the human imagination to foresee all possible interconnections, and, above all, the inability of humans to avoid bias resulting from their personal values. Indeed, personal values can distort the structuring of the assessment itself. This lack of finality is not unique to technology assessments; it is present in every scientific field. In physics, often regarded as an "exact," or "hard" field of science, research results published in the open literature are subject to criticism, to challenge, and lead to remeasurements and recalculations until, slowly, consensus arises about what is a verifiable description of nature. Technology assessors must expect that their predictions will be similarly challenged, contradicted and reargued. As the result of this process, society will be far more informed of the policy options available for its future than if the assessments were never begun.[8]

Finally, it is worth recording our perspective on what can be expected from the kind of work we have done. Obviously, one cannot expect to predict absolutely the future form and impact of the technology studied. Nor can one hope to eliminate all undesirable side effects. What we can achieve, however, is some foresight and a sense of anticipation, which might ultimately enhance the potential social benefits and reduce the more severe and foreseeable negative effects. This conviction has stimulated the recent surge of activity in the area of technology assessment—a process that needs to be continued as long as a new technology is in development. Above all, it must not be terminated at the conclusion of the first broad study; at that time, the critical questions may be clarified, but they will surely not be answered.

NOTES

1. Jeffrey Frey and Raymond Bowers, *Advances in Electronics and Electron Physics*, vol. 38 (New York: Academic Press, 1975), p. 147; Jeffrey Frey and Raymond Bowers, "What's Ahead for Microwaves," *IEEE Spectrum* 9 (March 1972):41-47; Raymond Bowers and Jeffrey Frey, "Technology Assessment and Microwave Diodes," *Scientific American* 226,

no. 2 (February 1972): 13-21; Edward M. Dickson and Raymond Bowers, *The Video Telephone* (New York: Praeger Publishers, 1973).

2. National Academy of Sciences, "Technology: Processes of Assessment and Choice," report prepared for the U.S. House of Representatives, Committee on Science and Astronautics (Washington, D.C.: U.S. Government Printing Office, 1969), p. 122; National Academy of Engineering, "A Study of Technology Assessment," report prepared for the U.S. House of Representatives, Committee on Science and Astronautics (Washington, D.C.: U.S. Government Printing Office, 1969), p. 15.

3. Alfred M. Lee, "Technology Assessment: A Structural Computer-Assisted Approach" (Masters thesis, Cornell University, 1975).

4. National Academy of Sciences, "Technology: Processes of Assessment and Choice," p. 43.

5. See, for example, Sidney H. Aronson, "The Sociology of the Telephone," *International Journal of Comparative Sociology* 12, no. 3 (September 1971): 153-67.

6. Alex Reid, "New Directions in Telecommunications Research," report prepared for the Sloan Commission on Cable Communications, Washington, D.C., June 1971.

7. National Academy of Engineering, "A Study of Technology Assessment," p. 13.

8. Dickson and Bowers, *The Video Telephone*, p. 4.

A NOTE ON THE CORNELL UNIVERSITY PROGRAM ON SCIENCE, TECHNOLOGY, AND SOCIETY

The Program on Science, Technology, and Society (STS) was established in 1969 to promote teaching and research on the interactions of science and technology with political and social institutions, as well as the effect of these interactions on individuals' lives. The program was conceived to contribute to understanding in these areas, utilizing a holistic approach that engages scholars from many disciplines, including the physical, biological, and social sciences; the humanities; engineering; business and public administration; and law. Topics of concern to the program include: political processes and large-scale technologies; the scope and limits of rational decision making; technology assessment of developments in telecommunication technologies; science policy, both national and international; biomedical and environmental ethics; energy policy; and science, technology, and the arms race.

The STS program is involved in the development of new interdisciplinary courses at both the graduate and undergraduate levels and in several major research activities. It also engages in programs aimed at promoting public understanding of the societal impact of science and technology.

STS receives substantial funds from Cornell, but most of its financial resources result from grants by private foundations, corporations, government agencies, and private donors. Organizations currently contributing funds for program activities include the Cogar Foundation, the Fleischmann Foundation, the General Electric Foundation, the IBM Corporation, the Henry R. Luce Foundation, the National Science Foundation, the Alfred P. Sloan Foundation, and the Xerox Corporation.

Details of STS activities are described in booklets that can be obtained from the Program on Science, Technology, and Society, Clark Hall, Cornell University, Ithaca, New York 14853 (telephone 607/256-3810).

INDEX

ABOUT THE AUTHORS

BETH BALDWIN was born in 1954 in Erie, Pennsylvania. She received a B.A. from Cornell University in 1976. Since graduating, she has worked as an administrative secretary for the University of Iowa Hospitals and Clinics and served as a volunteer at Planned Parenthood. She is currently assistant to the president of Baldwin Brothers, Inc., in Erie, Pennsylvania, while studying part time for an M.B.A. at Gannon College.

PHILIP L. BEREANO was born in 1940. He received the B.Ch.E. in 1962 and the Master's in Regional Planning in 1971 from Cornell University; in 1965, he received a J.D. from the Columbia University Law School. Since that time, he has served as a legislative assistant for the National Air Pollution Control Administration of the U.S. Public Health Service (1966–1970) and as an Assistant Professor at Cornell in Civil and Environmental Engineering and City and Regional Planning (1970–1975). While at Cornell he was executive secretary of the Program on Science, Technology, and Society. He has served as counsel for various environmental and citizens' groups and indigent criminal defendants. Current research is in the areas of environmental policy, technology assessment, "intermediate" (low impact) technologies, and women and technology. In 1976 he authored/edited *Technology as a Social and Political Phenomenon,* published by John Wiley and Sons. He is an Associate Professor in the Program in Social Management of Technology at the University of Washington.

ERWIN A. BLACKSTONE was born in 1942 in Utica, New York. He received the A.B. in economics from Syracuse University, and the M.A. and Ph.D. in economics from the University of Michigan. He has taught at Dartmouth College and Cornell University; currently, he is an Associate Professor of Economics at Temple University. Professor Blackstone specializes in industrial organization, medical economics, antitrust, and regulatory economics, and is now engaged in research on the telecommunications and health industries. He has published articles in many professional journals.

RAYMOND BOWERS was born in London, England, in 1927. He received a B.Sc. from London University in 1948 and a Ph.D. from Oxford University in 1951. From 1951-1953, he was a Postdoctoral Fellow in Physics at the University of Chicago, and from 1954-1960 a Research Physicist at Westinghouse Electric Corporation. Since 1960, he has been Professor of Physics at Cornell University, directing the research group on conduction processes in metals. In addition to his work in solid-state physics, Professor Bowers has also been actively involved in issues of science, technology, and society. In 1966-1967 he served as a member of the staff of the Office of Science and Technology, Executive Office of the President, and he has worked on national committees, including the National Academy of Science study on "Technology Assessment." He is currently director of Cornell University's Program on Science, Technology, and Society, and conducting research on the implications of future developments in telecommunications. Professor Bowers has been a consultant to a number of United States corporations, as well as to the National Science Foundation and the Department of State.

SARA EDMONDSON was born in 1946 in Arlington, Virginia. She received the A.B. in political science from Bryn Mawr College in 1969. After serving a year in the Peace Corps in Senegal, she taught high school for three years in Ann Arundel County, Maryland. She completed a Master's in Regional Planning in 1977 and is currently a candidate for the Ph.D. at Cornell University. While in attendance at Cornell, she has served as a research assistant on two major research projects, one studying the implementation problems of the Supplemental Security Income Program in New York State and the other analyzing the potential social effects of new mobile communication technologies. She also designed an evaluation for a human services program during an internship at the Southern Tier Regional Planning and Development Board.

JEFFREY FREY was born in New York City in 1939. He received the B.E.E. degree from Cornell University in 1960, the M.S. in 1963 and the Ph.D. in 1965 in electrical engineering from the University of California, Berkeley. His Ph.D. research was concerned with electron-stream instabilities, and in 1966-1967 he was awarded a NATO Postdoctoral Fellowship for experimental research on this problem at the Rutherford Laboratory, Chilton, England. In 1965 and 1966 he worked at the Watkins-Johnson Company in Palo Alto, California, on problems in millimeter wave generation and microwave integrated circuits. From 1967 to 1969 he was employed at the United Kingdom Atomic Energy Research Establishment, investigating device applications of ion implantation, and in 1976 he was a Fellow of the Japan Society for the Promotion of Science at the University of Tokyo. He is currently an Associate Professor at Cornell University, working in the fields of microwave semiconductor device theory and technology, and technology assessment. Professor Frey is the editor of the volume *Microwave Integrated Circuits,* published by Artech House in 1975. He has published numerous papers in the fields of plasma and semiconductor theory and technology and is the author, with Raymond Bowers, of the monograph "Technology Assessment of Microwave Semiconductor Devices," which appeared in *Advances in Electronics and Electron Physics* in 1975.

ROBERT E. GLANVILLE received the B.A. in social sciences from the State University of New York at Binghamton in 1972. In 1976 he received the J.D. from the Cornell University School of Law. Employed as a researcher with the Cornell University Program on Science, Technology, and Society in 1975, he is currently employed as a Law Assistant to the Appellate Division of the New York Supreme Court.

HOWARD HAMMERMAN received the B.A. from Antioch College and the M.S.W. and Ph.D. in sociology from the University of Michigan. He taught planning methodology, neighborhood theory, and the human ecology as an Assistant Professor in the Cornell University Department of City and Regional Planning from 1972 to 1977. At Michigan and Cornell he developed a systems approach to the study of man-environment interactions. He is currently a contracts officer in the Office of Policy Development and Research at the Department of Housing and Urban Development in Washington, D.C.

KURT L. HANSLOWE was born in 1926 in Vienna, Austria. He came to the United States in 1940, subsequently receiving the B.A. degree from Yale University in 1947 and the J.D. from the Harvard Law School in 1951. He was a member of the legal staff of the United Automobile Workers from 1951–1958, then joined the faculty of Cornell University, where he is Professor of Law and Professor of Industrial and Labor Relations. He has written numerous articles and several books in the fields of labor law, administrative law, and science, technology, and law. In 1977 he was appointed Senior Fulbright-Hays Scholar a d Visiting Professor of Law at the University of Vienna.

CARY HERSHEY was born in 1945 in New York City. He graduated from Cornell University in 1967 with a B.A. in government. He completed an M.P.A. at New York University in 1969 and a Ph.D. in public administration in 1971. He has been an Assistant Professor in the Department of City and Regional Planning at Cornell University since 1971, where he directed the Planning Department's program in Social Policy Planning and its HUD Internship Program. His primary research interests are in the areas of public services delivery, strategies for social change, and the social implications of communications, having written several articles in these areas, as well as *Protest in the Public Service,* published by D. C. Heath in 1973.

HEIDI KARGMAN graduated from Cornell University in 1975 with a B.A. in Technology and Society. While attending Cornell she served as editor of the Alternative Energy Newsletter, president of the Triphammer Women's Cooperative, and an announcer on WVBR-FM, a Cornell-affiliated radio station. During the fall of 1973 and 1974 she was an undergraduate teaching assistant in the Department of Computer Science. She was a research aide with the Cornell University Program on Science, Technology, and Society from the summer of 1975 to January 1977. Presently, she is a master's degree candidate at the Center for the Study of the Human Dimensions of Science and Technology at Renselaer Polytechnic Institute.

ALFRED M. LEE was born in 1951 in Bloomington, Indiana. He received the B.S. in electrical engineering from the University of Illinois, Urbana-Champaign, in 1973 and the M.S. in civil and environmental engineering from Cornell University in 1975. For over two years he was employed as a research specialist by the Cornell University Program on Science, Technology, and Society. He is currently a Ph.D. candidate at Cornell University in the School of Civil and Environmental Engineering.

ARNIM MEYBURG was born in Bremerhaven, West Germany, and attended the University of Hamburg and the Free University of Berlin. He received an M.S. in quantitative geography in 1968 and a Ph.D. in civil engineering (transportation) in 1971, both from Northwestern University. He has been a faculty member at Cornell University since 1969 and is currently Associate Professor of Transportation Planning and Engineering in the School of Civil and Environmental Engineering, and Acting Chairman of the Department of Environmental Engineering. In addition to being joint author with Peter R. Stopher of an earlier book, he has written a number of technical papers in the areas of travel-demand modeling, urban goods movement, and transportation-systems analysis.

MARK NADEL was formerly Assistant Professor of Government at Cornell University and a member of the Cornell Program on Science, Technology and Society. He is the author of *The Politics of Consumer Protection, Corporations and Political Accountability,* and numberous articles. Recently, he directed a study on public participation in the federal regulatory process for the Senate Committee on Governmental Affairs and is now a policy analyst for the General Accounting Office, Washington, D.C.

T.J.PEMPEL was born in 1942. He received the B.S. in 1966, the M.A. in 1969, and the Ph.D. in 1972, all from Columbia University, he is currently Associate Professor of Government at Cornell University. His most recent research has been on problems of Japanese public policy and policy making. Articles on this research have appeared in the *American Journal of Political Science, Journal of Asian Studies, Polity, International Organization,* and elsewhere. His book, *Policymaking in Contemporary Japan,* was recently published by Cornell University Press.

ERIC SHOTT received his B.A. in Geography from Chicago State University and attended the Cornell University master's program in City and Reg onal Planning. His areas of specialization are public policy analysis and application of quantitative techniques to planning problems. Presently, he is Regional Systems

Planner for the Northern Virginia Planning District Commission where he is involved in developing a regional impact assessment system.

RUSSELL THATCHER received the B.S. in Public Systems Planning and Analysis with a minor in Transportation Planning from the Cornell University School of Civil and Environmental Engineering. He is currently employed by Call-A-Ride of Barnstable County, Inc., a regional demand-responsive transportation system serving Cape Code, Massachusetts.

HAROLD WARE was born in 1947 in Newburg, New York. He received his B.A. from the State University of New York at Stony Brook in 1969. During his undergraduate career, he worked as a research assistant for the Department of Economics. He was employed as a high school mathematics teacher from 1969–1974. During that time he also attended Yeshiva University and received the M.S. degree in mathematics/education. He received the M.A. in economics in 1976 from Cornell University, and his Ph.D. dissertation was accepted in August 1977. While at Cornell, he worked as an instructor in economics, a teaching assistant, and did research for the Technology Assessment Project of the Program on Science, Technology, and Society. He is currently a member of National Economic Research Associates, Inc., an economics consulting firm in New York City.